COURSE OF THEORETICAL PHYSICS

VOLUME 7

THEORY OF ELASTICITY

Third Edition

Other titles in the COURSE OF THEORETICAL PHYSICS by L. D. Landau and E. M. Lifshitz

Vol. 1. Mechanics, 3rd edition

Vol. 2. The Classical Theory of Fields, 4th edition

Vol. 3. Quantum Mechanics—Non-Relativistic Theory, 3rd edition

Vol. 4. Quantum Electrodynamics, 2nd edition

Vol. 5. Statistical Physics, Part 1, 3rd edition

Vol. 6. Fluid Mechanics

Vol 8. Electrodynamics of Continuous Media, 2nd edition

Vol. 9. Statistical Physics, Part 2

Vol. 10. Physical Kinetics

THEORY OF ELASTICITY

by

L. D. LANDAU and E. M. LIFSHITZ
Institute of Physical Problems, U.S.S.R. Academy of Sciences

Volume 7 of Course of Theoretical Physics

Third English Edition, Revised and Enlarged

by E. M. LIFSHITZ, A. M. KOSEVICH and L. P. PITAEVSKIĬ

Translated from the Russian by

J. B. SYKES and W. H. REID

ELSEVIER
BUTTERWORTH
HEINEMANN

AMSTERDAM • BOSTON • HEIDELBERG • LONDON • NEW YORK • OXFORD
PARIS • SAN DIEGO • SAN FRANCISCO • SINGAPORE • SYDNEY • TOKYO

Butterworth-Heinemann is an imprint of Elsevier
The Boulevard, Langford Lane, Kidlington, Oxford, OX5 1GB
30 Corporate Drive, Suite 400, Burlington, MA 01803, USA

Translated from the fourth edition of *Teoria uprogosti* 'Nakuka', Moscow 1986

First published in English by Pergamon Press plc 1959
Second edition 1970
Reprinted 1975, 1981, 1984
Third edition 1986
Reprinted 1995, 1997, 1998, 2000, 2002, 2005, 2006, 2007, 2008

British Library Cataloguing in Publication Data
Landau, Lev Davidovich
Theory of elasticity – 3rd ed. (Course of theoretical physics; vol. 7)
1. Elasticity
I. Title II. Lifshitz, E. M.
III. Kosevich, A. M. IV. Pitaevskii, L. P. V. Teoria uprugosti
VI. Series
531′.3823 QA931

Library of Congress Cataloging-in-Publication Data
Landau, Lev Davidovich, 1908–1968
Theory of elasticity (Course of theoretical physics; vol. 7)
Translation of: Teoria uprugosti
Includes index
1. Elasticity 2. Elastic solids I. Lifshitz, E. M.
II. Kosevich, A. M. III. Pitaevskii, L. P IV. Title
V. Series: Landau, L. D. Course of theoretical physics vol. 7
QA931.L283 1986 531'.38 86-2450

ISBN: 978-0-7506-2633-0

Transferred to Digital Printing 2009

CONTENTS

PREFACE TO THE FIRST ENGLISH EDITION

THE present volume of our *Theoretical Physics* deals with the theory of elasticity.

Being written by physicists, and primarily for physicists, it naturally includes not only the ordinary theory of the deformation of solids, but also some topics not usually found in textbooks on the subject, such as thermal conduction and viscosity in solids, and various problems in the theory of elastic vibrations and waves. On the other hand, we have discussed only very briefly certain special matters, such as complex mathematical methods in the theory of elasticity and the theory of shells, which are outside the scope of this book.

Our thanks are due to Dr. Sykes and Dr. Reid for their excellent translation of the book.

Moscow L. D. LANDAU
E. M. LIFSHITZ

PREFACE TO THE SECOND ENGLISH EDITION

AS WELL as some minor corrections and additions, a chapter on the macroscopic theory of dislocations has been added in this edition. The chapter has been written jointly by myself and A. M. Kosevich.

A number of useful comments have been made by G. I. Barenblatt, V. L. Ginzburg, M. A. Isakovich, I. M. Lifshitz and I. M. Shmushkevich for the Russian edition, while the vigilance of Dr. Sykes and Dr. Reid has made it possible to eliminate some further errors from the English translation.

I should like to express here my sincere gratitude to all the above-named.

Moscow E. M. LIFSHITZ

PREFACE TO THE THIRD ENGLISH EDITION

THE major part of this book (Chapters I, II, III and V) is not very different from what was in the first two English editions (1959 and 1970). This is a natural result of the fact that the basic equations and conclusions of elasticity theory have long since been established.

The second edition included a chapter on the theory of dislocations in crystals, written jointly with A. M. Kosevich, which has been only slightly changed in the present edition.

This new edition contains a further chapter, on the mechanics of liquid crystals, written jointly with L. P. Pitaevskiĭ—a new branch of continuum mechanics which combines features of liquids and elastic solids, and whose proper position in the *Course of Theoretical Physics* is therefore after both fluid mechanics and elasticity of solids.

As always, I have derived much benefit from discussing with my friends and colleagues various topics dealt with in the book. I should like to mention with gratitude the names of V. L. Ginzburg, V. L. Indenbom, E. I. Kats, Yu. A. Kosevich, V. V. Lebedev, V. P. Mineev and G. E. Volovik for their various comments used in preparing the book.

Moscow E. M. LIFSHITZ

NOTATION

ρ density of matter

$\mathbf{u}$ displacement vector

$u_{ik} = \frac{1}{2}\left(\frac{\partial u_i}{\partial x_k} + \frac{\partial u_k}{\partial x_i}\right)$ strain tensor

σ_{ik} stress tensor

K modulus of compression

μ modulus of rigidity

E Young's modulus

σ Poisson's ratio

c_l longitudinal velocity of sound

c_t transverse velocity of sound

c_l and c_t are expressed in terms of K, μ or of E, σ by formulae given in §22.

The quantities K, μ, E and σ are related by

$E = 9K\mu/(3K + \mu)$

$\sigma = (3K - 2\mu)/2(3K + \mu)$

$K = E/3(1 - 2\sigma)$

$\mu = E/2(1 + \sigma)$

The summation convention always applies to suffixes occurring twice in vector and tensor expressions. In Chapter VI, ∂_i ($\equiv \partial/\partial x_i$) is used to denote differentiation with respect to a coordinate.

References to other volumes in the *Course of Theoretical Physics*:

Fields = Vol. 2 (*The Classical Theory of Fields*, fourth English edition, 1975).

SP 1 = Vol. 5 (*Statistical Physics*, Part 1, third English edition, 1980).

FM = Vol. 6 (*Fluid Mechanics*, Second English edition, 1987).

ECM = Vol. 8 (*Electrodynamics of Continuous Media*, second English edition, 1984).

All are published by Pergamon Press.

CHAPTER I

FUNDAMENTAL EQUATIONS

§1. The strain tensor

THE mechanics of solid bodies, regarded as continuous media, forms the content of the *theory of elasticity*.†

Under the action of applied forces, solid bodies exhibit deformation to some extent, i.e. they change in shape and volume. The deformation of a body is described mathematically in the following way. The position of any point in the body is defined by its position vector **r** (with components $x_1 = x$, $x_2 = y$, $x_3 = z$) in some coordinate system. When the body is deformed, every point in it is in general displaced. Let us consider some particular point; let its position vector before the deformation be **r**, and after the deformation have a different value **r**′ (with components x'_i). The displacement of this point due to the deformation is then given by the vector **r**′ − **r**, which we shall denote by **u**:

$$u_i = x'_i - x_i. \tag{1.1}$$

The vector **u** is called the *displacement vector*. The coordinates x'_i of the displaced point are, of course, functions of the coordinates x_i of the point before displacement. The displacement vector u_i is therefore also a function of the coordinates x_i. If the vector **u** is given as a function of x_i, the deformation of the body is entirely determined.

When a body is deformed, the distances between its points change. Let us consider two points very close together. If the radius vector joining them before the deformation is $\mathrm{d}x_i$, the radius vector joining the same two points in the deformed body is $\mathrm{d}x'_i = \mathrm{d}x_i + \mathrm{d}u_i$. The distance between the points is $\mathrm{d}l = \sqrt{(\mathrm{d}x_1{}^2 + \mathrm{d}x_2{}^2 + \mathrm{d}x_3{}^2)}$ before the deformation, and $\mathrm{d}l' = \sqrt{(\mathrm{d}x'_1{}^2 + \mathrm{d}x'_2{}^2 + \mathrm{d}x'_3{}^2)}$ after it. Using the general summation rule, we can write $\mathrm{d}l^2 = \mathrm{d}x_i{}^2$, $\mathrm{d}l'^2 = \mathrm{d}x'_i{}^2 = (\mathrm{d}x_i + \mathrm{d}u_i)^2$. Substituting $\mathrm{d}u_i = (\partial u_i/\partial x_k)\mathrm{d}x_k$, we can write

$$\mathrm{d}l'^2 = \mathrm{d}l^2 + 2\frac{\partial u_i}{\partial x_k}\mathrm{d}x_i\,\mathrm{d}x_k + \frac{\partial u_i}{\partial x_k}\frac{\partial u_i}{\partial x_l}\mathrm{d}x_k\,\mathrm{d}x_l.$$

Since the summation is taken over both suffixes i and k in the second term on the right, this term can be put in the explicitly symmetrical form

$$\left(\frac{\partial u_i}{\partial x_k} + \frac{\partial u_k}{\partial x_i}\right)\mathrm{d}x_i\,\mathrm{d}x_k.$$

In the third term, we interchange the suffixes i and l. Then $\mathrm{d}l'^2$ takes the final form

$$\mathrm{d}l'^2 = \mathrm{d}l^2 + 2u_{ik}\,\mathrm{d}x_i\,\mathrm{d}x_k, \tag{1.2}$$

† The basic equations of elasticity theory were established in the 1820s by Cauchy and by Poisson.

where the tensor u_{ik} is defined as

$$u_{ik} = \frac{1}{2}\left(\frac{\partial u_i}{\partial x_k} + \frac{\partial u_k}{\partial x_i} + \frac{\partial u_l}{\partial x_i}\frac{\partial u_l}{\partial x_k}\right). \tag{1.3}$$

These expressions give the change in an element of length when the body is deformed.

The tensor u_{ik} is called the *strain tensor*. We see from its definition that it is symmetrical, i.e.

$$u_{ik} = u_{ki}. \tag{1.4}$$

Like any symmetrical tensor, u_{ik} can be *diagonalized* at any given point. This means that, at any given point, we can choose coordinate axes (the *principal axes* of the tensor) in such a way that only the diagonal components u_{11}, u_{22}, u_{33} of the tensor u_{ik} are different from zero. These components, the *principal values* of the strain tensor, will be denoted by $u^{(1)}$, $u^{(2)}$, $u^{(3)}$. It should be remembered, of course, that, if the tensor u_{ik} is diagonalized at any point in the body, it will not in general be diagonal at any other point.

If the strain tensor is diagonalized at a given point, the element of length (1.2) near it becomes

$$\begin{aligned} dl'^2 &= (\delta_{ik} + 2u_{ik})\,dx_i\,dx_k \\ &= (1+2u^{(1)})\,dx_1{}^2 + (1+2u^{(2)})\,dx_2{}^2 + (1+2u^{(3)})\,dx_3{}^2. \end{aligned}$$

We see that the expression is the sum of three independent terms. This means that the strain in any volume element may be regarded as composed of independent strains in three mutually perpendicular directions, namely those of the principal axes of the strain tensor. Each of these strains is a simple extension (or compression) in the corresponding direction: the length dx_1 along the first principal axis becomes $dx'_1 = \sqrt{(1+2u^{(1)})}\,dx_1$, and similarly for the other two axes. The quantity $\sqrt{(1+2u^{(i)})}-1$ is consequently equal to the relative extension $(dx'_i - dx_i)/dx_i$ along the ith principal axis.

In almost all cases occurring in practice, the strains are small. This means that the change in any distance in the body is small compared with the distance itself. In other words, the relative extensions are small compared with unity. In what follows we shall suppose that all strains are small.

If a body is subjected to a small deformation, all the components of the strain tensor are small, since they give, as we have seen, the relative changes in lengths in the body. The displacement vector u_i, however, may sometimes be large, even for small strains. For example, let us consider a long thin rod. Even for a large deflection, in which the ends of the rod move a considerable distance, the extensions and compressions in the rod itself will be small.

Except in such special cases,† the displacement vector for a small deformation is itself small. For it is evident that a three-dimensional body (i.e. one whose dimension in no direction is small) cannot be deformed in such a way that parts of it move a considerable distance without the occurrence of considerable extensions and compressions in the body.

† Which include, besides deformations of thin rods, those of thin plates to form cylindrical surfaces. We have also to exclude the case where the deformation of a three-dimensional body is accompanied by a rotation through a finite angle.

Thin rods will be discussed in Chapter II. In other cases the u_i and their derivatives are small for small deformations, and we can therefore neglect the last term in the general expression (1.3), as being of the second order of smallness. Thus, for small deformations, the strain tensor is given by

$$u_{ik} = \frac{1}{2}\left(\frac{\partial u_i}{\partial x_k} + \frac{\partial u_k}{\partial x_i}\right). \tag{1.5}$$

The relative extensions of the elements of length along the principal axes of the strain tensor (at a given point) are, to within higher-order quantities, $\sqrt{(1+2u^{(i)})} - 1 \approx u^{(i)}$, i.e. they are the principal values of the tensor u_{ik}.

Let us consider an infinitesimal volume element $\mathrm{d}V$, and find its volume $\mathrm{d}V'$ after the deformation. To do so, we take the principal axes of the strain tensor, at the point considered, as the coordinate axes. Then the elements of length $\mathrm{d}x_1, \mathrm{d}x_2, \mathrm{d}x_3$ along these axes become, after the deformation, $\mathrm{d}x'_1 = (1+u^{(1)})\mathrm{d}x_1$, etc. The volume $\mathrm{d}V$ is the product $\mathrm{d}x_1\,\mathrm{d}x_2\,\mathrm{d}x_3$, while $\mathrm{d}V'$ is $\mathrm{d}x'_1\,\mathrm{d}x'_2\,\mathrm{d}x'_3$. Thus $\mathrm{d}V' = \mathrm{d}V(1+u^{(1)})(1+u^{(2)}) \times \times (1+u^{(3)})$. Neglecting higher-order terms, we therefore have $\mathrm{d}V' = \mathrm{d}V(1+u^{(1)}+u^{(2)}+ + u^{(3)})$. The sum $u^{(1)}+u^{(2)}+u^{(3)}$ of the principal values of a tensor is well known to be invariant, and is equal to the sum of the diagonal components $u_{ii} = u_{11}+u_{22}+u_{33}$ in any coordinate system. Thus

$$\mathrm{d}V' = \mathrm{d}V(1+u_{ii}). \tag{1.6}$$

We see that the sum of the diagonal components of the strain tensor is the relative volume change $(\mathrm{d}V' - \mathrm{d}V)/\mathrm{d}V$.

It is often convenient to use the components of the strain tensor in spherical polar or cylindrical polar coordinates. We give here, for reference, the corresponding formulae, which express the components in terms of the derivatives of the components of the displacement vector in the same coordinates. In spherical polar coordinates r, θ, ϕ, we have

$$\left.\begin{aligned}
&u_{rr} = \frac{\partial u_r}{\partial r}, \quad u_{\theta\theta} = \frac{1}{r}\frac{\partial u_\theta}{\partial \theta} + \frac{u_r}{r}, \quad u_{\phi\phi} = \frac{1}{r\sin\theta}\frac{\partial u_\phi}{\partial \phi} + \frac{u_\theta}{r}\cot\theta + \frac{u_r}{r},\\
&2u_{\theta\phi} = \frac{1}{r}\left(\frac{\partial u_\phi}{\partial \theta} - u_\phi\cot\theta\right) + \frac{1}{r\sin\theta}\frac{\partial u_\theta}{\partial \phi}, \quad 2u_{r\theta} = \frac{\partial u_\theta}{\partial r} - \frac{u_\theta}{r} + \frac{1}{r}\frac{\partial u_r}{\partial \theta},\\
&2u_{\phi r} = \frac{1}{r\sin\theta}\frac{\partial u_r}{\partial \phi} + \frac{\partial u_\phi}{\partial r} - \frac{u_\phi}{r}.
\end{aligned}\right\} \tag{1.7}$$

In cylindrical polar coordinates r, ϕ, z,

$$\left.\begin{aligned}
&u_{rr} = \frac{\partial u_r}{\partial r}, \quad u_{\phi\phi} = \frac{1}{r}\frac{\partial u_\phi}{\partial \phi} + \frac{u_r}{r}, \quad u_{zz} = \frac{\partial u_z}{\partial z},\\
&2u_{\phi z} = \frac{1}{r}\frac{\partial u_z}{\partial \phi} + \frac{\partial u_\phi}{\partial z}, \quad 2u_{rz} = \frac{\partial u_r}{\partial z} + \frac{\partial u_z}{\partial r},\\
&2u_{r\phi} = \frac{\partial u_\phi}{\partial r} - \frac{u_\phi}{r} + \frac{1}{r}\frac{\partial u_r}{\partial \phi}.
\end{aligned}\right\} \tag{1.8}$$

§2. The stress tensor

In a body that is not deformed, the arrangement of the molecules corresponds to a state of thermal equilibrium. All parts of the body are in mechanical equilibrium. This means that, if some portion of the body is considered, the resultant of the forces on that portion is zero.

When a deformation occurs, the arrangement of the molecules is changed, and the body ceases to be in its original state of equilibrium. Forces therefore arise which tend to return the body to equilibrium. These internal forces which occur when a body is deformed are called *internal stresses*. If no deformation occurs, there are no internal stresses.

The internal stresses are due to molecular forces, i.e. the forces of interaction between the molecules. An important fact in the theory of elasticity is that the molecular forces have a very short range of action. Their effect extends only to the neighbourhood of the molecule exerting them, over a distance of the same order as that between the molecules, whereas in the theory of elasticity, which is a macroscopic theory, the only distances considered are those large compared with the distances between the molecules. The range of action of the molecular forces should therefore be taken as zero in the theory of elasticity. We can say that the forces which cause the internal stresses are, as regards the theory of elasticity, "near-action" forces, which act from any point only to neighbouring points. Hence it follows that the forces exerted on any part of the body by surrounding parts act only on the surface of this part.

The following reservation should be made here. The above assertion is not valid in cases where the deformation of the body results in macroscopic electric fields in it (pyroelectric and piezoelectric bodies). Such bodies are discussed in *ECM*.

Let us consider the total force on some portion of the body. Firstly, this total force is equal to the sum of all the forces on all the volume elements in that portion of the body, i.e. it can be written as the volume integral $\int \mathbf{F}\,\mathrm{d}V$, where $\mathbf{F}$ is the force per unit volume and $\mathbf{F}\,\mathrm{d}V$ the force on the volume element $\mathrm{d}V$. Secondly, the forces with which various parts of the portion considered act on one another cannot give anything but zero in the total resultant force, since they cancel by Newton's third law. The required total force can therefore be regarded as the sum of the forces exerted on the given portion of the body by the portions surrounding it. From above, however, these forces act on the surface of that portion, and so the resultant force can be represented as the sum of forces acting on all the surface elements, i.e. as an integral over the surface.

Thus, for any portion of the body, each of the three components $\int F_i\,\mathrm{d}V$ of the resultant of all the internal stresses can be transformed into an integral over the surface. As we know from vector analysis, the integral of a scalar over an arbitrary volume can be transformed into an integral over the surface if the scalar is the divergence of a vector. In the present case we have the integral of a vector, and not of a scalar. Hence the vector F_i must be the divergence of a tensor of rank two, i.e. be of the form

$$F_i = \partial\sigma_{ik}/\partial x_k. \tag{2.1}$$

Then the force on any volume can be written as an integral over the closed surface bounding that volume:†

† The vector $\mathbf{df}$ is along the normal outward from the closed surface. The integral over a closed surface is transformed into one over the volume enclosed by the surface by replacing the surface element $\mathrm{d}f_i$ by the operator $\mathrm{d}V\partial/\partial x_i$.

Strictly speaking, to determine the total force on a deformed portion of the body we should integrate, not over the old coordinates x_i, but over the coordinates x'_i of the points of the deformed body. The derivatives (2.1) should therefore be taken with respect to x'_i. However, in view of the smallness of the deformation, the derivatives with respect to x_i and x'_i differ only by higher-order quantities, and so the derivatives can be taken with respect to the coordinates x_i.

$$\int F_i \, dV = \int \frac{\partial \sigma_{ik}}{\partial x_k} dV = \oint \sigma_{ik} \, df_k. \tag{2.2}$$

The tensor σ_{ik} is called the *stress tensor*. As we see from (2.2), $\sigma_{ik} df_k$ is the ith component of the force on the surface element $\mathbf{df}$. By taking elements of area in the planes of xy, yz, zx, we find that the component σ_{ik} of the stress tensor is the ith component of the force on unit area perpendicular to the x_k-axis. For instance, the force on unit area perpendicular to the x-axis, normal to the area (i.e. along the x-axis), is σ_{xx}, and the tangential forces (along the y and z axes) are σ_{yx} and σ_{zx}.

The following remark should be made concerning the sign of the force $\sigma_{ik} df_k$. The surface integral in (2.2) is the force exerted on the volume enclosed by the surface by the surrounding parts of the body. The force which this volume exerts on the surface surrounding it is the same with the opposite sign. Hence, for example, the force exerted by the internal stresses on the surface of the body itself is $-\oint \sigma_{ik} df_k$, where the integral is taken over the surface of the body and $\mathbf{df}$ is along the outward normal.

Let us determine the moment of the forces on a portion of the body. The moment of the force $\mathbf{F}$ can be written as an antisymmetrical tensor of rank two, whose components are $F_i x_k - F_k x_i$, where x_i are the coordinates of the point where the force is applied.† Hence the moment of the forces on the volume element dV is $(F_i x_k - F_k x_i) dV$, and the moment of the forces on the whole volume is $M_{ik} = \int (F_i x_k - F_k x_i) dV$. Like the total force on any volume, this moment can be expressed as an integral over the surface bounding the volume. Substituting the expression (2.1) for F_i, we find

$$M_{ik} = \int \left(\frac{\partial \sigma_{il}}{\partial x_l} x_k - \frac{\partial \sigma_{kl}}{\partial x_l} x_i \right) dV$$

$$= \int \frac{\partial (\sigma_{il} x_k - \sigma_{kl} x_i)}{\partial x_l} dV - \int \left(\sigma_{il} \frac{\partial x_k}{\partial x_l} - \sigma_{kl} \frac{\partial x_i}{\partial x_l} \right) dV.$$

In the second term we use the fact that the derivative of a coordinate with respect to itself is unity, and with respect to another coordinate is zero (since the three coordinates are independent variables); thus $\partial x_k / \partial x_l = \delta_{kl}$, where δ_{kl} is the unit tensor. In the first term, the integrand is the divergence of a tensor; the integral can be transformed into one over the surface. The result is

$$M_{ik} = \oint (\sigma_{il} x_k - \sigma_{kl} x_i) \, df_l + \int (\sigma_{ki} - \sigma_{ik}) \, dV. \tag{2.3}$$

The tensor M_{ik} will be an integral over the surface alone if the stress tensor is symmetrical:

$$\sigma_{ik} = \sigma_{ki}, \tag{2.4}$$

so that the volume integral vanishes; the basis for this important statement will be further

† The moment of the force $\mathbf{F}$ is defined as the vector product $\mathbf{F} \times \mathbf{r}$, and we know from vector analysis that the components of a vector product form an antisymmetrical tensor of rank two as written here.

discussed at the end of the section. The moment of the forces on a portion of the body can then be written simply as

$$M_{ik} = \int (F_i x_k - F_k x_i)\,dV = \oint (\sigma_{il} x_k - \sigma_{kl} x_i)\,df_l. \tag{2.5}$$

It is easy to find the stress tensor for a body undergoing uniform compression from all sides (*hydrostatic* compression). In this case a pressure of the same magnitude acts on every unit area on the surface of the body, and its direction is along the inward normal. If this pressure is denoted by p, a force $-p\,df_i$ acts on the surface element df_i. This force, in terms of the stress tensor, must be $\sigma_{ik}\,df_k$. Writing $-p\,df_i = -p\delta_{ik}\,df_k$, we see that the stress tensor in hydrostatic compression is

$$\sigma_{ik} = -p\delta_{ik}. \tag{2.6}$$

Its non-zero components are simply equal to the pressure.

In the general case of an arbitrary deformation, the non-diagonal components of the stress tensor are also non-zero. This means that not only a normal force but also tangential (shearing) stresses act on each surface element. These latter stresses tend to move the surface elements relative to each other.

In equilibrium the internal stresses in every volume element must balance, i.e. we must have $F_i = 0$. Thus the equations of equilibrium for a deformed body are

$$\partial\sigma_{ik}/\partial x_k = 0. \tag{2.7}$$

If the body is in a gravitational field, the sum $\mathbf{F} + \rho\mathbf{g}$ of the internal stresses and the force of gravity ($\rho\mathbf{g}$ per unit volume) must vanish; ρ is the density† and $\mathbf{g}$ the gravitational acceleration vector, directed vertically downwards. In this case the equations of equilibrium are

$$\partial\sigma_{ik}/\partial x_k + \rho g_i = 0. \tag{2.8}$$

The external forces applied to the surface of the body (which are the usual cause of deformation) appear in the boundary conditions on the equations of equilibrium. Let $\mathbf{P}$ be the external force on unit area of the surface of the body, so that a force $\mathbf{P}\,df$ acts on a surface element df. In equilibrium, this must be balanced by the force $-\sigma_{ik}\,df_k$ of the internal stresses acting on that element. Thus we must have $P_i\,df - \sigma_{ik}\,df_k = 0$. Writing $df_k = n_k\,df$, where $\mathbf{n}$ is a unit vector along the outward normal to the surface, we find

$$\sigma_{ik} n_k = P_i. \tag{2.9}$$

This is the condition which must be satisfied at every point on the surface of a body in equilibrium.

We shall derive also a formula giving the mean value of the stress tensor in a deformed body. To do so, we multiply equation (2.7) by x_k and integrate over the whole volume:

$$\int \frac{\partial\sigma_{il}}{\partial x_l} x_k\,dV = \int \frac{\partial(\sigma_{il} x_k)}{\partial x_l}\,dV - \int \sigma_{il}\frac{\partial x_k}{\partial x_l}\,dV = 0.$$

The first integral on the right is transformed into a surface integral; in the second integral we put $\partial x_k/\partial x_l = \delta_{kl}$. The result is $\oint \sigma_{il} x_k\,df_l - \int \sigma_{ik}\,dV = 0$. Substituting (2.9) in the first

† Strictly speaking, the density of a body changes when it is deformed. An allowance for this change, however, involves higher-order quantities in the case of small deformations, and is therefore unimportant.

integral, we find $\oint P_i x_k \, df = \int \sigma_{ik} \, dV = V\bar{\sigma}_{ik}$, where V is the volume of the body and $\bar{\sigma}_{ik}$ the mean value of the stress tensor. Since $\sigma_{ik} = \sigma_{ki}$, this formula can be written in the symmetrical form

$$\bar{\sigma}_{ik} = (1/2V)\oint (P_i x_k + P_k x_i)\, df. \tag{2.10}$$

Thus the mean value of the stress tensor can be found immediately from the external forces acting on the body, without solving the equations of equilibrium.

Let us now go back to the proof given above that the stress tensor is symmetrical, since it is in need of refinement. The physical condition imposed, that the tensor M_{ik} be representable as an integral over the surface alone, is satisfied not only if the antisymmetrical part of the tensor σ_{ik} (that is, the integrand in the volume integral in (2.3)) is zero, but also if it is a divergence, i.e. if

$$\sigma_{ik} - \sigma_{ki} = 2\partial\phi_{ikl}/\partial x_l, \quad \phi_{ikl} = -\phi_{kil}, \tag{2.11}$$

where ϕ_{ikl} is any tensor antisymmetrical in the first pair of suffixes. In the present case, this tensor is to be expressed in terms of the derivatives $\partial u_i/\partial x_k$, and accordingly the stress tensor contains terms in higher derivatives of the displacement vector. Within the theory of elasticity as described here, all such terms should be regarded as higher-order small quantities and omitted.

It is, however, important in principle that the stress tensor can be reduced to a symmetrical form even if these terms are not neglected.† The reason is that the definition (2.1) of this tensor is not unique: any transformation is possible that is of the form

$$\tilde{\sigma}_{ik} - \sigma_{ik} = \partial\chi_{ikl}/\partial x_l, \quad \chi_{ikl} = -\chi_{ilk}, \tag{2.12}$$

where χ_{ikl} is any tensor antisymmetrical in the last pair of suffixes. Evidently, the derivatives $\partial\sigma_{ik}/\partial x_k$ and $\partial\tilde{\sigma}_{ik}/\partial x_k$, which determine the force F, are identically equal. If the antisymmetrical part of σ_{ik} has the form (2.11), then an unsymmetrical σ_{ik} can be made symmetrical by a transformation of this type. The symmetrical tensor is

$$\tilde{\sigma}_{ik} = \tfrac{1}{2}(\sigma_{ik} + \sigma_{ki}) + \partial(\phi_{ilk} + \phi_{kli})/\partial x_l; \tag{2.13}$$

it is easy to see that $\tilde{\sigma}_{ik} - \sigma_{ik}$ has the form (2.12) with

$$\chi_{ikl} = \phi_{kli} + \phi_{ilk} - \phi_{ikl} \tag{2.14}$$

(P. C. Martin, O. Parodi and P. S. Pershan 1972).

§3. The thermodynamics of deformation

Let us consider some deformed body, and suppose that the deformation is changed in such a way that the displacement vector u_i changes by a small amount δu_i; and let us determine the work done by the internal stresses in this change. Multiplying the force $F_i = \partial\sigma_{ik}/\partial x_k$ by the displacement δu_i and integrating over the volume of the body, we have $\int \delta R \, dV = \int (\partial\sigma_{ik}/\partial x_k)\delta u_i \, dV$, where δR denotes the work done by the internal stresses per unit volume. We integrate by parts, obtaining

$$\int \delta R \, dV = \oint \sigma_{ik}\delta u_i \, df_k - \int \sigma_{ik}\frac{\partial \delta u_i}{\partial x_k} dV.$$

† In accordance with the general results of the microscopic theory (*Fields*, §32).

By considering an infinite medium which is not deformed at infinity, we make the surface of integration in the first integral tend to infinity; then $\sigma_{ik} = 0$ on the surface, and the integral is zero. The second integral can, by virtue of the symmetry of the tensor σ_{ik}, be written

$$\int \delta R \, dV = -\frac{1}{2}\int \sigma_{ik}\left(\frac{\partial \delta u_i}{\partial x_k} + \frac{\partial \delta u_k}{\partial x_i}\right) dV$$

$$= -\frac{1}{2}\int \sigma_{ik}\delta\left(\frac{\partial u_i}{\partial x_k} + \frac{\partial u_k}{\partial x_i}\right) dV$$

$$= -\int \sigma_{ik}\delta u_{ik} \, dV.$$

Thus we find

$$\delta R = -\sigma_{ik}\delta u_{ik}. \tag{3.1}$$

This formula gives the work δR in terms of the change in the strain tensor.

If the deformation of the body is fairly small, it returns to its original undeformed state when the external forces causing the deformation cease to act. Such deformations are said to be *elastic*. For large deformations, the removal of the external forces does not result in the total disappearance of the deformation; a *residual deformation* remains, so that the state of the body is not that which existed before the forces were applied. Such deformations are said to be *plastic*. In what follows (except in Chapter IV) we shall consider only elastic deformations.

We shall also suppose that the process of deformation occurs so slowly that thermodynamic equilibrium is established in the body at every instant, in accordance with the external conditions. This assumption is almost always justified in practice. The process will then be thermodynamically reversible.

In what follows we shall take all such thermodynamic quantities as the entropy S, the internal energy $\mathscr{E}$, etc., relative to unit volume of the body, and not relative to unit mass as in fluid mechanics, and denote them by the corresponding capital letters.

The following remark should be made here. Strictly speaking, the unit volumes before and after the deformation should be distinguished, since they in general contain different amounts of matter. We shall always (except in Chapter VI) relate the thermodynamic quantities to unit volume of the undeformed body, i.e. to the amount of matter therein, which may occupy a different volume after the deformation. Accordingly, the total energy of the body, for example, is obtained by integrating $\mathscr{E}$ over the volume of the undeformed body.

An infinitesimal change $d\mathscr{E}$ in the internal energy is equal to the difference between the heat acquired by the unit volume considered and the work dR done by the internal stresses. The amount of heat is, for a reversible process, TdS, where T is the temperature. Thus $d\mathscr{E} = TdS - dR$; with dR given by (3.1), we obtain

$$d\mathscr{E} = TdS + \sigma_{ik} \, du_{ik}. \tag{3.2}$$

This is the fundamental thermodynamic relation for deformed bodies.

In hydrostatic compression, the stress tensor is $\sigma_{ik} = -p\delta_{ik}$ (2.6). Then $\sigma_{ik} du_{ik} = -p\delta_{ik} du_{ik} = -p \, du_{ii}$. We have seen, however (cf. (1.6)), that the sum u_{ii} is the relative volume change due to the deformation. If we consider unit volume, therefore, u_{ii} is simply the change in that volume, and du_{ii} is the volume element dV. The thermodynamic relation then takes its usual form

$$d\mathscr{E} = TdS - pdV. \tag{3.2a}$$

Introducing the (Helmholtz) free energy of the body, $F = \mathscr{E} - TS$, we find the form

$$dF = -S\,dT + \sigma_{ik}\,du_{ik} \tag{3.3}$$

of the relation (3.2). Finally, the thermodynamic potential (Gibbs free energy) Φ is defined as

$$\Phi = \mathscr{E} - TS - \sigma_{ik}u_{ik} = F - \sigma_{ik}u_{ik}. \tag{3.4}$$

This is a generalization of the usual expression $\Phi = \mathscr{E} - TS + pV$.† Substituting (3.4) in (3.3), we find

$$d\Phi = -S\,dT - u_{ik}\,d\sigma_{ik}. \tag{3.5}$$

The independent variables in (3.2) and (3.3) are respectively S, u_{ik} and T, u_{ik}. The components of the stress tensor can be obtained by differentiating $\mathscr{E}$ or F with respect to the components of the strain tensor, for constant entropy S or temperature T respectively:

$$\sigma_{ik} = (\partial\mathscr{E}/\partial u_{ik})_S = (\partial F/\partial u_{ik})_T. \tag{3.6}$$

Similarly, by differentiating Φ with respect to the components σ_{ik}, we can obtain the components u_{ik}:

$$u_{ik} = -(\partial\Phi/\partial\sigma_{ik})_T. \tag{3.7}$$

§4. Hooke's law

In order to be able to apply the general formulae of thermodynamics to any particular case, we must know the free energy F of the body as a function of the strain tensor. This expression is easily obtained by using the fact that the deformation is small and expanding the free energy in powers of u_{ik}. We shall at present consider only isotropic bodies. The corresponding results for crystals will be obtained in §10.

In considering a deformed body at some temperature (constant throughout the body), we shall take the undeformed state to be the state of the body in the absence of external forces and at the same temperature; this last condition is necessary on account of the thermal expansion (see §6). Then, for $u_{ik} = 0$, the internal stresses are zero also, i.e. $\sigma_{ik} = 0$. Since $\sigma_{ik} = \partial F/\partial u_{ik}$, it follows that there is no linear term in the expansion of F in powers of u_{ik}.

Next, since the free energy is a scalar, each term in the expansion of F must be a scalar also. Two independent scalars of the second degree can be formed from the components of the symmetrical tensor u_{ik}: they can be taken as the squared sum of the diagonal components (u_{ii}^2) and the sum of the squares of all the components (u_{ik}^2). Expanding F in powers of u_{ik}, we therefore have as far as terms of the second order

$$F = F_0 + \tfrac{1}{2}\lambda u_{ii}^2 + \mu u_{ik}^2. \tag{4.1}$$

This is the general expression for the free energy of a deformed isotropic body. The quantities λ and μ are called *Lamé coefficients*.

We have seen in §1 that the change in volume in the deformation is given by the sum u_{ii}. If this sum is zero, then the volume of the body is unchanged by the deformation, only its shape being altered. Such a deformation is called a *pure shear*.

† For hydrostatic compression, the expression (3.4) becomes $\Phi = F + pu_{ii} = F + p(V - V_0)$, where $V - V_0$ is the volume change resulting from the deformation. Hence we see that the definition of Φ used here differs by a term $-pV_0$ from the usual definition $\Phi = F + pV$.

The opposite case is that of a deformation which causes a change in the volume of the body but no change in its shape. Each volume element of the body retains its shape also. We have seen in §1 that the tensor of such a deformation is $u_{ik} = \text{constant} \times \delta_{ik}$. Such a deformation is called a *hydrostatic compression.*

Any deformation can be represented as the sum of a pure shear and a hydrostatic compression. To do so, we need only use the identity

$$u_{ik} = (u_{ik} - \tfrac{1}{3}\delta_{ik}u_{ll}) + \tfrac{1}{3}\delta_{ik}u_{ll}. \tag{4.2}$$

The first term on the right is evidently a pure shear, since the sum of its diagonal terms is zero ($\delta_{ii} = 3$). The second term is a hydrostatic compression.

As a general expression for the free energy of a deformed isotropic body, it is convenient to replace (4.1) by another formula, using this decomposition of an arbitrary deformation into a pure shear and a hydrostatic compression. We take as the two independent scalars of the second degree the sums of the squared components of the two terms in (4.2). Then F becomes†

$$F = \mu(u_{ik} - \tfrac{1}{3}\delta_{ik}u_{ll})^2 + \tfrac{1}{2}Ku_{ll}^2. \tag{4.3}$$

The quantities K and μ are called respectively the *bulk modulus* or *modulus of hydrostatic compression* (or simply the *modulus of compression*) and the *shear modulus* or *modulus of rigidity*. K is related to the Lamé coefficients by

$$K = \lambda + \tfrac{2}{3}\mu. \tag{4.4}$$

In a state of thermodynamic equilibrium, the free energy is a minimum. If no external forces act on the body, then F as a function of u_{ik} must have a minimum for $u_{ik} = 0$. This means that the quadratic form (4.3) must be positive. If the tensor u_{ik} is such that $u_{ll} = 0$, only the first term remains in (4.3); if, on the other hand, the tensor is of the form $u_{ik} = \text{constant} \times \delta_{ik}$, then only the second term remains. Hence it follows that a necessary (and evidently sufficient) condition for the form (4.3) to be positive is that each of the coefficients K and μ be positive. Thus we conclude that the moduli of compression and rigidity are always positive:

$$K > 0, \ \mu > 0. \tag{4.5}$$

We now use the general thermodynamic relation (3.6) to determine the stress tensor. To calculate the derivatives $\partial F/\partial u_{ik}$, we write the total differential $\mathrm{d}F$ (for constant temperature):

$$\mathrm{d}F = Ku_{ll}\,\mathrm{d}u_{ll} + 2\mu(u_{ik} - \tfrac{1}{3}u_{ll}\delta_{ik})\,\mathrm{d}(u_{ik} - \tfrac{1}{3}u_{ll}\delta_{ik}).$$

In the second term, multiplication of the first parenthesis by δ_{ik} gives zero, leaving $\mathrm{d}F = Ku_{ll}\,\mathrm{d}u_{ll} + 2\mu(u_{ik} - \tfrac{1}{3}u_{ll}\delta_{ik})\,\mathrm{d}u_{ik}$, or writing $\mathrm{d}u_{ll} = \delta_{ik}\,\mathrm{d}u_{ik}$,

$$\mathrm{d}F = [Ku_{ll}\delta_{ik} + 2\mu(u_{ik} - \tfrac{1}{3}u_{ll}\delta_{ik})]\,\mathrm{d}u_{ik}.$$

Hence the stress tensor is

$$\sigma_{ik} = Ku_{ll}\delta_{ik} + 2\mu(u_{ik} - \tfrac{1}{3}\delta_{ik}u_{ll}). \tag{4.6}$$

This expression determines the stress tensor in terms of the strain tensor for an isotropic body. It shows that, if the deformation is a pure shear or a pure hydrostatic compression, the relation between σ_{ik} and u_{ik} is determined only by the modulus of rigidity or of hydrostatic compression respectively.

† The constant term F_0 is the free energy of the undeformed body, and is of no further interest. We shall therefore omit it, for brevity, taking F to be only the free energy of the deformation (the elastic free energy, as it is called).

It is not difficult to obtain the converse formula which expresses u_{ik} in terms of σ_{ik}. To do so, we find the sum σ_{ii} of the diagonal terms. Since this sum is zero for the second term of (4.6), we have $\sigma_{ii} = 3Ku_{ii}$, or

$$u_{ii} = \sigma_{ii}/3K. \tag{4.7}$$

Substituting this expression in (4.6) and so determining u_{ik}, we find

$$u_{ik} = \delta_{ik}\sigma_{ll}/9K + (\sigma_{ik} - \tfrac{1}{3}\delta_{ik}\sigma_{ll})/2\mu, \tag{4.8}$$

which gives the strain tensor in terms of the stress tensor.

Equation (4.7) shows that the relative change in volume (u_{ii}) in any deformation of an isotropic body depends only on the sum σ_{ii} of the diagonal components of the stress tensor, and the relation between u_{ii} and σ_{ii} is determined only by the modulus of hydrostatic compression. In hydrostatic compression of a body, the stress tensor is $\sigma_{ik} = -p\delta_{ik}$. Hence we have in this case, from (4.7),

$$u_{ii} = -p/K. \tag{4.9}$$

Since the deformations are small, u_{ii} and p are small quantities, and we can write the ratio u_{ii}/p of the relative volume change to the pressure in the differential form $(1/V)(\partial V/\partial p)_T$. Thus

$$\frac{1}{K} = -\frac{1}{V}\left(\frac{\partial V}{\partial p}\right)_T.$$

The quantity $1/K$ is called the *coefficient of hydrostatic compression* (or simply the *coefficient of compression*).

We see from (4.8) that the strain tensor u_{ik} is a linear function of the stress tensor σ_{ik}. That is, the deformation is proportional to the applied forces. This law, valid for small deformations, is called *Hooke's law*.†

We may give also a useful form of the expression for the free energy of a deformed body, which is obtained immediately from the fact that F is quadratic in the strain tensor. According to Euler's theorem, $u_{ik}\partial F/\partial u_{ik} = 2F$, whence, since $\partial F/\partial u_{ik} = \sigma_{ik}$, we have

$$F = \tfrac{1}{2}\sigma_{ik}u_{ik}. \tag{4.10}$$

If we substitute in this formula the u_{ik} as linear combinations of the components σ_{ik}, the elastic energy will be represented as a quadratic function of the σ_{ik}. Again applying Euler's theorem, we obtain $\sigma_{ik}\partial F/\partial\sigma_{ik} = 2F$, and a comparison with (4.10) shows that

$$u_{ik} = \partial F/\partial\sigma_{ik}. \tag{4.11}$$

It should be emphasized, however, that, whereas the formula $\sigma_{ik} = \partial F/\partial u_{ik}$ is a general relation of thermodynamics, the inverse formula (4.11) is applicable only if Hooke's law is valid.

§5. Homogeneous deformations

Let us consider some simple cases of what are called *homogeneous deformations*, i.e. those in which the strain tensor is constant throughout the volume of the body.‡ For

† Hooke's law is actually applicable to almost all elastic deformations. The reason is that deformations usually cease to be elastic when they are still so small that Hooke's law is a good approximation. Substances such as rubber form an exception.

‡ The six components of the tensor u_{ik} are not entirely independent, since they are expressed in terms of the derivatives of only three independent functions, the components of the vector **u** (see §7, Problem 9). But the six constants u_{ik} can in principle be specified arbitrarily.

example, the hydrostatic compression already considered is a homogeneous deformation.

We first consider a *simple extension* (or compression) of a rod. Let the rod be along the z-axis, and let forces be applied to its ends which stretch it in both directions. These forces act uniformly over the end surfaces of the rod; let the force on unit area be p.

Since the deformation is homogeneous, i.e. u_{ik} is constant through the body, the stress tensor σ_{ik} is also constant, and so it can be determined at once from the boundary conditions (2.8). There is no external force on the sides of the rod, and therefore $\sigma_{ik}n_k = 0$. Since the unit vector **n** on the side of the rod is perpendicular to the z-axis, i.e. $n_z = 0$, it follows that all the components σ_{ik} except σ_{zz} are zero. On the end surface we have $\sigma_{zi}n_i = p$, or $\sigma_{zz} = p$.

From the general expression (4.8) which relates the components of the strain and stress tensors, we see that all the components u_{ik} with $i \neq k$ are zero. For the remaining components we find

$$u_{xx} = u_{yy} = -\frac{1}{3}\left(\frac{1}{2\mu} - \frac{1}{3K}\right)p, \quad u_{zz} = \frac{1}{3}\left(\frac{1}{3K} + \frac{1}{\mu}\right)p. \tag{5.1}$$

The component u_{zz} gives the relative lengthening of the rod. The coefficient of p is called the *coefficient of extension*, and its reciprocal is the *modulus of extension* or *Young's modulus*, E:

$$u_{zz} = p/E, \tag{5.2}$$

where

$$E = 9K\mu/(3K + \mu). \tag{5.3}$$

The components u_{xx} and u_{yy} give the relative compression of the rod in the transverse direction. The ratio of the transverse compression to the longitudinal extension is called *Poisson's ratio*, σ:†

$$u_{xx} = -\sigma u_{zz}, \tag{5.4}$$

where

$$\sigma = \tfrac{1}{2}(3K - 2\mu)/(3K + \mu). \tag{5.5}$$

Since K and μ are always positive, Poisson's ratio can vary between -1 (for $K = 0$) and $\frac{1}{2}$ (for $\mu = 0$). Thus‡

$$-1 \leqslant \sigma \leqslant \tfrac{1}{2}. \tag{5.6}$$

Finally, the relative increase in the volume of the rod is

$$u_{ii} = p/3K. \tag{5.7}$$

The free energy of a stretched rod can be obtained immediately from formula (4.10). Since only the component σ_{zz} is not zero, we have $F = \frac{1}{2}\sigma_{zz}u_{zz}$, whence

$$F = p^2/2E. \tag{5.8}$$

† The use of σ to denote Poisson's ratio and σ_{ik} to denote the components of the stress tensor cannot lead to ambiguity, since the latter always have suffixes.

‡ In practice, Poisson's ratio varies only between 0 and $\frac{1}{2}$. There are no substances known for which $\sigma < 0$, i.e. which would expand transversely when stretched longitudinally. It may be mentioned that the inequality $\sigma > 0$ corresponds to $\lambda > 0$, where λ is the Lamé coefficient appearing in (4.1); in other words, both terms in (4.1), as well as in (4.3), are always positive in practice, although this is not thermodynamically necessary. Values of σ close to $\frac{1}{2}$ (e.g. for rubber) correspond to a modulus of rigidity which is small compared with the modulus of compression.

In what follows we shall, as is customary, use E and σ instead of K and μ. These and the second Lamé coefficient are given in terms of E and σ by

$$\left.\begin{aligned} \lambda &= E\sigma/(1-2\sigma)(1+\sigma), \\ \mu &= E/2(1+\sigma),\ K = E/3(1-2\sigma). \end{aligned}\right\} \tag{5.9}$$

We shall write out here the general formulae of §4, with the coefficients expressed in terms of E and σ. The free energy is

$$F = \frac{E}{2(1+\sigma)}\left(u_{ik}{}^2 + \frac{\sigma}{1-2\sigma}u_{ll}{}^2\right). \tag{5.10}$$

The stress tensor is given in terms of the strain tensor by

$$\sigma_{ik} = \frac{E}{1+\sigma}\left(u_{ik} + \frac{\sigma}{1-2\sigma}u_{ll}\delta_{ik}\right). \tag{5.11}$$

Conversely,

$$u_{ik} = [(1+\sigma)\sigma_{ik} - \sigma\sigma_{ll}\delta_{ik}]/E. \tag{5.12}$$

Since formulae (5.11) and (5.12) are in frequent use, we shall give them also in component form:

$$\left.\begin{aligned} \sigma_{xx} &= \frac{E}{(1+\sigma)(1-2\sigma)}[(1-\sigma)u_{xx} + \sigma(u_{yy}+u_{zz})], \\ \sigma_{yy} &= \frac{E}{(1+\sigma)(1-2\sigma)}[(1-\sigma)u_{yy} + \sigma(u_{xx}+u_{zz})], \\ \sigma_{zz} &= \frac{E}{(1+\sigma)(1-2\sigma)}[(1-\sigma)u_{zz} + \sigma(u_{xx}+u_{yy})], \\ \sigma_{xy} &= \frac{E}{1+\sigma}u_{xy},\ \sigma_{xz} = \frac{E}{1+\sigma}u_{xz},\ \sigma_{yz} = \frac{E}{1+\sigma}u_{yz}, \end{aligned}\right\} \tag{5.13}$$

and conversely

$$\left.\begin{aligned} u_{xx} &= \frac{1}{E}[\sigma_{xx} - \sigma(\sigma_{yy}+\sigma_{zz})], \\ u_{yy} &= \frac{1}{E}[\sigma_{yy} - \sigma(\sigma_{xx}+\sigma_{zz})], \\ u_{zz} &= \frac{1}{E}[\sigma_{zz} - \sigma(\sigma_{xx}+\sigma_{yy})], \\ u_{xy} &= \frac{1+\sigma}{E}\sigma_{xy},\ u_{xz} = \frac{1+\sigma}{E}\sigma_{xz},\ u_{yz} = \frac{1+\sigma}{E}\sigma_{yz}. \end{aligned}\right\} \tag{5.14}$$

Let us now consider the compression of a rod whose sides are fixed in such a way that they cannot move. The external forces which cause the compression of the rod are applied to its ends and act along its length, which we again take to be along the z-axis. Such a deformation is called a *unilateral compression*. Since the rod is deformed only in the z-direction, only the component u_{zz} of u_{ik} is not zero. Then we have from (5.11)

$$\sigma_{xx} = \sigma_{yy} = \frac{E}{(1+\sigma)(1-2\sigma)}u_{zz},\quad \sigma_{zz} = \frac{E(1-\sigma)}{(1+\sigma)(1-2\sigma)}u_{zz}.$$

Again denoting the compressing force by p ($\sigma_{zz} = p$, which is negative for a compression), we have

$$u_{zz} = p(1+\sigma)(1-2\sigma)/E(1-\sigma). \tag{5.15}$$

The coefficient of p is called the *coefficient of unilateral compression*. For the transverse stresses we have

$$\sigma_{xx} = \sigma_{yy} = p\sigma/(1-\sigma). \tag{5.16}$$

Finally, the free energy of the rod is

$$F = p^2(1+\sigma)(1-2\sigma)/2E(1-\sigma). \tag{5.17}$$

§6. Deformations with change of temperature

Let us now consider deformations which are accompanied by a change in the temperature of the body; this can occur either as a result of the deformation process itself, or from external causes.

We shall regard as the undeformed state the state of the body in the absence of external forces at some given temperature T_0. If the body is at a temperature T different from T_0, then, even if there are no external forces, it will in general be deformed, on account of thermal expansion. In the expansion of the free energy $F(T)$, there will therefore be terms linear, as well as quadratic, in the strain tensor. From the components of the tensor u_{ik}, of rank two, we can form only one linear scalar quantity, the sum u_{ii} of its diagonal components. We shall also assume that the temperature change $T - T_0$ which accompanies the deformation is small. We can then suppose that the coefficient of u_{ii} in the expansion of F (which must vanish for $T = T_0$) is simply proportional to the difference $T - T_0$. Thus we find the free energy to be (instead of (4.3))

$$F(T) = F_0(T) - K\alpha(T-T_0)u_{ll} + \mu(u_{ik} - \tfrac{1}{3}\delta_{ik}u_{ll})^2 + \tfrac{1}{2}Ku_{ll}^2, \tag{6.1}$$

where the coefficient of $T - T_0$ has been written as $-K\alpha$. The quantities μ, K and α can here be supposed constant; an allowance for their temperature dependence would lead to terms of higher order.

Differentiating F with respect to u_{ik}, we obtain the stress tensor:

$$\sigma_{ik} = -K\alpha(T-T_0)\delta_{ik} + Ku_{ll}\delta_{ik} + 2\mu(u_{ik} - \tfrac{1}{3}\delta_{ik}u_{ll}). \tag{6.2}$$

The first term gives the additional stresses caused by the change in temperature. In free thermal expansion of the body (external forces being absent), there can be no internal stresses. Equating σ_{ik} to zero, we find that u_{ik} is of the form constant $\times\ \delta_{ik}$, and

$$u_{ll} = \alpha(T-T_0). \tag{6.3}$$

But u_{ll} is the relative change in volume caused by the deformation. Thus α is just the *thermal expansion coefficient* of the body.

Among the various (thermodynamic) types of deformation, isothermal and adiabatic deformations are of importance. In isothermal deformations, the temperature of the body does not change. Accordingly, we must put $T = T_0$ in (6.1), returning to the usual formulae; the coefficients K and μ may therefore be called *isothermal moduli*.

A deformation is adiabatic if there is no exchange of heat between the various parts of the body (or, of course, between the body and the surrounding medium). The entropy S remains constant. It is the derivative $-\partial F/\partial T$ of the free energy with respect to temperature. Differentiating the expression (6.1), we have as far as terms of the first order in u_{ik}

$$S(T) = S_0(T) + K\alpha u_{ll}. \tag{6.4}$$

Putting S constant, we can determine the change of temperature $T - T_0$ due to the deformation, which is therefore proportional to u_{ll}:

$$C_v(T - T_0)/T_0 = -K\alpha u_{ll}. \tag{6.5}$$

Substituting this expression for $T - T_0$ in (6.2), we obtain for σ_{ik} an expression of the usual kind,

$$\sigma_{ik} = K_{ad} u_{ll}\delta_{ik} + 2\mu(u_{ik} - \tfrac{1}{3}\delta_{ik}u_{ll}), \tag{6.6}$$

with the same modulus of rigidity μ but a different modulus of compression K_{ad}. The relation between the adiabatic modulus K_{ad} and the ordinary isothermal modulus K can also be found directly from the thermodynamic formula

$$\left(\frac{\partial V}{\partial p}\right)_S = \left(\frac{\partial V}{\partial p}\right)_T + \frac{T(\partial V/\partial T)_p^{\,2}}{C_p},$$

where C_p is the specific heat per unit volume at constant pressure. If V is taken to be the volume occupied by matter which before the deformation occupied unit volume, the derivatives $\partial V/\partial T$ and $\partial V/\partial p$ give the relative volume changes in heating and compression respectively. That is,

$$(\partial V/\partial T)_p = \alpha, \quad (\partial V/\partial p)_S = -1/K_{ad}, \quad (\partial V/\partial p)_T = -1/K.$$

Thus we find the relation between the adiabatic and isothermal moduli to be†

$$1/K_{ad} = 1/K - T\alpha^2/C_p, \quad \mu_{ad} = \mu. \tag{6.7}$$

For the adiabatic Young's modulus and Poisson's ratio we easily obtain

$$E_{ad} = \frac{E}{1 - ET\alpha^2/9C_p}, \qquad \sigma_{ad} = \frac{\sigma + ET\alpha^2/9C_p}{1 - ET\alpha^2/9C_p}. \tag{6.8}$$

In practice, $ET\alpha^2/C_p$ is usually small, and it is therefore sufficiently accurate to put

$$E_{ad} = E + E^2T\alpha^2/9C_p, \qquad \sigma_{ad} = \sigma + (1+\sigma)ET\alpha^2/9C_p. \tag{6.9}$$

In isothermal deformation, the stress tensor is given in terms of the derivatives of the free energy:

$$\sigma_{ik} = (\partial F/\partial u_{ik})_T.$$

For constant entropy, on the other hand, we have (see (3.6))

$$\sigma_{ik} = (\partial \mathscr{E}/\partial u_{ik})_S,$$

where $\mathscr{E}$ is the internal energy. Accordingly, the expression analogous to (4.3) determines, for adiabatic deformations, not the free energy but the internal energy per unit volume:

$$\mathscr{E} = \tfrac{1}{2}K_{ad}u_{ll}^{\,2} + \mu(u_{ik} - \tfrac{1}{3}u_{ll}\delta_{ik})^2. \tag{6.10}$$

---

† To derive these formulae from (6.5) and (6.6), we should have to use also the thermodynamic formula $C_p - C_v = T\alpha^2K$.

§7. The equations of equilibrium for isotropic bodies

Let us now derive the equations of equilibrium for isotropic solid bodies. To do so, we substitute in the general equations (2.7)

$$\partial\sigma_{ik}/\partial x_k + \rho g_i = 0$$

the expression (5.11) for the stress tensor. We have

$$\frac{\partial\sigma_{ik}}{\partial x_k} = \frac{E\sigma}{(1+\sigma)(1-2\sigma)}\frac{\partial u_{ll}}{\partial x_i} + \frac{E}{1+\sigma}\frac{\partial u_{ik}}{\partial x_k}.$$

Substituting

$$u_{ik} = \frac{1}{2}\left(\frac{\partial u_i}{\partial x_k} + \frac{\partial u_k}{\partial x_i}\right),$$

we obtain the equations of equilibrium in the form

$$\frac{E}{2(1+\sigma)}\frac{\partial^2 u_i}{\partial x_k{}^2} + \frac{E}{2(1+\sigma)(1-2\sigma)}\frac{\partial^2 u_l}{\partial x_i \partial x_l} + \rho g_i = 0. \tag{7.1}$$

These equations can be conveniently rewritten in vector notation. The quantities $\partial^2 u_i/\partial x_k{}^2$ are components of the vector $\triangle\mathbf{u}$, and $\partial u_l/\partial x_l \equiv \operatorname{div}\mathbf{u}$. Thus the equations of equilibrium become

$$\triangle\mathbf{u} + \frac{1}{1-2\sigma}\,\mathbf{grad}\operatorname{div}\mathbf{u} = -\rho\mathbf{g}\frac{2(1+\sigma)}{E}. \tag{7.2}$$

It is sometimes useful to transform this equation by using the vector identity $\mathbf{grad}\operatorname{div}\mathbf{u} = \triangle\mathbf{u} + \mathbf{curl}\,\mathbf{curl}\,\mathbf{u}$. Then (7.2) becomes

$$\mathbf{grad}\operatorname{div}\mathbf{u} - \frac{1-2\sigma}{2(1-\sigma)}\,\mathbf{curl}\,\mathbf{curl}\,\mathbf{u} = -\rho\mathbf{g}\frac{(1+\sigma)(1-2\sigma)}{E(1-\sigma)}. \tag{7.3}$$

We have written the equations of equilibrium for a uniform gravitational field, since this is the body force most usually encountered in the theory of elasticity. If there are other body forces, the vector $\rho\mathbf{g}$ on the right-hand side of the equation must be replaced accordingly.

A very important case is that where the deformation of the body is caused, not by body forces, but by forces applied to its surface. The equation of equilibrium then becomes

$$(1-2\sigma)\triangle\mathbf{u} + \mathbf{grad}\operatorname{div}\mathbf{u} = 0 \tag{7.4}$$

or

$$2(1-\sigma)\,\mathbf{grad}\operatorname{div}\mathbf{u} - (1-2\sigma)\,\mathbf{curl}\,\mathbf{curl}\,\mathbf{u} = 0. \tag{7.5}$$

The external forces appear in the solution only through the boundary conditions.

Taking the divergence of equation (7.4) and using the identity

$$\operatorname{div}\mathbf{grad} \equiv \triangle,$$

we find

$$\triangle\operatorname{div}\mathbf{u} = 0, \tag{7.6}$$

i.e. $\operatorname{div}\mathbf{u}$ (which determines the volume change due to the deformation) is a harmonic function. Taking the Laplacian of equation (7.4), we then obtain

$$\triangle\triangle \mathbf{u} = 0, \tag{7.7}$$

i.e. in equilibrium the displacement vector satisfies the *biharmonic equation*. These results remain valid in a uniform gravitational field (since the right-hand side of equation (7.2) gives zero on differentiation), but not in the general case of external forces which vary through the body.

The fact that the displacement vector satisfies the biharmonic equation does not, of course, mean that the general integral of the equations of equilibrium (in the absence of body forces) is an arbitrary biharmonic vector; it must be remembered that the function $\mathbf{u}(x, y, z)$ also satisfies the lower-order differential equation (7.4). It is possible, however, to express the general integral of the equations of equilibrium in terms of the derivatives of an arbitrary biharmonic vector (see Problem 10).

If the body is non-uniformly heated, an additional term appears in the equation of equilibrium. The stress tensor must include the term

$$-K\alpha(T - T_0)\delta_{ik}$$

(see (6.2)), and $\partial\sigma_{ik}/\partial x_k$ accordingly contains a term

$$-K\alpha\partial T/\partial x_i = -[E\alpha/3(1-2\sigma)]\partial T/\partial x_i.$$

The equation of equilibrium thus takes the form

$$\frac{3(1-\sigma)}{1+\sigma}\,\mathbf{grad}\,\mathrm{div}\,\mathbf{u} - \frac{3(1-2\sigma)}{2(1+\sigma)}\,\mathbf{curl}\,\mathbf{curl}\,\mathbf{u} = \alpha\,\mathbf{grad}\,T. \tag{7.8}$$

Let us consider the particular case of a *plane deformation*, in which one component of the displacement vector (u_z) is zero throughout the body, while the components u_x, u_y depend only on x and y. The components u_{zz}, u_{xz}, u_{yz} of the strain tensor then vanish identically, and therefore so do the components σ_{xz}, σ_{yz} of the stress tensor (but not the longitudinal stress σ_{zz}, the existence of which is implied by the constancy of the length of the body in the z-direction).†

Since all quantities are independent of the coordinate z, the equations of equilibrium (in the absence of external body forces) $\partial\sigma_{ik}/\partial x_k = 0$ reduce in this case to two equations:

$$\frac{\partial\sigma_{xx}}{\partial x} + \frac{\partial\sigma_{xy}}{\partial y} = 0, \quad \frac{\partial\sigma_{yx}}{\partial x} + \frac{\partial\sigma_{yy}}{\partial y} = 0. \tag{7.9}$$

The most general functions σ_{xx}, σ_{xy}, σ_{yy} satisfying these equations are of the form

$$\sigma_{xx} = \partial^2\chi/\partial y^2, \qquad \sigma_{xy} = -\partial^2\chi/\partial x\partial y, \qquad \sigma_{yy} = \partial^2\chi/\partial x^2, \tag{7.10}$$

where χ is an arbitrary function of x and y. It is easy to obtain an equation which must be satisfied by this function. Such an equation must exist, since the three quantities σ_{xx}, σ_{xy}, σ_{yy} can be expressed in terms of the two quantities u_x, u_y, and are therefore not independent. Using formulae (5.13), we find, for a plane deformation,

$$\sigma_{xx} + \sigma_{yy} = E(u_{xx} + u_{yy})/(1+\sigma)(1-2\sigma).$$

† The use of the theory of functions of a complex variable provides very powerful methods of solving plane problems in the theory of elasticity. See N. I. Muskhelishvili, *Some Basic Problems of the Mathematical Theory of Elasticity*, 2nd English ed., P. Noordhoff, Groningen 1963.

But

$$\sigma_{xx}+\sigma_{yy}=\triangle\chi, \qquad u_{xx}+u_{yy}=\frac{\partial u_x}{\partial x}+\frac{\partial u_y}{\partial y}\equiv \mathrm{div}\,\mathbf{u},$$

and, since by (7.6) div **u** is harmonic, we conclude that the function χ satisfies the equation

$$\triangle\triangle\chi=0, \tag{7.11}$$

i.e. it is biharmonic. This function is called the *stress function*. When the plane problem has been solved and the function χ is known, the longitudinal stress σ_{zz} is determined at once from the formula

$$\sigma_{zz}=\sigma E(u_{xx}+u_{yy})/(1+\sigma)(1-2\sigma)=\sigma(\sigma_{xx}+\sigma_{yy}),$$

or

$$\sigma_{zz}=\sigma\triangle\chi. \tag{7.12}$$

PROBLEMS

PROBLEM 1. Determine the deformation of a long rod (with length l) standing vertically in a gravitational field.

SOLUTION. We take the z-axis along the axis of the rod, and the xy-plane in the plane of its lower end. The equations of equilibrium are $\partial\sigma_{xi}/\partial x_i=\partial\sigma_{yi}/\partial x_i=0$, $\partial\sigma_{zi}/\partial x_i=\rho g$. On the sides of the rod all the components σ_{ik} except σ_{zz} must vanish, and on the upper end ($z=l$) $\sigma_{xz}=\sigma_{yz}=\sigma_{zz}=0$. The solution of the equations of equilibrium satisfying these conditions is $\sigma_{zz}=-\rho g(l-z)$, with all other σ_{ik} zero. From σ_{ik} we find u_{ik} to be $u_{xx}=u_{yy}=\sigma\rho g(l-z)/E$, $u_{zz}=-\rho g(l-z)/E$, $u_{xy}=u_{xz}=u_{yz}=0$, and hence by integration we have the components of the displacement vector, $u_x=\sigma\rho g(l-z)x/E$, $u_y=\sigma\rho g(l-z)y/E$, $u_z=-(\rho g/2E)\{l^2-(l-z)^2-\sigma(x^2+y^2)\}$. The expression for u_z satisfies the boundary condition $u_z=0$ only at one point on the lower end of the rod. Hence the solution obtained is not valid near the lower end.

PROBLEM 2. Determine the deformation of a hollow sphere (with external and internal radii R_2 and R_1) with a pressure p_1 inside and p_2 outside.

SOLUTION. We use spherical polar coordinates, with the origin at the centre of the sphere. The displacement vector **u** is everywhere radial, and is a function of r alone. Hence **curl u** $=0$, and equation (7.5) becomes **grad** div **u** $=0$. Hence

$$\mathrm{div}\,\mathbf{u}=\frac{1}{r^2}\frac{\mathrm{d}(r^2u)}{\mathrm{d}r}=\text{constant}\equiv 3a,$$

or $u=ar+b/r^2$. The components of the strain tensor are (see formulae (1.7)) $u_{rr}=a-2b/r^3$, $u_{\theta\theta}=u_{\phi\phi}=a+b/r^3$. The radial stress is

$$\sigma_{rr}=\frac{E}{(1+\sigma)(1-2\sigma)}\{(1-\sigma)u_{rr}+2\sigma u_{\theta\theta}\}=\frac{E}{1-2\sigma}a-\frac{2E}{1+\sigma}\frac{b}{r^3}.$$

The constants a and b are determined from the boundary conditions: $\sigma_{rr}=-p_1$ at $r=R_1$, and $\sigma_{rr}=-p_2$ at $r=R_2$. Hence we find

$$a=\frac{p_1R_1^3-p_2R_2^3}{R_2^3-R_1^3}\cdot\frac{1-2\sigma}{E}, \qquad b=\frac{R_1^3R_2^3(p_1-p_2)}{R_2^3-R_1^3}\cdot\frac{1+\sigma}{2E}.$$

For example, the stress distribution in a spherical shell with a pressure $p_1=p$ inside and $p_2=0$ outside is given by

$$\sigma_{rr}=\frac{pR_1^3}{R_2^3-R_1^3}\left(1-\frac{R_2^3}{r^3}\right), \qquad \sigma_{\theta\theta}=\sigma_{\phi\phi}=\frac{pR_1^3}{R_2^3-R_1^3}\left(1+\frac{R_2^3}{2r^3}\right).$$

For a thin spherical shell with thickness $h=R_2-R_1\ll R$ we have approximately

$$u=pR^2(1-\sigma)/2Eh, \qquad \sigma_{\theta\theta}=\sigma_{\phi\phi}=\tfrac{1}{2}pR/h, \qquad \bar{\sigma}_{rr}=\tfrac{1}{2}p,$$

where $\bar{\sigma}_{rr}$ is the mean value of the radial stress over the thickness of the shell.

The stress distribution in an infinite elastic medium with a spherical cavity (with radius R) subjected to hydrostatic compression is obtained by putting $R_1 = R$, $R_2 = \infty$, $p_1 = 0$, $p_2 = p$:

$$\sigma_{rr} = -p\left(1-\frac{R^3}{r^3}\right), \qquad \sigma_{\theta\theta} = \sigma_{\phi\phi} = -p\left(1+\frac{R^3}{2r^3}\right).$$

At the surface of the cavity the tangential stresses $\sigma_{\theta\theta} = \sigma_{\phi\phi} = -3p/2$, i.e. they exceed the pressure at infinity.

Problem 3. Determine the deformation of a solid sphere (with radius R) in its own gravitational field.

Solution. The force of gravity on unit mass in a spherical body is $-\mathbf{g}r/R$. Substituting this expression in place of $\mathbf{g}$ in equation (7.3), we obtain the following equation for the radial displacement:

$$\frac{E(1-\sigma)}{(1+\sigma)(1-2\sigma)}\frac{\mathrm{d}}{\mathrm{d}r}\left(\frac{1}{r^2}\frac{\mathrm{d}(r^2u)}{\mathrm{d}r}\right) = \rho g\frac{r}{R}.$$

The solution finite for $r = 0$ which satisfies the condition $\sigma_{rr} = 0$ for $r = R$ is

$$u = -\frac{g\rho R(1-2\sigma)(1+\sigma)}{10E(1-\sigma)}r\left(\frac{3-\sigma}{1+\sigma}-\frac{r^2}{R^2}\right).$$

It should be noticed that the substance is compressed ($u_{rr} < 0$) inside a spherical surface of radius $R\sqrt{\{(3-\sigma)/3(1+\sigma)\}}$ and stretched outside it ($u_{rr} > 0$). The pressure at the centre of the sphere is $(3-\sigma)g\rho R/10(1-\sigma)$.

Problem 4. Determine the deformation of a cylindrical pipe (with external and internal radii R_2 and R_1), with a pressure p inside and no pressure outside.†

Solution. We use cylindrical polar coordinates, with the z-axis along the axis of the pipe. When the pressure is uniform along the pipe, the deformation is a purely radial displacement $u_r = u(r)$. Similarly to Problem 2, we have

$$\operatorname{div}\mathbf{u} = \frac{1}{r}\frac{\mathrm{d}(ru)}{\mathrm{d}r} = \text{constant} \equiv 2a.$$

Hence $u = ar + b/r$. The non-zero components of the strain tensor are (see formulae (1.8)) $u_{rr} = \mathrm{d}u/\mathrm{d}r = a - b/r^2$, $u_{\phi\phi} = u/r = a + b/r^2$. From the conditions $\sigma_{rr} = 0$ at $r = R_2$, and $\sigma_{rr} = -p$ at $r = R_1$, we find

$$a = \frac{pR_1^2}{R_2^2-R_1^2}\cdot\frac{(1+\sigma)(1-2\sigma)}{E}, \qquad b = \frac{pR_1^2R_2^2}{R_2^2-R_1^2}\cdot\frac{1+\sigma}{E}.$$

The stress distribution is given by the formulae

$$\sigma_{rr} = \frac{pR_1^2}{R_2^2-R_1^2}\left(1-\frac{R_2^2}{r^2}\right), \qquad \sigma_{\phi\phi} = \frac{pR_1^2}{R_2^2-R_1^2}\left(1+\frac{R_2^2}{r^2}\right),$$

$$\sigma_{zz} = 2p\sigma R_1^2/(R_2^2-R_1^2).$$

Problem 5. Determine the deformation of a cylinder rotating uniformly about its axis.

Solution. Replacing the gravitational force in (7.3) by the centrifugal force $\rho\Omega^2 r$ (where Ω is the angular velocity), we have in cylindrical polar coordinates the following equation for the displacement $u_r = u(r)$:

$$\frac{E(1-\sigma)}{(1+\sigma)(1-2\sigma)}\frac{\mathrm{d}}{\mathrm{d}r}\left(\frac{1}{r}\frac{\mathrm{d}(ru)}{\mathrm{d}r}\right) = -\rho\Omega^2 r.$$

The solution which is finite for $r = 0$ and satisfies the condition $\sigma_{rr} = 0$ for $r = R$ is

$$u = \frac{\rho\Omega^2(1+\sigma)(1-2\sigma)}{8E(1-\sigma)}r[(3-2\sigma)R^2 - r^2].$$

Problem 6. Determine the deformation of a non-uniformly heated sphere with a spherically symmetrical temperature distribution.

† In Problems 4, 5 and 7 it is assumed that the length of the cylinder is maintained constant, so that there is no longitudinal deformation.

SOLUTION. In spherical polar coordinates, equation (7.8) for a purely radial deformation is

$$\frac{d}{dr}\left(\frac{1}{r^2}\frac{d(r^2u)}{dr}\right) = \alpha\frac{1+\sigma}{3(1-\sigma)}\frac{dT}{dr}.$$

The solution which is finite for $r = 0$ and satisfies the condition $\sigma_{rr} = 0$ for $r = R$ is

$$u = \alpha\frac{1+\sigma}{3(1-\sigma)}\left\{\frac{1}{r^2}\int_0^r T(r)r^2\,dr + \frac{2(1-2\sigma)}{1+\sigma}\frac{r}{R^3}\int_0^R T(r)r^2\,dr\right\}.$$

The temperature $T(r)$ is measured from the value for which the sphere, if uniformly heated, is regarded as undeformed. In the above formula the temperature in question is taken as that of the outer surface of the sphere, so that $T(R) = 0$.

PROBLEM 7. The same as Problem 6, but for a non-uniformly heated cylinder with an axially symmetrical temperature distribution.

SOLUTION. We similarly have in cylindrical polar coordinates

$$u = \alpha\frac{1+\sigma}{3(1-\sigma)}\left\{\frac{1}{r}\int_0^r T(r)r\,dr + (1-2\sigma)\frac{r}{R^2}\int_0^R T(r)r\,dr\right\}.$$

PROBLEM 8. Determine the deformation of an infinite elastic medium with a given temperature distribution $T(x, y, z)$ which is such that the temperature tends to a constant value T_0 at infinity, there being no deformation there.

SOLUTION. Equation (7.8) has an obvious solution for which $\mathbf{curl}\,\mathbf{u} = 0$ and

$$\operatorname{div}\mathbf{u} = \alpha(1+\sigma)[T(x, y, z) - T_0]/3(1-\sigma).$$

The vector $\mathbf{u}$, whose divergence is a given function defined in all space and vanishing at infinity, and whose curl is zero identically, can be written, as we know from vector analysis, in the form

$$\mathbf{u}(x, y, z) = -\frac{1}{4\pi}\,\mathbf{grad}\int\frac{\operatorname{div}'\mathbf{u}(x', y', z')}{r}\,dV',$$

where

$$r = \sqrt{\{(x-x')^2 + (y-y')^2 + (z-z')^2\}}.$$

We therefore obtain the general solution of the problem in the form

$$\mathbf{u} = -\frac{\alpha(1+\sigma)}{12\pi(1-\sigma)}\,\mathbf{grad}\int\frac{T'-T_0}{r}\,dV', \tag{1}$$

where $T' \equiv T(x', y', z')$.

If a finite quantity of heat q is evolved in a very small volume at the origin, the temperature distribution can be written $T - T_0 = (q/C)\delta(x)\delta(y)\delta(z)$, where C is the specific heat of the medium. The integral in (1) is then q/Cr, and the deformation is given by

$$\mathbf{u} = \frac{\alpha(1+\sigma)q}{12\pi(1-\sigma)C}\cdot\frac{\mathbf{r}}{r^3}.$$

PROBLEM 9. Derive the equations of equilibrium for an isotropic body (in the absence of body forces) in terms of the components of the stress tensor.

SOLUTION. The required system of equations contains the three equations

$$\partial\sigma_{ik}/\partial x_k = 0 \tag{1}$$

and also the equations resulting from the fact that the six different components of u_{ik} are not independent quantities. To derive these equations, we first write down the system of differential relations satisfied by the components of the tensor u_{ik}. It is easy to see that the quantities

$$u_{ik} = \frac{1}{2}\left(\frac{\partial u_i}{\partial x_k} + \frac{\partial u_k}{\partial x_i}\right)$$

satisfy identically the relations

$$\frac{\partial^2 u_{ik}}{\partial x_l \partial x_m} + \frac{\partial^2 u_{lm}}{\partial x_i \partial x_k} = \frac{\partial^2 u_{il}}{\partial x_k \partial x_m} + \frac{\partial^2 u_{km}}{\partial x_i \partial x_l}.$$

Here there are only six essentially different relations, namely those corresponding to the following values of i, k, l, m: 1122, 1133, 2233, 1123, 2213, 3312. All these are retained if the above tensor equation is contracted with respect to l and m:

$$\triangle u_{ik} + \frac{\partial^2 u_{ll}}{\partial x_i \partial x_k} = \frac{\partial^2 u_{il}}{\partial x_k \partial x_l} + \frac{\partial^2 u_{kl}}{\partial x_i \partial x_l}. \tag{2}$$

Substituting here u_{ik} in terms of σ_{ik} according to (5.12) and using (1), we obtain the required equations:

$$(1+\sigma)\triangle \sigma_{ik} + \frac{\partial^2 \sigma_{ll}}{\partial x_i \partial x_k} = 0. \tag{3}$$

These equations remain valid in the presence of external forces constant throughout the body.

Contracting equation (3) with respect to the suffixes i and k, we find that $\triangle \sigma_{ll} = 0$, i.e. σ_{ll} is a harmonic function. Taking the Laplacian of equation (3), we then find that $\triangle\triangle \sigma_{ik} = 0$, i.e. the components σ_{ik} are biharmonic functions. These results follow also from (7.6) and (7.7), since σ_{ik} and u_{ik} are linearly related.

PROBLEM 10. Express the general integral of the equations of equilibrium (in the absence of body forces) in terms of an arbitrary biharmonic vector. (B. G. Galerkin 1930).

SOLUTION. It is natural to seek a solution of equation (7.4) in the form

$$\mathbf{u} = \triangle \mathbf{f} + A\, \mathbf{grad}\, \mathrm{div}\, \mathbf{f}.$$

Hence $\mathrm{div}\, \mathbf{u} = (1+A)\mathrm{div}\, \triangle \mathbf{f}$. Substituting in (7.4), we obtain

$$(1-2\sigma)\triangle\triangle \mathbf{f} + [2(1-\sigma)A + 1]\, \mathbf{grad}\, \mathrm{div}\, \triangle \mathbf{f} = 0.$$

From this we see that, if $\mathbf{f}$ is an arbitrary biharmonic vector ($\triangle\triangle \mathbf{f} = 0$), then

$$\mathbf{u} = \triangle \mathbf{f} - \frac{1}{2(1-\sigma)}\mathbf{grad}\, \mathrm{div}\, \mathbf{f}.$$

PROBLEM 11. Express the stresses σ_{rr}, $\sigma_{\phi\phi}$, $\sigma_{r\phi}$ for a plane deformation (in polar coordinates r, ϕ) as derivatives of the stress function.

SOLUTION. Since the required expressions cannot depend on the choice of the initial line of ϕ, they do not contain ϕ explicitly. Hence we can proceed as follows: we transform the Cartesian derivatives (7.10) into derivatives with respect to r, ϕ, and use the results that $\sigma_{rr} = (\sigma_{xx})_{\phi=0}$, $\sigma_{\phi\phi} = (\sigma_{yy})_{\phi=0}$, $\sigma_{r\phi} = (\sigma_{xy})_{\phi=0}$, the angle ϕ being measured from the x-axis. Thus

$$\sigma_{rr} = \frac{1}{r}\frac{\partial \chi}{\partial r} + \frac{1}{r^2}\frac{\partial^2 \chi}{\partial \phi^2}, \qquad \sigma_{\phi\phi} = \frac{\partial^2 \chi}{\partial r^2}, \qquad \sigma_{r\phi} = -\frac{\partial}{\partial r}\left(\frac{1}{r}\frac{\partial \chi}{\partial \phi}\right).$$

PROBLEM 12. Determine the stress distribution in an infinite elastic medium containing a spherical cavity and subjected to a homogeneous deformation at infinity.

SOLUTION. A general homogeneous deformation can be represented as a combination of a homogeneous hydrostatic extension (or compression) and a homogeneous shear. The former has been considered in Problem 2, so that we need only consider a homogeneous shear.

Let $\sigma_{ik}^{(0)}$ be the homogeneous stress field which would be found in all space if the cavity were absent: in a pure shear $\sigma_{ii}^{(0)} = 0$. The corresponding displacement vector is denoted by $\mathbf{u}^{(0)}$, and we seek the required solution in the form $\mathbf{u} = \mathbf{u}^{(0)} + \mathbf{u}^{(1)}$, where the function $\mathbf{u}^{(1)}$ arising from the presence of the cavity is zero at infinity.

Any solution of the biharmonic equation can be written as a linear combination of centrally symmetrical solutions and their spatial derivatives of various orders. The functions r^2, r, 1, $1/r$ are independent centrally symmetrical solutions. Hence the most general form of a biharmonic vector $\mathbf{u}^{(1)}$, depending only on the components of the constant tensor $\sigma_{ik}^{(0)}$ as parameters and vanishing at infinity, is

$$u_i^{(1)} = A\sigma_{ik}^{(0)}\frac{\partial}{\partial x_k}\left(\frac{1}{r}\right) + B\sigma_{kl}^{(0)}\frac{\partial^3}{\partial x_i \partial x_k \partial x_l}\left(\frac{1}{r}\right) + C\sigma_{kl}^{(0)}\frac{\partial^3}{\partial x_i \partial x_k \partial x_l} r. \tag{1}$$

Substituting this expression in equation (7.4), we obtain

$$(1-2\sigma)\frac{\partial^2 u_i}{\partial x_l^2} + \frac{\partial}{\partial x_i}\frac{\partial u_l}{\partial x_l} = [2(1-2\sigma)C + (A+2C)]\sigma_{kl}^{(0)}\frac{\partial^3}{\partial x_i \partial x_k \partial x_l}\frac{1}{r} = 0,$$

whence $A = -4C(1-\sigma)$. Two further relations between the constants A, B, C are obtained from the condition at the surface of the cavity: $(\sigma_{ik}^{(0)} + \sigma_{ik}^{(1)})n_k = 0$ for $r = R$ (R being the radius of the cavity, the origin at its centre, and $\mathbf{n}$ a unit vector parallel to $\mathbf{r}$). A somewhat lengthy calculation, using (1), gives the following values:

$$B = CR^2/5, \qquad C = 5R^3(1+\sigma)/2E(7-5\sigma).$$

The final expression for the stress distribution is

$$\sigma_{ik} = \sigma_{ik}^{(0)}\left\{1 + \frac{5(1-2\sigma)}{7-5\sigma}\left(\frac{R}{r}\right)^3 + \frac{3}{7-5\sigma}\left(\frac{R}{r}\right)^5\right\} +$$
$$+ \frac{15}{7-5\sigma}\left(\frac{R}{r}\right)^3\left\{\sigma - \left(\frac{R}{r}\right)^2\right\}(\sigma_{il}^{(0)}n_k n_l + \sigma_{kl}^{(0)}n_i n_l) +$$
$$+ \frac{15}{2(7-5\sigma)}\left(\frac{R}{r}\right)^3\left\{-5 + 7\left(\frac{R}{r}\right)^2\right\}\sigma_{lm}^{(0)}n_l n_m n_i n_k +$$
$$+ \frac{15}{2(7-5\sigma)}\left(\frac{R}{r}\right)^3\left\{1 - 2\sigma - \left(\frac{R}{r}\right)^2\right\}\delta_{ik}\sigma_{lm}^{(0)}n_l n_m.$$

In order to obtain the stress distribution for arbitrary $\sigma_{ik}^{(0)}$ (not a pure shear), $\sigma_{ik}^{(0)}$ in this expression must be replaced by $\sigma_{ik}^{(0)} - \frac{1}{3}\delta_{ik}\sigma_{ll}^{(0)}$, and the expression

$$\tfrac{1}{3}\sigma_{ll}^{(0)}\left[\delta_{ik} + \frac{R^3}{2r^3}(\delta_{ik} - 3n_i n_k)\right]$$

corresponding to a deformation homogeneous at infinity (cf. Problem 2) must be added. We may give here the general formula for the stresses at the surface of the cavity:

$$\sigma_{ik} = \frac{15}{7-5\sigma}\left\{(1-\sigma)(\sigma_{ik}^{(0)} - \sigma_{il}^{(0)}n_l n_k - \sigma_{kl}^{(0)}n_l n_i) +\right.$$
$$\left.+ \sigma_{lm}^{(0)}n_l n_m n_i n_k - \sigma\sigma_{lm}^{(0)}n_l n_m \delta_{ik} + \frac{5\sigma - 1}{10}\sigma_{ll}^{(0)}(\delta_{ik} - n_i n_k)\right\}.$$

Near the cavity, the stresses considerably exceed the stresses at infinity, but this extends over only a short distance (the *concentration of stresses*). For example, if the medium is subjected to a homogeneous extension (only $\sigma_{zz}^{(0)}$ different from zero), the greatest stress occurs on the equator of the cavity, where

$$\sigma_{zz} = \frac{27 - 15\sigma}{2(7-5\sigma)}\sigma_{zz}^{(0)}.$$

§8. Equilibrium of an elastic medium bounded by a plane

Let us consider an elastic medium occupying a half-space, i.e. bounded on one side by an infinite plane, and determine the deformation of the medium caused by forces applied to its free surface.† The distribution of these forces need satisfy only one condition: they must vanish at infinity in such a way that there is no deformation at infinity. In such a case the equations of equilibrium can be integrated in a general form (J. Boussinesq 1885).

The equation of equilibrium (7.4) holds throughout the space occupied by the medium:

$$\mathbf{grad}\,\mathrm{div}\,\mathbf{u} + (1-2\sigma)\triangle\mathbf{u} = 0. \tag{8.1}$$

We seek a solution of this equation in the form

$$\mathbf{u} = \mathbf{f} + \mathbf{grad}\,\phi, \tag{8.2}$$

where ϕ is some scalar and the vector $\mathbf{f}$ satisfies Laplace's equation:

$$\triangle\mathbf{f} = 0. \tag{8.3}$$

† The most direct and regular method of solving this problem is to use Fourier's method on equation (8.1). In that case, however, some fairly complicated integrals have to be calculated. The method given below is based on a number of artificial devices, but the calculations are simpler.

Substituting (8.2) in (8.1), we then obtain the following equation for ϕ:

$$2(1-\sigma)\triangle\phi = -\operatorname{div}\mathbf{f}. \tag{8.4}$$

We take the free surface of the elastic medium as the xy-plane; the medium is in $z > 0$. We write the functions f_x and f_y as the z-derivatives of some functions g_x and g_y:

$$f_x = \partial g_x/\partial z, \qquad f_y = \partial g_y/\partial z. \tag{8.5}$$

Since f_x and f_y are harmonic functions, we can always choose the functions g_x and g_y so as to satisfy Laplace's equation:

$$\triangle g_x = 0, \qquad \triangle g_y = 0. \tag{8.6}$$

Equation (8.4) then becomes

$$2(1-\sigma)\triangle\phi = -\frac{\partial}{\partial z}\left(\frac{\partial g_x}{\partial x}+\frac{\partial g_y}{\partial y}+f_z\right).$$

Since g_x, g_y and f_z are harmonic functions, we easily see that a function ϕ which satisfies this equation can be written as

$$\phi = -\frac{z}{4(1-\sigma)}\left(f_z+\frac{\partial g_x}{\partial x}+\frac{\partial g_y}{\partial y}\right)+\psi, \tag{8.7}$$

where ψ is again a harmonic function:

$$\triangle\psi = 0. \tag{8.8}$$

Thus the problem of determining the displacement $\mathbf{u}$ reduces to that of finding the functions g_x, g_y, f_z, ψ, all of which satisfy Laplace's equation.

We shall now write out the boundary conditions which must be satisfied at the free surface of the medium (the plane $z = 0$). Since the unit outward normal vector $\mathbf{n}$ is in the negative z-direction, it follows from the general formula (2.9) that $\sigma_{iz} = -P_i$. Using for σ_{ik} the general expression (5.11) and expressing the components of the vector $\mathbf{u}$ in terms of the auxiliary quantities g_x, g_y, f_z and ψ, we obtain after a simple calculation the boundary conditions

$$\left.\begin{aligned}
&\left[\frac{\partial^2 g_x}{\partial z^2}\right]_{z=0}+\left[\frac{\partial}{\partial x}\left\{\frac{1-2\sigma}{2(1-\sigma)}f_z-\frac{1}{2(1-\sigma)}\left(\frac{\partial g_x}{\partial x}+\frac{\partial g_y}{\partial y}\right)+2\frac{\partial\psi}{\partial z}\right\}\right]_{z=0}\\
&\qquad = -2(1+\sigma)P_x/E,\\
&\left[\frac{\partial^2 g_y}{\partial z^2}\right]_{z=0}+\left[\frac{\partial}{\partial v}\left\{\frac{1-2\sigma}{2(1-\sigma)}f_z-\frac{1}{2(1-\sigma)}\left(\frac{\partial g_x}{\partial x}+\frac{\partial g_y}{\partial y}\right)+2\frac{\partial\psi}{\partial z}\right\}\right]_{z=0}\\
&\qquad = -2(1+\sigma)P_y/E,
\end{aligned}\right\} \tag{8.9}$$

$$\left[\frac{\partial}{\partial z}\left\{f_z-\left(\frac{\partial g_x}{\partial x}+\frac{\partial g_y}{\partial y}\right)+2\frac{\partial\psi}{\partial z}\right\}\right]_{z=0} = -2(1+\sigma)P_z/E. \tag{8.10}$$

The components P_x, P_y, P_z of the external forces applied to the surface are given functions of the coordinates x and y, and vanish at infinity.

The formulae by which the auxiliary quantities g_x, g_y, f_z and ψ were defined do not determine them uniquely. We can therefore impose an arbitrary additional condition on these quantities, and it is convenient to make the quantity in the braces in equations (8.9) vanish:†

$$(1-2\sigma)f_z - \left(\frac{\partial g_x}{\partial x} + \frac{\partial g_y}{\partial y}\right) + 4(1-\sigma)\frac{\partial \psi}{\partial z} = 0. \tag{8.11}$$

Then the conditions (8.9) become simply

$$\left[\frac{\partial^2 g_x}{\partial z^2}\right]_{z=0} = -\frac{2(1+\sigma)}{E}P_x, \quad \left[\frac{\partial^2 g_y}{\partial z^2}\right]_{z=0} = -\frac{2(1+\sigma)}{E}P_y. \tag{8.12}$$

Equations (8.10)–(8.12) suffice to determine completely the harmonic functions g_x, g_y, f_z and ψ.

For simplicity, we shall consider the case where the free surface of an elastic half-space is subjected to a concentrated force **F**, i.e. one which is applied to an area so small that it can be regarded as a point. The effect of this force is the same as that of surface forces given by $\mathbf{P} = \mathbf{F}\delta(x)\delta(y)$, the origin being at the point of application of the force. If we know the solution for a concentrated force, we can immediately find the solution for any force distribution $\mathbf{P}(x, y)$. For, if

$$u_i = G_{ik}(x, y, z)F_k \tag{8.13}$$

is the displacement due to the action of a concentrated force **F** applied at the origin, then the displacement caused by forces $\mathbf{P}(x, y)$ is given by the integral‡

$$u_i = \iint G_{ik}(x-x', y-y', z)\, P_k(x', y')\,\mathrm{d}x'\,\mathrm{d}y'. \tag{8.14}$$

We know from potential theory that a harmonic function f which is zero at infinity and has a given normal derivative $\partial f/\partial z$ on the plane $z = 0$ is given by the formula

$$f(x, y, z) = -\frac{1}{2\pi}\iint\left[\frac{\partial f(x', y', z)}{\partial z}\right]_{z=0}\frac{\mathrm{d}x'\,\mathrm{d}y'}{r},$$

where

$$r = \sqrt{\{(x-x')^2 + (y-y')^2 + z^2\}}.$$

Since the quantities $\partial g_x/\partial z$, $\partial g_y/\partial z$ and that in the braces in equation (8.10) satisfy Laplace's equation, while equations (8.10) and (8.12) determine the values of their normal derivatives on the plane $z = 0$, we have

$$\begin{aligned} f_z - \left(\frac{\partial g_x}{\partial x} + \frac{\partial g_y}{\partial y}\right) + 2\frac{\partial \psi}{\partial z} &= \frac{1+\sigma}{\pi E}\iint\frac{P_z(x', y')}{r}\,\mathrm{d}x'\,\mathrm{d}y' \\ &= \frac{1+\sigma}{\pi E}\cdot\frac{F_z}{r}, \end{aligned} \tag{8.15}$$

$$\frac{\partial g_x}{\partial z} = \frac{1+\sigma}{\pi E}\cdot\frac{F_x}{r}, \quad \frac{\partial g_y}{\partial z} = \frac{1+\sigma}{\pi E}\cdot\frac{F_y}{r}, \tag{8.16}$$

where now $r = \sqrt{(x^2 + y^2 + z^2)}$.

† We shall not prove here that this condition can in fact be imposed; this follows from the absence of contradiction in the result.

‡ In mathematical terms, G_{ik} is the Green's tensor for the equations of equilibrium of a semi-infinite medium.

The expressions for the components of the required vector **u** involve the derivatives of g_x, g_y with respect to x, y, z, but not g_x, g_y themselves. To calculate $\partial g_x/\partial x$, $\partial g_y/\partial y$, we differentiate equations (8.16) with respect to x and y respectively:

$$\frac{\partial^2 g_x}{\partial x \partial z} = -\frac{1+\sigma}{\pi E}\cdot\frac{F_x x}{r^3}, \qquad \frac{\partial^2 g_y}{\partial y \partial z} = -\frac{1+\sigma}{\pi E}\cdot\frac{F_y y}{r^3}.$$

Now, integrating over z from ∞ to z, we obtain

$$\left.\begin{aligned}\frac{\partial g_x}{\partial x} &= \frac{1+\sigma}{\pi E}\cdot\frac{F_x x}{r(r+z)},\\ \frac{\partial g_y}{\partial y} &= \frac{1+\sigma}{\pi E}\cdot\frac{F_y y}{r(r+z)}.\end{aligned}\right\} \qquad (8.17)$$

We shall not pause to complete the remaining calculations, which are elementary but laborious. We determine f_z and $\partial\psi/\partial z$ from equations (8.11), (8.15) and (8.17). Knowing $\partial\psi/\partial z$, it is easy to calculate $\partial\psi/\partial x$ and $\partial\psi/\partial y$ by integrating with respect to z and then differentiating with respect to x and y. We thus obtain all the quantities needed to calculate the displacement vector from (8.2), (8.5) and (8.7). The following are the final formulae:

$$\left.\begin{aligned}u_x &= \frac{1+\sigma}{2\pi E}\left\{\left[\frac{xz}{r^3} - \frac{(1-2\sigma)x}{r(r+z)}\right]F_z + \frac{2(1-\sigma)r+z}{r(r+z)}F_x + \right.\\ &\qquad\left. + \frac{[2r(\sigma r+z)+z^2]x}{r^3(r+z)^2}(xF_x+yF_y)\right\},\\ u_y &= \frac{1+\sigma}{2\pi E}\left\{\left[\frac{yz}{r^3} - \frac{(1-2\sigma)y}{r(r+z)}\right]F_z + \frac{2(1-\sigma)r+z}{r(r+z)}F_y + \right.\\ &\qquad\left. + \frac{[2r(\sigma r+z)+z^2]y}{r^3(r+z)^2}(xF_x+yF_y)\right\},\\ u_z &= \frac{1+\sigma}{2\pi E}\left\{\left[\frac{2(1-\sigma)}{r} + \frac{z^2}{r^3}\right]F_z + \left[\frac{1-2\sigma}{r(r+z)} + \frac{z}{r^3}\right](xF_x+yF_y)\right\}.\end{aligned}\right\} \qquad (8.18)$$

In particular, the displacement of points on the surface of the medium is given by putting $z = 0$:

$$\left.\begin{aligned}u_x &= \frac{1+\sigma}{2\pi E}\cdot\frac{1}{r}\left\{-\frac{(1-2\sigma)x}{r}F_z + 2(1-\sigma)F_x + \frac{2\sigma x}{r^2}(xF_x+yF_y)\right\},\\ u_y &= \frac{1+\sigma}{2\pi E}\cdot\frac{1}{r}\left\{-\frac{(1-2\sigma)y}{r}F_z + 2(1-\sigma)F_y + \frac{2\sigma y}{r^2}(xF_x+yF_y)\right\},\\ u_z &= \frac{1+\sigma}{2\pi E}\cdot\frac{1}{r}\left\{2(1-\sigma)F_z + (1-2\sigma)\frac{1}{r}(xF_x+yF_y)\right\}.\end{aligned}\right\} \qquad (8.19)$$

PROBLEM

Determine the deformation of an infinite elastic medium when a force **F** is applied to a small region in it (W. Thomson 1848).†

† The corresponding problem for an arbitrary infinite anisotropic medium has been solved by I. M. Lifshitz and L. N. Rozentsveĭg (*Zhurnal eksperimental'noĭ i teoreticheskoĭ fiziki* **17**, 783, 1947).

SOLUTION. If we consider the deformation at distances r which are large compared with the dimension of the region where the force is applied, we can suppose that the force is applied at a point. The equation of equilibrium is (cf. (7.2)).

$$\triangle \mathbf{u} + \frac{1}{1-2\sigma}\,\mathbf{grad}\,\mathrm{div}\,\mathbf{u} = -\frac{2(1+\sigma)}{E}\mathbf{F}\delta(\mathbf{r}), \tag{1}$$

where $\delta(\mathbf{r}) \equiv \delta(x)\delta(y)\delta(z)$, the origin being at the point where the force is applied. We seek the solution in the form $\mathbf{u} = \mathbf{u}_0 + \mathbf{u}_1$, where $\mathbf{u}_0$ satisfies the Poisson-type equation

$$\triangle \mathbf{u}_0 = -\frac{2(1+\sigma)}{E}\mathbf{F}\delta(\mathbf{r}). \tag{2}$$

We then have for $\mathbf{u}_1$ the equation

$$\mathbf{grad}\,\mathrm{div}\,\mathbf{u}_1 + (1-2\sigma)\triangle \mathbf{u}_1 = -\mathbf{grad}\,\mathrm{div}\,\mathbf{u}_0. \tag{3}$$

The solution of equation (2) which vanishes at infinity is $\mathbf{u}_0 = (1+\sigma)\mathbf{F}/2\pi E r$. Taking the curl of equation (3), we have $\triangle\,\mathbf{curl}\,\mathbf{u}_1 = 0$. At infinity we must have $\mathbf{curl}\,\mathbf{u}_1 = 0$. But a function harmonic in all space and zero at infinity must be zero identically. Thus $\mathbf{curl}\,\mathbf{u}_1 = 0$, and we can therefore write $\mathbf{u}_1 = \mathbf{grad}\,\phi$. From (3) we obtain $\mathbf{grad}\,\{2(1-\sigma)\triangle\phi + \mathrm{div}\,\mathbf{u}_0\} = 0$. Hence it follows that the quantity in braces is a constant, and it must be zero at infinity; we therefore have in all space

$$\triangle\phi = -\frac{\mathrm{div}\,\mathbf{u}_0}{2(1-\sigma)} = -\frac{1+\sigma}{4\pi E(1-\sigma)}\mathbf{F}\cdot\mathbf{grad}\left(\frac{1}{r}\right).$$

If ψ is a solution of the equation $\triangle\psi = 1/r$, then

$$\phi = -\frac{1+\sigma}{4\pi E(1-\sigma)}\mathbf{F}\cdot\mathbf{grad}\,\psi.$$

Taking the solution $\psi = \frac{1}{2}r$, which has no singularities, we obtain

$$\mathbf{u}_1 = \mathbf{grad}\,\phi = \frac{1+\sigma}{8\pi E(1-\sigma)}\,\frac{(\mathbf{F}\cdot\mathbf{n})\mathbf{n}-\mathbf{F}}{r},$$

where $\mathbf{n}$ is a unit vector parallel to the position vector $\mathbf{r}$. The final result is

$$\mathbf{u} = \frac{1+\sigma}{8\pi E(1-\sigma)}\cdot\frac{(3-4\sigma)\mathbf{F}+\mathbf{n}(\mathbf{n}\cdot\mathbf{F})}{r}.$$

On putting this formula into the form (8.13) we obtain the Green's tensor for the equations of equilibrium of an infinite isotropic medium:†

$$G_{ik} = \frac{1+\sigma}{8\pi E(1-\sigma)}[(3-4\sigma)\delta_{ik} + n_i n_k]\frac{1}{r}$$
$$= \frac{1}{4\pi\mu}\left[\frac{\delta_{ik}}{r} - \frac{1}{4(1-\sigma)}\frac{\partial^2 r}{\partial x_i \partial x_k}\right].$$

§9. Solid bodies in contact

Let two solid bodies be in contact at a point which is not a singular point on either surface. Fig. 1a shows a cross-section of the two surfaces near the point of contact O. The surfaces have a common tangent plane at O, which we take as the xy-plane. We regard the positive z-direction as being into either body (i.e. in opposite directions for the two bodies) and denote the corresponding coordinates by z and z'.

† The fact that the components of the tensor G_{ik} are first-order homogeneous functions of the coordinates x, y, z is evident from arguments of homogeneity applied to the form of equation (1), where the left-hand side is a linear combination of the second derivatives of the components of the vector $\mathbf{u}$, and the right-hand side is a third-order homogeneous function ($\delta(a\mathbf{r}) = a^{-3}\delta(\mathbf{r})$).

This property remains valid in the general case of an arbitrary anisotropic medium.

Near a point of ordinary contact with the xy-plane, the equation of the surface can be written

$$z = \kappa_{\alpha\beta} x_\alpha x_\beta, \tag{9.1}$$

where summation is understood over the values 1, 2, of the repeated suffixes α, β ($x_1 = x$, $x_2 = y$), and $\kappa_{\alpha\beta}$ is a symmetrical tensor of rank two, which characterizes the curvature of the surface: the principal values of the tensor $\kappa_{\alpha\beta}$ are $1/2R_1$ and $1/2R_2$, where R_1 and R_2 are the principal radii of curvature of the surface at the point of contact. A similar relation for the surface of the other body near the point of contact can be written

$$z' = \kappa'_{\alpha\beta} x_\alpha x_\beta. \tag{9.2}$$

Let us now assume that the two bodies are pressed together by applied forces, and approach a short distance h.† Then a deformation occurs near the original point of contact, and the two bodies will be in contact over a small but finite portion of their surfaces. Let u_z and u'_z be the components (along the z and z' axes respectively) of the corresponding displacement vectors for points on the surfaces of the two bodies. The broken lines in Fig. 1b show the surfaces as they would be in the absence of any deformation, while the continuous lines show the surfaces of the deformed bodies; the letters z and z' denote the distances given by equations (9.1) and (9.2). It is seen at once from the figure that the equation

$$(z + u_z) + (z' + u'_z) = h,$$

or

$$(\kappa_{\alpha\beta} + \kappa'_{\alpha\beta}) x_\alpha x_\beta + u_z + u'_z = h, \tag{9.3}$$

holds everywhere in the region of contact. At points outside the region of contact, we have

$$z + z' + u_z + u'_z < h.$$

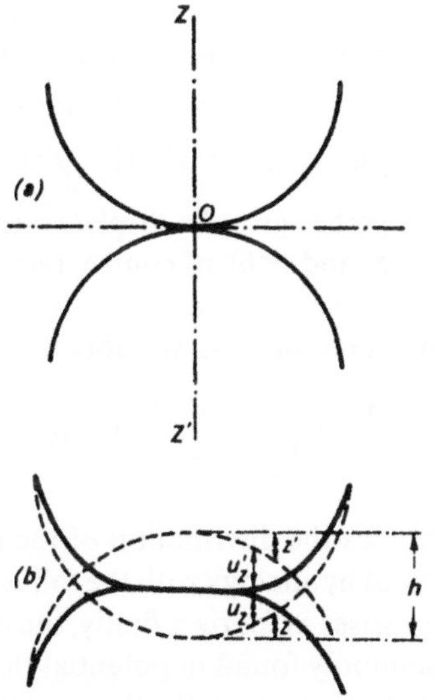

FIG. 1

† This *contact problem* in the theory of elasticity was first solved by H. Hertz (1882).

We choose the x and y axes to be the principal axes of the tensor $\kappa_{\alpha\beta}+\kappa'_{\alpha\beta}$. Denoting the principal values of this tensor by A and B, we can rewrite equation (9.3) as

$$Ax^2+By^2+u_z+u'_z=h. \tag{9.4}$$

The quantities A and B are related to the radii of curvature R_1, R_2 and R'_1, R'_2 by formulae which will be given without proof:

$$2(A+B)=\frac{1}{R_1}+\frac{1}{R_2}+\frac{1}{R'_1}+\frac{1}{R'_2},$$

$$4(A-B)^2=\left(\frac{1}{R_1}-\frac{1}{R_2}\right)^2+\left(\frac{1}{R'_1}-\frac{1}{R'_2}\right)^2+2\cos 2\phi\left(\frac{1}{R_1}-\frac{1}{R_2}\right)\left(\frac{1}{R'_1}-\frac{1}{R'_2}\right),$$

where ϕ is the angle between the normal sections whose radii of curvature are R_1 and R'_1. The radii of curvature are regarded as positive if the centre of curvature lies within the body concerned, and negative in the contrary case.

We denote by $P_z(x, y)$ the pressure between the two deformed bodies at points in the region of contact; outside this region, of course, $P_z=0$. To determine the relation between P_z and the displacements u_z, u'_z, we can with sufficient accuracy regard the surfaces as plane and use the formulae obtained in §8. According to the third of formulae (8.19) and (8.14), the displacement u_z under the action of normal forces $P_z(x, y)$ is given by

$$\left.\begin{aligned}u_z&=\frac{1-\sigma^2}{\pi E}\iint\frac{P_z(x',y')}{r}\,dx'\,dy',\\ u'_z&=\frac{1-\sigma'^2}{\pi E'}\iint\frac{P_z(x',y')}{r}\,dx'\,dy',\end{aligned}\right\} \tag{9.5}$$

where σ, σ' and E, E' are the Poisson's ratios and the Young's moduli of the two bodies. Since $P_z=0$ outside the region of contact, the integration extends only over this region. It may be noted that, from these formulae, the ratio u_z/u'_z is constant:

$$u_z/u'_z=(1-\sigma^2)E'/(1-\sigma'^2)E. \tag{9.6}$$

The relations (9.4) and (9.6) together give the displacements u_z, u'_z at every point of the region of contact (although (9.5) and (9.6), of course, relate to points outside that region also).

Substituting the expressions (9.5) in (9.4), we obtain

$$\frac{1}{\pi}\left(\frac{1-\sigma^2}{E}+\frac{1-\sigma'^2}{E'}\right)\iint\frac{P_z(x',y')}{r}\,dx'\,dy'=h-Ax^2-By^2. \tag{9.7}$$

This integral equation determines the distribution of the pressure P_z over the region of contact. Its solution can be found by analogy with the following results of potential theory. The idea of using this analogy arises as follows: firstly, the integral on the left-hand side of equation (9.7) is of a type commonly found in potential theory, where such integrals give the potential of a charge distribution; secondly, the potential inside a uniformly charged ellipsoid is a quadratic function of the coordinates.

If the ellipsoid $x^2/a^2+y^2/b^2+z^2/c^2=1$ is uniformly charged (with volume charge density ρ), the potential in the ellipsoid is given by

$\phi(x, y, z)$

$$= \pi\rho abc \int_0^\infty \left\{1 - \frac{x^2}{a^2+\xi} - \frac{y^2}{b^2+\xi} - \frac{z^2}{c^2+\xi}\right\} \frac{d\xi}{\sqrt{\{(a^2+\xi)(b^2+\xi)(c^2+\xi)\}}}.$$

In the limiting case of an ellipsoid which is very much flattened in the z-direction ($c \to 0$), we have

$$\phi(x, y) = \pi\rho abc \int_0^\infty \left\{1 - \frac{x^2}{a^2+\xi} - \frac{y^2}{b^2+\xi}\right\} \frac{d\xi}{\sqrt{\{(a^2+\xi)(b^2+\xi)\xi\}}};$$

in passing to the limit $c \to 0$ we must, of course, put $z = 0$ for points inside the ellipsoid. The potential $\phi(x, y, z)$ can also be written as

$$\phi(x, y, z) = \iiint \frac{\rho\, dx'\, dy'\, dz'}{\sqrt{\{(x-x')^2 + (y-y')^2 + (z-z')^2\}}},$$

where the integration is over the volume of the ellipsoid. In passing to the limit $c \to 0$, we must put $z = z' = 0$ in the radicand; integrating over z' between the limits

$$\pm c\sqrt{\{1 - (x'^2/a^2) - (y'^2/b^2)\}},$$

we obtain

$$\phi(x, y) = 2\rho c \iint \frac{dx'\, dy'}{r} \sqrt{\left(1 - \frac{x'^2}{a^2} - \frac{y'^2}{b^2}\right)},$$

where

$$r = \sqrt{\{(x-x')^2 + (y-y')^2\}},$$

and the integration is over the area inside the ellipse

$$x'^2/a^2 + y'^2/b^2 = 1.$$

Equating the two expressions for $\phi(x, y)$, we obtain the identity

$$\iint \frac{dx'\, dy'}{r} \sqrt{\left(1 - \frac{x'^2}{a^2} - \frac{y'^2}{b^2}\right)}$$

$$= \tfrac{1}{2}\pi ab \int_0^\infty \left(1 - \frac{x^2}{a^2+\xi} - \frac{y^2}{b^2+\xi}\right) \frac{d\xi}{\sqrt{\{(a^2+\xi)(b^2+\xi)\xi\}}}. \qquad (9.8)$$

Comparing this relation with equation (9.7), we see that the right-hand sides are quadratic functions of x and y of the same form, and the left-hand sides are integrals of the same form. We can therefore deduce immediately that the region of contact (i.e. the region of integration in (9.7)) is bounded by an ellipse of the form

$$\frac{x^2}{a^2} + \frac{y^2}{b^2} = 1 \qquad (9.9)$$

and that the function $P_z(x, y)$ must be of the form

$$P_z(x, y) = \text{constant} \times \sqrt{\left(1 - \frac{x^2}{a^2} - \frac{y^2}{b^2}\right)}.$$

Taking the constant such that the integral $\iint P_z \, dx \, dy$ over the region of contact is equal to the given total force F which moves the bodies together, we obtain

$$P_z(x, y) = \frac{3F}{2\pi ab}\sqrt{\left(1 - \frac{x^2}{a^2} - \frac{y^2}{b^2}\right)}. \tag{9.10}$$

This formula gives the distribution of pressure over the area of the region of contact. It may be pointed out that the pressure at the centre of this region is $\frac{3}{2}$ times the mean pressure $F/\pi ab$.

Substituting (9.10) in equation (9.7) and replacing the resulting integral in accordance with (9.8), we obtain

$$\frac{FD}{\pi}\int_0^\infty \left(1 - \frac{x^2}{a^2+\xi} - \frac{y^2}{b^2+\xi}\right) d\xi / \sqrt{\{(a^2+\xi)(b^2+\xi)\xi\}}$$

$$= h - Ax^2 - By^2,$$

where

$$D = \frac{3}{4}\left(\frac{1-\sigma^2}{E} + \frac{1-\sigma'^2}{E'}\right).$$

This equation must hold identically for all values of x and y inside the ellipse (9.9); the coefficients of x and y and the free terms must therefore be respectively equal on each side. Hence we find

$$h = \frac{FD}{\pi}\int_0^\infty \frac{d\xi}{\sqrt{\{(a^2+\xi)(b^2+\xi)\xi\}}}, \tag{9.11}$$

$$\left.\begin{aligned} A &= \frac{FD}{\pi}\int_0^\infty \frac{d\xi}{(a^2+\xi)\sqrt{\{(a^2+\xi)(b^2+\xi)\xi\}}}, \\ B &= \frac{FD}{\pi}\int_0^\infty \frac{d\xi}{(b^2+\xi)\sqrt{\{(a^2+\xi)(b^2+\xi)\xi\}}}. \end{aligned}\right\} \tag{9.12}$$

Equations (9.12) determine the semi-axes a and b of the region of contact from the given force F (A and B being known for given bodies). The relation (9.11) then gives the distance of approach h as a function of the force F. The right-hand sides of these equations involve elliptic integrals.

Thus the problem of bodies in contact can be regarded as completely solved. The form of the surfaces (i.e. the displacements u_z, u'_z) outside the region of contact is determined by the same formulae (9.5) and (9.10); the values of the integrals can be found immediately from the analogy with the potential outside a charged ellipsoid. Finally, the formulae of §8 enable us to find also the deformation at various points in the bodies (but only, of course, at distances small compared with the dimensions of the bodies).

Let us apply these formulae to the case of contact between two spheres with radii R and R'. Here $A = B = 1/2R + 1/2R'$. It is clear from symmetry that $a = b$, i.e. the region of contact is a circle. From (9.12) we find the radius a of this circle to be

$$a = F^{1/3}\{DRR'/(R+R')\}^{1/3}. \tag{9.13}$$

h is in this case the difference between the sum $R+R'$ and the distance between the centres of the spheres. From (9.10) we obtain the following relation between F and h:

$$h = F^{2/3}\left[D^2\left(\frac{1}{R}+\frac{1}{R'}\right)\right]^{1/3}. \tag{9.14}$$

It should be noticed that h is proportional to $F^{2/3}$; conversely, the force F varies as $h^{3/2}$. We can write down also the potential energy U of the spheres in contact. Since $-F = -\partial U/\partial h$, we have

$$U = h^{5/2}\frac{2}{5D}\sqrt{\frac{RR'}{R+R'}}. \tag{9.15}$$

Finally, it may be mentioned that a relation of the form $h = \text{constant}\times F^{2/3}$, or $F = \text{constant}\times h^{3/2}$, holds not only for spheres but also for other finite bodies in contact. This is easily seen from similarity arguments. If we make the substitution

$$a^2 \to \alpha a^2, \qquad b^2 \to \alpha b^2, \qquad F \to \alpha^{3/2}F,$$

where α is an arbitrary constant, equations (9.12) remain unchanged. In equation (9.11), the right-hand side is multiplied by α, and so h must be replaced by αh if this equation is to remain unchanged. Hence it follows that F must be proportional to $h^{3/2}$.

PROBLEMS

Problem 1. Determine the time for which two colliding elastic spheres remain in contact.

Solution. In a system of coordinates in which the centre of mass of the two spheres is at rest, the energy before the collision is equal to the kinetic energy of the relative motion $\frac{1}{2}\mu v^2$, where v is the relative velocity of the colliding spheres and $\mu = m_1m_2/(m_1+m_2)$ their reduced mass. During the collision, the total energy is the sum of the kinetic energy, which may be written $\frac{1}{2}\mu\dot{h}^2$, and the potential energy (9.15). By the law of conservation of energy we have

$$\mu\left(\frac{dh}{dt}\right)^2 + kh^{5/2} = \mu v^2, \qquad k = \frac{4}{5D}\sqrt{\frac{RR'}{R+R'}}.$$

The maximum approach h_0 of the spheres corresponds to the time when their relative velocity $\dot{h} = 0$, and is $h_0 = (\mu/k)^{2/5}v^{4/5}$.

The time τ during which the collision takes place (i.e. h varies from 0 to h_0 and back) is

$$\tau = 2\int_0^{h_0}\frac{dh}{\sqrt{(v^2-kh^{5/2}/\mu)}} = 2\left(\frac{\mu^2}{k^2v}\right)^{1/5}\int_0^1\frac{dx}{\sqrt{(1-x^{2/5})}},$$

or

$$\tau = \frac{4\sqrt{\pi}\,\Gamma(2/5)}{5\Gamma(9/10)}\left(\frac{\mu^2}{k^2v}\right)^{1/5} = 2{\cdot}94\left(\frac{\mu^2}{k^2v}\right)^{1/5}.$$

By using the statical formulae obtained in the text to solve this problem, we have neglected elastic oscillations of the spheres resulting from the collision. If this is legitimate, the velocity v must be small compared with the velocity of sound. In practice, however, the validity of the theory is limited by the still more stringent requirement that the resulting deformations should not exceed the elastic limit of the substance.

Problem 2. Determine the dimensions of the region of contact and the pressure distribution when two cylinders are pressed together along a generator.

Solution. In this case the region of contact is a narrow strip along the length of the cylinders. Its width $2a$ and the pressure distribution across it can be found from the formulae in the text by going to the limit $b/a \to \infty$. The pressure distribution will be of the form $P_z(x) = \text{constant}\times\sqrt{(1-x^2/a^2)}$, where x is the coordinate across the strip; normalizing the pressure to give a force F per unit length, we obtain

$$P_z(x) = \frac{2F}{\pi a}\sqrt{\left(1-\frac{x^2}{a^2}\right)}.$$

Substituting this expression in (9.7) and effecting the integration by means of (9.8), we have

$$A = \frac{4DF}{3\pi}\int_0^\infty \frac{d\xi}{(a^2+\xi)^{3/2}\xi} = \frac{8DF}{3\pi a^2}.$$

One of the radii of curvature of a cylindrical surface is infinite, and the other is the radius of the cylinder; in this case, therefore, $A = 1/2R + 1/2R'$, $B = 0$. We have finally for the width of the region of contact

$$a = \sqrt{\left(\frac{16DF}{3\pi}\cdot\frac{RR'}{R+R'}\right)}.$$

§10. The elastic properties of crystals

The change in the free energy in isothermal compression of a crystal is, as with isotropic bodies, a quadratic function of the strain tensor. Unlike what happens for isotropic bodies, however, this function contains not just two coefficients, but a larger number of them. The general form of the free energy of a deformed crystal is

$$F = \tfrac{1}{2}\lambda_{iklm}u_{ik}u_{lm}, \tag{10.1}$$

where λ_{iklm} is a tensor of rank four, called the *elastic modulus tensor*. Since the strain tensor is symmetrical, the product $u_{ik}u_{lm}$ is unchanged when the suffixes *i*, *k*, or *l*, *m*, or *i*, *l* and *k*, *m*, are interchanged. Hence we see that the tensor λ_{iklm} can be defined so that it has the same symmetry properties:

$$\lambda_{iklm} = \lambda_{kilm} = \lambda_{ikml} = \lambda_{lmik}. \tag{10.2}$$

A simple calculation shows that the number of different components of a tensor of rank four having these symmetry properties is in general 21.†

In accordance with the expression (10.1) for the free energy, the stress tensor for a crystal is given in terms of the strain tensor by

$$\sigma_{ik} = \partial F/\partial u_{ik} = \lambda_{iklm}u_{lm}; \tag{10.3}$$

cf. also the last footnote to this section.

If the crystal possesses symmetry, relations exist between the various components of the tensor λ_{iklm}, so that the number of independent components is less than 21.

We shall discuss these relations for each possible type of macroscopic symmetry of crystals, i.e. for each of the crystal classes, dividing these into the corresponding crystal systems (see *SP* 1, §§130, 131).

(1) *Triclinic system.* Triclinic symmetry (classes $\mathbf{C}_1$ and $\mathbf{C}_i$) does not place any restrictions on the components of the tensor λ_{iklm}, and the system of coordinates may be chosen arbitrarily as regards the symmetry. All the 21 moduli of elasticity are non-zero and independent. However, the arbitrariness of the choice of coordinate system enables us to impose additional conditions on the components of the tensor λ_{iklm}. Since the orientation of the coordinate system relative to the body is defined by three quantities (angles of rotation), there can be three such conditions; for example, three of the components may be taken as zero. Then the independent quantities which describe the elastic properties of the crystal will be 18 non-zero moduli and 3 angles defining the orientation of the axes in the crystal.

† Another notation used for λ_{iklm} in the literature is $\lambda_{\alpha\beta}$, with α and β taking values from 1 to 6 in correspondence with *xx*, *yy*, *zz*, *yz*, *zx*, *xy*.

(2) *Monoclinic system.* Let us consider the class C_s; we take a coordinate system with the xy-plane as the plane of symmetry. On reflection in this plane, the coordinates undergo the transformation $x \to x$, $y \to y$, $z \to -z$. The components of a tensor are transformed as the products of the corresponding coordinates. It is therefore clear that, in the transformation mentioned, all components λ_{iklm} whose suffixes include z an odd number of times (1 or 3) will change sign, while the other components will remain unchanged. By the symmetry of the crystal, however, all quantities characterizing its properties (including all components λ_{iklm}) must remain unchanged on reflection in the plane of symmetry. Hence it is evident that all components with an odd number of suffixes z must be zero. Accordingly, the general expression for the elastic free energy of a crystal belonging to the monoclinic system is

$$F = \tfrac{1}{2}\lambda_{xxxx}u_{xx}^2 + \tfrac{1}{2}\lambda_{yyyy}u_{yy}^2 + \tfrac{1}{2}\lambda_{zzzz}u_{zz}^2 + \lambda_{xxyy}u_{xx}u_{yy} + \lambda_{xxzz}u_{xx}u_{zz} + \\ + \lambda_{yyzz}u_{yy}u_{zz} + 2\lambda_{xyxy}u_{xy}^2 + 2\lambda_{xzxz}u_{xz}^2 + 2\lambda_{yzyz}u_{yz}^2 + 2\lambda_{xxxy}u_{xx}u_{xy} + \\ + 2\lambda_{yyyx}u_{yy}u_{yx} + 2\lambda_{xyzz}u_{xy}u_{zz} + 4\lambda_{xzyz}u_{xz}u_{yz}. \tag{10.4}$$

This contains 13 independent coefficients. A similar expression is obtained for the class C_2, and also for the class C_{2h}, which contains both symmetry elements (C_2 and σ_h). In the argument given, however, the direction of only one coordinate axis (that of z) is fixed; those of x and y can have arbitrary directions in the perpendicular plane. This arbitrariness can be used to make one coefficient, say λ_{xyzz}, vanish by a suitable choice of axes. Then the 13 quantities which describe the elastic properties of the crystal will be 12 non-zero moduli and one angle defining the orientation of the axes in the xy-plane.

(3) *Orthorhombic system.* In all the classes of this system (C_{2v}, D_2, D_{2h}) the choice of coordinate axes is determined by the symmetry, and the expression obtained for the free energy is the same for each class.

Let us consider, for example, the class D_{2h}; we take the three planes of symmetry as the coordinate planes. Reflections in each of these planes are transformations in which one coordinate changes sign and the other two remain unchanged. It is evident therefore that the only non-zero components λ_{iklm} are those whose suffixes contain each of x, y, z an even number of times; the other components would have to change sign on reflection in some plane of symmetry. Thus the general expression for the free energy in the orthorhombic system is

$$F = \tfrac{1}{2}\lambda_{xxxx}u_{xx}^2 + \tfrac{1}{2}\lambda_{yyyy}u_{yy}^2 + \tfrac{1}{2}\lambda_{zzzz}u_{zz}^2 + \lambda_{xxyy}u_{xx}u_{yy} + \lambda_{xxzz}u_{xx}u_{zz} + \\ + \lambda_{yyzz}u_{yy}u_{zz} + 2\lambda_{xyxy}u_{xy}^2 + 2\lambda_{xzxz}u_{xz}^2 + 2\lambda_{yzyz}u_{yz}^2. \tag{10.5}$$

It contains nine moduli of elasticity.

(4) *Tetragonal system.* Let us consider the class C_{4v}; we take the axis C_4 as the z-axis, and the x and y axes perpendicular to two of the vertical planes of symmetry. Reflections in these two planes signify tranformations

$$x \to -x, \qquad y \to y, \qquad z \to z$$

and

$$x \to x, \qquad y \to -y, \qquad z \to z;$$

all components λ_{iklm} with an odd number of like suffixes therefore vanish. Furthermore, a rotation through an angle $\tfrac{1}{2}\pi$ about the axis C_4 is the transformation

$$x \to y, \qquad y \to -x, \qquad z \to z.$$

Hence we have

$$\lambda_{xxxx} = \lambda_{yyyy}, \qquad \lambda_{xxzz} = \lambda_{yyzz}, \qquad \lambda_{xzxz} = \lambda_{yzyz}.$$

The remaining transformations in the class C_{4v} do not give any further conditions. Thus the free energy of crystals in the tetragonal system is

$$F = \tfrac{1}{2}\lambda_{xxxx}(u_{xx}{}^2 + u_{yy}{}^2) + \tfrac{1}{2}\lambda_{zzzz}u_{zz}{}^2 + \lambda_{xxzz}(u_{xx}u_{zz} + u_{yy}u_{zz}) + \\ + \lambda_{xxyy}u_{xx}u_{yy} + 2\lambda_{xyxy}u_{xy}{}^2 + 2\lambda_{xzxz}(u_{xz}{}^2 + u_{yz}{}^2). \qquad (10.6)$$

It contains six moduli of elasticity.

A similar result is obtained for those other classes of the tetragonal system where the natural choice of the coordinate axes is determined by symmetry (D_{2d}, D_4, D_{4h}). In the classes C_4, S_4, C_{4h}, on the other hand, only the choice of the z-axis is unique (along the axis C_4 or S_4). The requirements of symmetry then allow a further component $\lambda_{xxxy} = -\lambda_{yyyx}$ in addition to those which appear in (10.6). These components may be made to vanish by suitably choosing the directions of the x and y axes, and F then reduces to the form (10.6).

(5) ***Rhombohedral system.*** Let us consider the class C_{3v}; we take the third-order axis as the z-axis, and the y-axis perpendicular to one of the vertical planes of symmetry. In order to find the restrictions imposed on the components of the tensor λ_{iklm} by the presence of the axis C_3, it is convenient to make a formal transformation using the complex coordinates $\xi = x + iy$, $\eta = x - iy$, the z coordinate remaining unchanged. We transform the tensor λ_{iklm} to the new coordinate system also, so that its suffixes take the values ξ, η, z. It is easy to see that, in a rotation through $2\pi/3$ about the axis C_3, the new coordinates undergo the transformation $\xi \to \xi e^{2\pi i/3}$, $\eta \to \eta e^{-2\pi i/3}$, $z \to z$. By symmetry only those components λ_{iklm} which are unchanged by this transformation can be different from zero. These components are evidently the ones whose suffixes contain ξ three times, or η three times (since $(e^{2\pi i/3})^3 = e^{2\pi i} = 1$), or ξ and η the same number of times (since $e^{2\pi i/3}e^{-2\pi i/3} = 1$), i.e. λ_{zzzz}, $\lambda_{\xi\eta\xi\eta}$, $\lambda_{\xi\xi\eta\eta}$, $\lambda_{\xi\eta zz}$, $\lambda_{\xi z\eta z}$, $\lambda_{\xi\xi\xi z}$, $\lambda_{\eta\eta\eta z}$. Furthermore, a reflection in the symmetry plane perpendicular to the y-axis gives the transformation $x \to x$, $y \to -y$, $z \to z$, or $\xi \to \eta$, $\eta \to \xi$. Since $\lambda_{\xi\xi\xi z}$ becomes $\lambda_{\eta\eta\eta z}$ in this transformation, these two components must be equal. Thus crystals of the rhombohedral system have only six moduli of elasticity. In order to obtain an expression for the free energy, we must form the sum $\frac{1}{2}\lambda_{iklm}u_{ik}u_{lm}$, in which the suffixes take the values ξ, η, z; since F is to be expressed in terms of the components of the strain tensor in the coordinates x, y, z, we must express in terms of these the components in the coordinates ξ, η, z. This is easily done by using the fact that the components of the tensor u_{ik} transform as the products of the corresponding coordinates. For example, since

$$\xi^2 = (x + iy)^2 = x^2 - y^2 + 2ixy,$$

it follows that

$$u_{\xi\xi} = u_{xx} - u_{yy} + 2iu_{xy}.$$

Consequently, the expression for F is found to be

$$F = \tfrac{1}{2}\lambda_{zzzz}u_{zz}{}^2 + 2\lambda_{\xi\eta\xi\eta}(u_{xx} + u_{yy})^2 + \lambda_{\xi\xi\eta\eta}\{(u_{xx} - u_{yy})^2 + 4u_{xy}{}^2\} + \\ + 2\lambda_{\xi\eta zz}(u_{xx} + u_{yy})u_{zz} + 4\lambda_{\xi z\eta z}(u_{xz}{}^2 + u_{yz}{}^2) + 4\lambda_{\xi\xi\xi z}\{(u_{xx} - u_{yy})u_{xz} - 2u_{xy}u_{yz}\}. \qquad (10.7)$$

This contains 6 independent coefficients. A similar result is obtained for the classes D_3 and D_{3d}, but in the classes C_3 and S_6, where the choice of the x and y axes remains arbitrary, requirements of symmetry allow also a non-zero value of the difference $\lambda_{\xi\xi\xi z}-\lambda_{\eta\eta\eta z}$. This, however, can be made to vanish by a suitable choice of the x and y axes.

(6) *Hexagonal system.* Let us consider the class C_6; we take the sixth-order axis as the z-axis, and again use the coordinates $\xi = x+iy$, $\eta = x-iy$. In a rotation through an angle $\frac{1}{3}\pi$ about the z-axis, the coordinates ξ, η undergo the transformation $\xi \to \xi e^{\pi i/3}$, $\eta \to \eta e^{-\pi i/3}$. Hence we see that only those components λ_{iklm} are non-zero which contain the same number of suffixes ξ and η. These are λ_{zzzz}, $\lambda_{\xi\eta\xi\eta}$, $\lambda_{\xi\xi\eta\eta}$, $\lambda_{\xi\eta zz}$, $\lambda_{\xi z\eta z}$. Other symmetry elements in the hexagonal system give no further restrictions. There are therefore only five moduli of elasticity. The free energy is

$$F = \tfrac{1}{2}\lambda_{zzzz}u_{zz}^2 + 2\lambda_{\xi\eta\xi\eta}(u_{xx}+u_{yy})^2 + \lambda_{\xi\xi\eta\eta}[(u_{xx}-u_{yy})^2+4u_{xy}^2] + \\ + 2\lambda_{\xi\eta zz}u_{zz}(u_{xx}+u_{yy}) + 4\lambda_{\xi z\eta z}(u_{xz}^2+u_{yz}^2). \tag{10.8}$$

It should be noticed that a deformation in the xy-plane (for which u_{xx}, u_{yy} and u_{xy} are non-zero) is determined by only two moduli of elasticity, as for an isotropic body; that is, the elastic properties of a hexagonal crystal are isotropic in the plane perpendicular to the sixth-order axis.

For this reason the choice of axis directions in this plane is unimportant and does not affect the form of F. The expression (10.8) therefore applies to all classes of the hexagonal system.

(7) *Cubic system.* We take the axes along the three fourth-order axes of the cubic system. Since there is tetragonal symmetry (with the fourth-order axis in the z-direction), the number of different components of the tensor λ_{iklm} is limited to at most the following six: λ_{xxxx}, λ_{zzzz}, λ_{xxzz}, λ_{xxyy}, λ_{xyxy}, λ_{xzxz}.† Rotations through $\frac{1}{2}\pi$ about the x and y axes give respectively the transformations $x\to x$, $y\to -z$, $z\to y$, and $x\to z$, $y\to y$, $z\to -x$. The components listed are therefore equal in successive pairs. Thus there remain only three different moduli of elasticity. The free energy of crystals of the cubic system is

$$F = \tfrac{1}{2}\lambda_{xxxx}(u_{xx}^2+u_{yy}^2+u_{zz}^2) + \lambda_{xxyy}(u_{xx}u_{yy}+u_{xx}u_{zz}+u_{yy}u_{zz}) + \\ + 2\lambda_{xyxy}(u_{xy}^2+u_{xz}^2+u_{yz}^2). \tag{10.9}$$

We may recapitulate the number of independent parameters (elastic moduli or angles defining the orientation of axes in the crystal) for the classes of the various systems:

Triclinic	21	Rhombohedral (C_3, S_6)	7
Monoclinic	13	Rhombohedral (C_{3v}, D_3, D_{3d})	6
Orthorhombic	9	Hexagonal	5
Tetragonal (C_4, S_4, C_{4h})	7	Cubic	3
Tetragonal (C_{4v}, D_{2d}, D_4, D_{4h})	6		

The least number of non-zero moduli that is possible by suitable choice of the coordinate axes is the same for all the classes in each system:

Triclinic	18	Rhombohedral	6
Monoclinic	12	Hexagonal	5
Orthorhombic	9	Cubic	3
Tetragonal	6		

† In the cubic classes T and T_d there are no fourth-order axes. The same result is, however, obtained in these cases by considering the third-order axes, rotations about which convert the x, y, z axes into one another.

All the above discussion relates, of course, to single crystals. Polycrystalline bodies whose component crystallites (grains) are sufficiently small may be regarded as isotropic bodies (since we are concerned with deformations in regions large compared with the dimensions of the crystallites). Like any isotropic body, a polycrystal has only two moduli of elasticity. It might be thought at first sight that these moduli could be obtained from those of the individual crystallites by simple averaging. This is not so, however. If we regard the deformation of a polycrystal as the result of a deformation of its component crystallites, it would in principle be necessary to solve the equations of equilibrium for every crystallite, taking into account the appropriate boundary conditions at their surfaces of separation. Hence we see that the relation between the elastic properties of the whole crystal and those of its component crystallites depends on the actual form of the latter and the amount of correlation of their mutual orientations. There is therefore no general relation between the moduli of elasticity of a polycrystal and those of a single crystal of the same substance.

The moduli of an isotropic polycrystal can be calculated with fair accuracy from those of a single crystal only when the elastic properties of the single crystal are nearly isotropic.† In a first approximation, the moduli of elasticity of the polycrystal can then simply be put equal to the "isotropic part" of the moduli of the single crystal. In the next approximation, terms appear which are quadratic in the small "anisotropic part" of these moduli. It is found‡ that these correction terms are independent of the shape of the crystallites and of the correlation of their orientations, and can be calculated in a general form.

Finally, let us consider the thermal expansion of crystals. In isotropic bodies, the thermal expansion is the same in every direction, so that the strain tensor in free thermal expansion is (see §6) $u_{ik} = \frac{1}{3}\alpha(T-T_0)\delta_{ik}$, where α is the thermal expansion coefficient. In crystals, however, we must put

$$u_{ik} = \tfrac{1}{3}\alpha_{ik}(T-T_0), \tag{10.10}$$

where α_{ik} is a tensor of rank two, symmetrical in the suffixes i and k. Let us calculate the number of independent components of this tensor in crystals of the various systems. The simplest way of doing this is to use the result of tensor algebra that to every symmetrical tensor of rank two there corresponds a *tensor ellipsoid*§. It follows at once from considerations of symmetry that, for triclinic, monoclinic and orthorhombic symmetry, the tensor ellipsoid has three axes of different length. For tetragonal, rhombohedral and hexagonal symmetry, on the other hand, we have an ellipsoid of revolution (with its axis of symmetry along the axes C_4, C_3 and C_6 respectively). Finally, for cubic symmetry the ellipsoid becomes a sphere. An ellipsoid of three axes is determined by three quantities, an ellipsoid of revolution by two, and a sphere by one (the radius). Thus the number of independent components of the tensor α_{ik} in crystals of the various systems is as follows: triclinic, monoclinic and orthorhombic, 3; tetragonal, rhombohedral and hexagonal, 2; cubic, 1.

Crystals of the first three systems are said to be *biaxial*, and those of the second three systems *uniaxial*. It should be noticed that the thermal expansion of crystals of the cubic system is determined by one quantity only, i.e. they behave in this respect as isotropic bodies.

---

† For example, a measure of the anisotropy of the elastic properties of a cubic crystal is the difference $\lambda_{xxxx} - \lambda_{xxyy} - 2\lambda_{xyxy}$; if this is zero, then (10.9) reduces to the expression (4.3) for the elastic energy of an isotropic body.

‡ I. M. Lifshitz and L. N. Rozentsveĭg, *Zhurnal èksperimental'noĭ i teoreticheskoĭ fiziki* **16**, 967, 1946.

§ Determined by the equation $\alpha_{ik}x_ix_k = 1$.

PROBLEMS

PROBLEM 1. Express the elastic energy of a hexagonal crystal in terms of the elastic moduli λ_{iklm} in the coordinates x, y, z (the x-axis being the sixth-order axis).

SOLUTION. For a general (not orthogonal) transformation of the coordinates, we have to distinguish the contravariant and covariant components of vectors and tensors, which are respectively transformed as the coordinates x^i themselves (and denoted by superscripts) and as the differentiation operators $\partial/\partial x^i$ (and denoted by subscripts). The scalar (10.1) is then to be written as

$$F = \tfrac{1}{2}\lambda_{iklm}u^{ik}u^{lm}.$$

In the expressions (10.8) and (10.9), the components u_{ik} are transformed as contravariant ones. To establish the relation between the components λ_{iklm} in the coordinates ξ, η, z and x, y, z, they are to be regarded as covariant; in Cartesian coordinates the two sets are of course the same. For the transformation of (10.7),

$$\frac{\partial}{\partial x} = \frac{\partial}{\partial \xi} + \frac{\partial}{\partial \eta}, \qquad \frac{\partial}{\partial y} = i\left(\frac{\partial}{\partial \xi} - \frac{\partial}{\partial \eta}\right).$$

Transforming the λ_{iklm} as products of these operators gives

$$\lambda_{xxxx} = \lambda_{yyyy} = 4\lambda_{\xi\eta\xi\eta} + 2\lambda_{\xi\xi\eta\eta},$$
$$\lambda_{xyxy} = 2\lambda_{\xi\xi\eta\eta},$$
$$\lambda_{xxyy} = 4\lambda_{\xi\eta\xi\eta} - 2\lambda_{\xi\xi\eta\eta},$$
$$\lambda_{xxzz} = \lambda_{yyzz} = 2\lambda_{\xi\eta zz},$$
$$\lambda_{xzxz} = \lambda_{yzyz} = 2\lambda_{\xi z\eta z}.$$

The free energy (10.8) in terms of these moduli is

$$F = \tfrac{1}{2}\lambda_{xxxx}(u_{xx}+u_{yy})^2 + \tfrac{1}{2}\lambda_{zzzz}u_{zz}^2 + \lambda_{xxzz}(u_{xx}+u_{yy})u_{zz} + 2\lambda_{xzxz}(u_{xz}^2+u_{yz}^2) + (\lambda_{xxxx}-\lambda_{xxyy})(u_{xy}^2 - u_{xx}u_{yy}).$$

PROBLEM 2. Find the conditions for the elastic energy of a cubic crystal to be positive.

SOLUTION. The first two terms in (10.9) constitute a quadratic form in the three independent variables u_{xx}, u_{yy}, u_{zz}. The conditions for this to be positive are that the determinant of its coefficients, one of the minors, and λ_{xxxx} be positive. The third term in (10.9) must be positive also. These conditions give the inequalities

$$\lambda_1 > 0, \quad \lambda_3 > 0, \quad -\tfrac{1}{2}\lambda_1 < \lambda_2 < \lambda_1,$$

where $\lambda_1 = \lambda_{xxxx}$, $\lambda_2 = \lambda_{xxyy}$, $\lambda_3 = \lambda_{xyxy}$.

PROBLEM 3. Determine the dependence of the Young's modulus of a cubic crystal on the direction in it.

SOLUTION. We take coordinate axes along the edges of the cube. Let the axis of a rod cut from the crystal be along a unit vector $\mathbf{n}$. The stress tensor in the stretched rod must satisfy the following conditions: (1) $\sigma_{ik}n_k = pn_i$, where p is the tensile force on unit area of the ends of the rod (condition at the ends of the rod); (2) for direction perpendicular to $\mathbf{n}$, $\sigma_{ik}t_k = 0$ (condition on the sides of the rod). Such a tensor must have the form $\sigma_{ik} = pn_in_k$. Calculating the components σ_{ik} by differentiating (10.9),† and comparing them with $\sigma_{ik} = pn_in_k$, we find the components of the strain tensor to be

$$u_{xx} = p\frac{(\lambda_1+2\lambda_2)n_x^2 - \lambda_2}{(\lambda_1-\lambda_2)(\lambda_1+2\lambda_2)}, \qquad u_{xy} = p\frac{n_xn_y}{2\lambda_3},$$

and similarly for the remaining components.

The relative elongation of the rod is $u = (dl' - dl)/dl$, where dl' is given by (1.2) and $dx_i/dl = n_i$. For small deformations this gives $u = u_{ik}n_in_k$. The Young's modulus is determined as the proportionality factor in $p = Eu$, and is found to be given by

$$\frac{1}{E} = \frac{\lambda_1+\lambda_2}{(\lambda_1+2\lambda_2)(\lambda_1-\lambda_2)} + \left(\frac{1}{\lambda_3} - \frac{2}{\lambda_1-\lambda_2}\right)(n_x^2n_y^2 + n_x^2n_z^2 + n_y^2n_z^2).$$

E has extremum values in the directions of the edges (the x, y, z axes) and the body diagonals of the cube. Along the edges,

$$E = (\lambda_1+2\lambda_2)(\lambda_1-\lambda_2)/(\lambda_1+\lambda_2).$$

The transverse compression of the rod is $u_{xx} = u_{yy} = -\sigma u_{zz} = -\sigma u$, where $\sigma = \lambda_2/(\lambda_1+\lambda_2)$ acts as Poisson's ratio. According to the inequalities derived in Problem 2, $-1 < \sigma < \tfrac{1}{2}$.

† If σ_{ik} is calculated not directly from $\sigma_{ik} = \lambda_{iklm}u_{lm}$ but by differentiating a specific expression for F, the derivatives with respect to u_{ik} with $i \neq k$ give doubled values for σ_{ik}. This is because the formula $\sigma_{ik} = \partial F/\partial u_{ik}$ is meaningful only as an expression of the fact that $dF = \sigma_{ik}du_{ik}$, and in the sum $\sigma_{ik}du_{ik}$ the terms in the differentials du_{ik} of each component with $i \neq k$ of the symmetrical tensor u_{ik} appear twice.

CHAPTER II

THE EQUILIBRIUM OF RODS AND PLATES

§11. The energy of a bent plate

In this chapter we shall study some particular cases of the equilibrium of deformed bodies, and we begin with that of thin deformed plates. When we speak of a *thin plate*, we mean that its thickness is small compared with its dimensions in the other two directions. The deformations themselves are supposed small, as before. In the present case the deformation is small if the displacements of points in the plate are small compared with its thickness.

The general equations of equilibrium are considerably simplified when applied to thin plates. It is more convenient, however, not to derive these simplified equations directly from the general ones, but to calculate afresh the free energy of a bent plate and then vary that energy.

When a plate is bent, it is stretched at some points and compressed at others: on the convex side there is evidently an extension, which decreases as we penetrate into the plate, finally becoming zero, after which a gradually increasing compression is found. The plate therefore contains a *neutral surface*, on which there is no extension or compression, and on opposite sides of which the deformation has oppoiste signs. The neutral surface clearly lies midway through the plate.

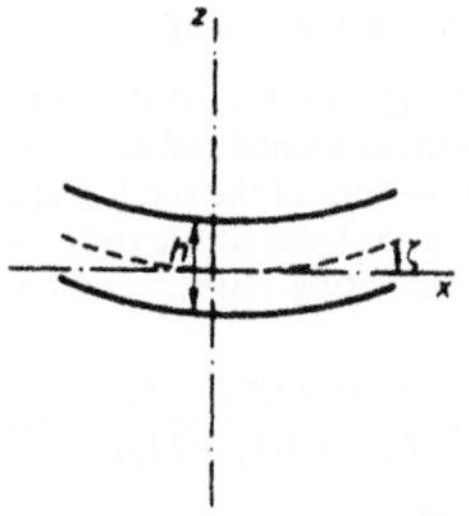

FIG. 2

We take a coordinate system with the origin on the neutral surface and the z-axis normal to the surface. The xy-plane is that of the undeformed plate. We denote by ζ the vertical displacement of a point on the neutral surface, i.e. its z coordinate (Fig. 2). The components of its displacement in the xy-plane are evidently of the second order of smallness relative to ζ, and can therefore be put equal to zero. Thus the displacement vector for points on the neutral surface is

$$u_x^{(0)} = u_y^{(0)} = 0, \qquad u_z^{(0)} = \zeta(x,y). \tag{11.1}$$

For further calculations it is necessary to note the following property of the stresses in a deformed plate. Since the plate is thin, comparatively small forces on its surface are needed to bend it. These forces are always considerably less than the internal stresses caused in the deformed plate by the extension and compression of its parts. We can therefore neglect the forces P_i in the boundary condition (2.8), leaving $\sigma_{ik}n_k = 0$. Since the plate is only slightly bent, we can suppose that the normal vector **n** is along the z-axis. Thus we must have on both surfaces of the plate $\sigma_{xz} = \sigma_{yz} = \sigma_{zz} = 0$. Since the plate is thin, however, these quantities must be small within the plate if they are zero on each surface. We therefore conclude that the components σ_{xz}, σ_{yz}, σ_{zz} are small compared with the remaining components of the stress tensor everywhere in the plate. We can therefore equate them to zero and use this condition to determine the components of the strain tensor.

By the general formulae (5.13), we have

$$\left.\begin{aligned} \sigma_{zx} &= \frac{E}{1+\sigma}u_{zx}, \qquad \sigma_{zy} = \frac{E}{1+\sigma}u_{zy}, \\ \sigma_{zz} &= \frac{E}{(1+\sigma)(1-2\sigma)}\{(1-\sigma)u_{zz} + \sigma(u_{xx}+u_{yy})\}. \end{aligned}\right\} \tag{11.2}$$

Equating these expressions to zero, we obtain $\partial u_x/\partial z = -\partial u_z/\partial x$, $\partial u_y/\partial z = -\partial u_z/\partial y$, $u_{zz} = -\sigma(u_{xx}+u_{yy})/(1-\sigma)$. In the first two of these equations u_z can, with sufficient accuracy, be replaced by $\zeta(x, y)$: $\partial u_x/\partial z = -\partial\zeta/\partial x$, $\partial u_y/\partial z = -\partial\zeta/\partial y$, whence

$$u_x = -z\partial\zeta/\partial x, \qquad u_y = -z\partial\zeta/\partial y. \tag{11.3}$$

The constants of integration are put equal to zero in order to make

$$u_x = u_y = 0 \quad \text{for} \quad z = 0.$$

Knowing u_x and u_y, we can determine all the components of the strain tensor:

$$\begin{gathered} u_{xx} = -z\partial^2\zeta/\partial x^2, \qquad u_{yy} = -z\partial^2\zeta/\partial y^2, \qquad u_{xy} = -z\partial^2\zeta/\partial x\partial y, \\ u_{xz} = u_{yz} = 0, \qquad u_{zz} = \frac{\sigma}{1-\sigma}z\left(\frac{\partial^2\zeta}{\partial x^2} + \frac{\partial^2\zeta}{\partial y^2}\right). \end{gathered} \tag{11.4}$$

We can now calculate the free energy F per unit volume of the plate, using the general formula (5.10). A simple calculation gives the expression

$$F = z^2\frac{E}{1+\sigma}\left\{\frac{1}{2(1-\sigma)}\left(\frac{\partial^2\zeta}{\partial x^2} + \frac{\partial^2\zeta}{\partial y^2}\right)^2 + \left[\left(\frac{\partial^2\zeta}{\partial x\partial y}\right)^2 - \frac{\partial^2\zeta}{\partial x^2}\frac{\partial^2\zeta}{\partial y^2}\right]\right\}. \tag{11.5}$$

The total free energy of the plate is obtained by integrating over the volume. The integration over z is from $-\frac{1}{2}h$ to $+\frac{1}{2}h$, where h is the thickness of the plate, and that over x, y is over the surface of the plate. The result is that the total free energy $F_{\text{pl}} = \int F\,\mathrm{d}V$ of a deformed plate is

$$F_{\text{pl}} = \frac{Eh^3}{24(1-\sigma^2)}\iint\left[\left(\frac{\partial^2\zeta}{\partial x^2} + \frac{\partial^2\zeta}{\partial y^2}\right)^2 + 2(1-\sigma)\left\{\left(\frac{\partial^2\zeta}{\partial x\partial y}\right)^2 - \frac{\partial^2\zeta}{\partial x^2}\frac{\partial^2\zeta}{\partial y^2}\right\}\right]\mathrm{d}x\,\mathrm{d}y; \tag{11.6}$$

the element of area can with sufficient accuracy be written as $\mathrm{d}x\ \mathrm{d}y$ simply, since the deformation is small.

Having obtained the expression for the free energy, we can regard the plate as being of infinitesimal thickness, i.e. as being a geometrical surface, since we are interested only in the form which it takes under the action of the applied forces, and not in the distribution of deformations inside it. The quantity ζ is then the displacement of points on the plate, regarded as a surface, when it is bent.

§12. The equation of equilibrium for a plate

The equation of equilibrium for a plate can be derived from the condition that its free energy be a minimum. To do so, we must calculate the variation of the expression (11.6).

We divide the integral in (11.6) into two, and vary the two parts separately. The first integral can be written in the form $\int (\triangle\zeta)^2\,\mathrm{d}f$, where $\mathrm{d}f = \mathrm{d}x\,\mathrm{d}y$ is a surface element and $\triangle \equiv \partial^2/\partial x^2 + \partial^2/\partial y^2$ is here (and in §§13, 14) the two-dimensional Laplacian. Varying this integral, we have

$$\delta\tfrac{1}{2}\int(\triangle\zeta)^2\,\mathrm{d}f = \int\triangle\zeta\,\triangle\delta\zeta\,\mathrm{d}f$$

$$= \int\triangle\zeta\,\mathrm{div}\,\mathbf{grad}\,\delta\zeta\,\mathrm{d}f$$

$$= \int\mathrm{div}(\triangle\zeta\,\mathbf{grad}\,\delta\zeta)\,\mathrm{d}f - \int\mathbf{grad}\,\delta\zeta\cdot\mathbf{grad}\,\triangle\zeta\,\mathrm{d}f.$$

All the vector operators, of course, relate to the two-dimensional coordinate system (x, y). The first integral on the right can be transformed into an integral along a closed contour enclosing the plate:†

$$\int\mathrm{div}(\triangle\zeta\,\mathbf{grad}\,\delta\zeta)\,\mathrm{d}f = \oint\triangle\zeta(\mathbf{n}\cdot\mathbf{grad}\,\delta\zeta)\,\mathrm{d}l$$

$$= \oint\triangle\zeta\frac{\partial\delta\zeta}{\partial n}\,\mathrm{d}l,$$

where $\partial/\partial n$ denotes differentiation along the outward normal to the contour.

In the second integral we use the same transformation to obtain

$$\int\mathbf{grad}\,\delta\zeta\cdot\mathbf{grad}\,\triangle\zeta\,\mathrm{d}f = \int\mathrm{div}(\delta\zeta\,\mathbf{grad}\,\triangle\zeta)\,\mathrm{d}f - \int\delta\zeta\,\triangle^2\zeta\,\mathrm{d}f$$

$$= \oint\delta\zeta(\mathbf{n}\cdot\mathbf{grad}\,\triangle\zeta)\,\mathrm{d}l - \int\delta\zeta\,\triangle^2\zeta\,\mathrm{d}f$$

$$= \oint\delta\zeta\frac{\partial\triangle\zeta}{\partial n}\,\mathrm{d}l - \int\delta\zeta\,\triangle^2\zeta\,\mathrm{d}f.$$

Substituting these results, we find that

$$\delta\tfrac{1}{2}\int(\triangle\zeta)^2\,\mathrm{d}f = \int\delta\zeta\,\triangle^2\zeta\,\mathrm{d}f - \oint\delta\zeta\frac{\partial\triangle\zeta}{\partial n}\,\mathrm{d}l + \oint\triangle\zeta\frac{\partial\delta\zeta}{\partial n}\,\mathrm{d}l. \tag{12.1}$$

---

† The transformation formula for two-dimensional integrals is exactly analogous to the one for three dimensions. The volume element $\mathrm{d}V$ is replaced by the surface element $\mathrm{d}f$ (a scalar), and the surface element $\mathbf{df}$ is replaced by a contour element $\mathrm{d}l$ multiplied by the vector $\mathbf{n}$ along the outward normal to the contour. The integral over $\mathrm{d}f$ is converted into one over $\mathrm{d}l$ by replacing $\mathrm{d}f\,\partial/\partial x_i$ by $n_i\,\mathrm{d}l$. For instance, if ϕ is a scalar, we have $\int\mathbf{grad}\,\phi\,\mathrm{d}f = \oint\phi\,\mathbf{n}\,\mathrm{d}l$.

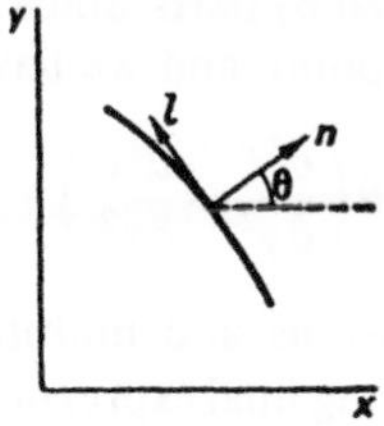

FIG. 3

The transformation of the variation of the second integral in (11.6) is somewhat more lengthy. This transformation is conveniently effected in components, and not in vector form. We have

$$\delta\iint\left\{\left(\frac{\partial^2\zeta}{\partial x\partial y}\right)^2-\frac{\partial^2\zeta}{\partial x^2}\frac{\partial^2\zeta}{\partial y^2}\right\}\mathrm{d}f$$
$$=\iint\left\{2\frac{\partial^2\zeta}{\partial x\partial y}\frac{\partial^2\delta\zeta}{\partial x\partial y}-\frac{\partial^2\zeta}{\partial x^2}\frac{\partial^2\delta\zeta}{\partial y^2}-\frac{\partial^2\delta\zeta}{\partial x^2}\frac{\partial^2\zeta}{\partial y^2}\right\}\mathrm{d}f.$$

The integrand can be written

$$\frac{\partial}{\partial x}\left(\frac{\partial\delta\zeta}{\partial y}\frac{\partial^2\zeta}{\partial x\delta y}-\frac{\partial\delta\zeta}{\partial x}\frac{\partial^2\zeta}{\partial y^2}\right)+\frac{\partial}{\partial y}\left(\frac{\partial\delta\zeta}{\partial x}\frac{\partial^2\zeta}{\partial x\partial y}-\frac{\partial\delta\zeta}{\partial y}\frac{\partial^2\zeta}{\partial x^2}\right),$$

i.e. as the (two-dimensional) divergence of a certain vector. The variation can therefore be written as a contour integral:

$$\delta\iint\left\{\left(\frac{\partial^2\zeta}{\partial x\partial y}\right)^2-\frac{\partial^2\zeta}{\partial x^2}\frac{\partial^2\zeta}{\partial y^2}\right\}\mathrm{d}f=\oint\mathrm{d}l\sin\theta\left\{\frac{\partial\delta\zeta}{\partial x}\frac{\partial^2\zeta}{\partial x\partial y}-\frac{\partial\delta\zeta}{\partial y}\frac{\partial^2\zeta}{\partial x^2}\right\}+$$
$$+\oint\mathrm{d}l\cos\theta\left\{\frac{\partial\delta\zeta}{\partial y}\frac{\partial^2\zeta}{\partial x\partial y}-\frac{\partial\delta\zeta}{\partial x}\frac{\partial^2\zeta}{\partial y^2}\right\},\qquad(12.2)$$

where θ is the angle between the x-axis and the normal to the contour (Fig. 3).

The derivatives of $\delta\zeta$ with respect to x and y are expressed in terms of its derivatives along the normal **n** and the tangent **l** to the contour:

$$\frac{\partial}{\partial x}=\cos\theta\frac{\partial}{\partial n}-\sin\theta\frac{\partial}{\partial l},$$
$$\frac{\partial}{\partial y}=\sin\theta\frac{\partial}{\partial n}+\cos\theta\frac{\partial}{\partial l}.$$

Then formula (12.2) becomes

$$\delta\iint\left\{\left(\frac{\partial^2\zeta}{\partial x\partial y}\right)^2-\frac{\partial^2\zeta}{\partial x^2}\frac{\partial^2\zeta}{\partial y^2}\right\}\mathrm{d}f$$
$$=\oint\mathrm{d}l\frac{\partial\delta\zeta}{\partial n}\left\{2\sin\theta\cos\theta\frac{\partial^2\zeta}{\partial x\partial y}-\sin^2\theta\frac{\partial^2\zeta}{\partial x^2}-\cos^2\theta\frac{\partial^2\zeta}{\partial y^2}\right\}+$$
$$+\oint\mathrm{d}l\frac{\partial\delta\zeta}{\partial l}\left\{\sin\theta\cos\theta\left(\frac{\partial^2\zeta}{\partial y^2}-\frac{\partial^2\zeta}{\partial x^2}\right)+(\cos^2\theta-\sin^2\theta)\frac{\partial^2\zeta}{\partial x\partial y}\right\}.$$

The second integral may be integrated by parts. Since it is taken along a closed contour, the limits of integration are the same point, and we have simply

$$-\oint \mathrm{d}l\,\delta\zeta \frac{\partial}{\partial l}\left\{\sin\theta\cos\theta\left(\frac{\partial^2\zeta}{\partial y^2}-\frac{\partial^2\zeta}{\partial x^2}\right)+(\cos^2\theta-\sin^2\theta)\frac{\partial^2\zeta}{\partial x\partial y}\right\}.$$

Collecting all the above expressions and multiplying by the coefficients shown in formula (11.6), we obtain the following final expression for the variation of the free energy:

$$\begin{aligned}\delta F_{\mathrm{pl}} = D\Bigg(&\int \triangle^2\zeta\,\delta\zeta\,\mathrm{d}f - \\ &-\oint \delta\zeta\,\mathrm{d}l\left[\frac{\partial\triangle\zeta}{\partial n}+(1-\sigma)\frac{\partial}{\partial l}\left\{\sin\theta\cos\theta\left(\frac{\partial^2\zeta}{\partial y^2}-\frac{\partial^2\zeta}{\partial x^2}\right)+\right.\right.\\ &\left.\left.+(\cos^2\theta-\sin^2\theta)\frac{\partial^2\zeta}{\partial x\partial y}\right\}\right]+\\ &+\oint \frac{\partial\delta\zeta}{\partial n}\,\mathrm{d}l\left\{\triangle\zeta+(1-\sigma)\left(2\sin\theta\cos\theta\frac{\partial^2\zeta}{\partial x\partial y}-\right.\right.\\ &\left.\left.-\sin^2\theta\frac{\partial^2\zeta}{\partial x^2}-\cos^2\theta\frac{\partial^2\zeta}{\partial y^2}\right)\right\}\Bigg),\end{aligned} \tag{12.3}$$

with

$$D = Eh^3/12(1-\sigma^2). \tag{12.4}$$

In order to derive from this the equation of equilibrium for the plate, we must equate to zero the sum of the variation δF and the variation δU of the potential energy of the plate due to the external forces acting on it. This latter variation is minus the work done by the external forces in deforming the plate. Let P be the external force acting on the plate, per unit area† and normal to the surface. Then the work done by the external forces when the points on the plate are displaced a distance $\delta\zeta$ is $\int P\delta\zeta\,\mathrm{d}f$. Thus the condition for the total free energy of the plate to be a minimum is

$$\delta F_{\mathrm{pl}} - \int P\delta\zeta\,\mathrm{d}f = 0.$$

On the left-hand side of this equation we have both surface and contour integrals. The surface integral is

$$\int\{D\triangle^2\zeta - P\}\delta\zeta\,\mathrm{d}f.$$

The variation $\delta\zeta$ in this integral is arbitrary. The integral can therefore vanish only if the coefficient of $\delta\zeta$ is zero, i.e.

$$D\triangle^2\zeta - P = 0. \tag{12.5}$$

This is the equation of equilibrium for a plate bent by external forces acting on it. The coefficient D is called the *flexural rigidity* or *cylindrical rigidity* of the plate.

† The force P may be the result of body forces (e.g. the force of gravity), and is then equal to the integral of the body force over the thickness of the plate.

The boundary conditions for this equation are obtained by equating to zero the contour integrals in (12.3). Here various particular cases have to be considered. Let us suppose that part of the edge of the plate is free, i.e. no external forces act on it. Then the variations $\delta\zeta$ and $\delta\partial\zeta/\partial n$ on this part of the edge are arbitrary, and their coefficients in the contour integrals must be zero. This gives the equations

$$-\frac{\partial\triangle\zeta}{\partial n}+(1-\sigma)\frac{\partial}{\partial l}\left\{\cos\theta\sin\theta\left(\frac{\partial^2\zeta}{\partial x^2}-\frac{\partial^2\zeta}{\partial y^2}\right)+\right.$$

$$\left.+(\sin^2\theta-\cos^2\theta)\frac{\partial^2\zeta}{\partial x\partial y}\right\}=0, \tag{12.6}$$

$$\triangle\zeta+(1-\sigma)\left\{2\sin\theta\cos\theta\frac{\partial^2\zeta}{\partial x\partial y}-\sin^2\theta\frac{\partial^2\zeta}{\partial x^2}-\cos^2\theta\frac{\partial^2\zeta}{\partial y^2}\right\}=0, \tag{12.7}$$

which must hold at all free points on the edge of the plate.

The boundary conditions (12.6) and (12.7) are very complex. Considerable simplifications occur when the edge of the plate is clamped or supported. If it is clamped (Fig. 4a), no vertical displacement is possible, and moreover no bending is possible at the edge. The angle through which a given part of the edge turns from its initial position is (for small displacements ζ) the derivative $\partial\zeta/\partial n$. Thus the variations $\delta\zeta$ and $\delta\partial\zeta/\partial n$ must be zero at clamped edges, so that the contour integrals in (12.3) are zero identically. The boundary conditions have in this case the simple form

$$\zeta=0, \qquad \partial\zeta/\partial n=0. \tag{12.8}$$

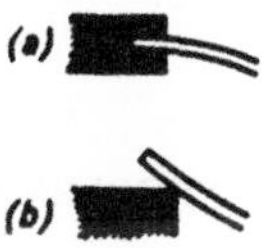

FIG. 4

The first of these expresses the fact that the edge of the plate undergoes no vertical displacement in the deformation, and the second that it remains horizontal.

It is easy to determine the reaction forces on a plate at a point where it is clamped. These are equal and opposite to the forces exerted by the plate on its support. As we know from mechanics, the force in any direction is equal to the space derivative, in that direction, of the energy. In particular, the force exerted by the plate on its support is given by minus the derivative of the energy with respect to the displacement ζ of the edge of the plate, and the reaction force by this derivative itself. The derivative in question, however, is just the coefficient of $\delta\zeta$ in the second integral in (12.3). Thus the reaction force per unit length is equal to the expression on the left of equation (12.6) (which of course, is not now zero), multiplied by D.

Similarly, the moment of the reaction forces is given by the expression on the left of equation (12.7), multiplied by the same factor. This follows at once from the result of mechanics that the moment of the force is equal to the derivative of the energy with respect to the angle through which the body turns. This angle is $\partial\zeta/\partial n$, so that the corresponding moment is given by the coefficient of $\partial\delta\zeta/\partial n$ in the third integral in (12.3). Both these expressions (that for the force and that for the moment) can be very much simplified by virtue of the conditions (12.8). Since ζ and $\partial\zeta/\partial n$ are zero everywhere on the edge of the plate, their tangential derivatives of all orders are zero also. Using this and converting the derivatives with respect to x and y in (12.6) and (12.7) into those in the directions of **n** and **l**, we obtain the following simple expressions for the reaction force F and the reaction moment M:

$$F = -D\left[\frac{\partial^3\zeta}{\partial n^3} + \frac{d\theta}{dl}\frac{\partial^2\zeta}{\partial n^2}\right], \tag{12.9}$$

$$M = D\frac{\partial^2\zeta}{\partial n^2}. \tag{12.10}$$

Another important case is that where the plate is supported (Fig. 4b), i.e. the edge rests on a fixed support, but is not clamped to it. In this case there is again no vertical displacement at the edge of the plate (i.e. on the line where it rests on the support), but its direction can vary. Accordingly, we have in (12.3) $\delta\zeta = 0$ in the contour integral, but $\partial\delta\zeta/\partial n \neq 0$. Hence only the condition (12.7) remains valid, and not (12.6). The expression on the left of (12.6) gives as before the reaction force at the points where the plate is supported; the moment of this force is zero in equilibrium. The boundary condition (12.7) can be simplified by converting to the derivatives in the direction of **n** and **l** and using the fact that, since $\zeta = 0$ everywhere on the edge, the derivatives $\partial\zeta/\partial l$ and $\partial^2\zeta/\partial l^2$ are also zero. We then have the boundary conditions in the form

$$\zeta = 0, \qquad \frac{\partial^2\zeta}{\partial n^2} + \sigma\frac{d\theta}{dl}\frac{\partial\zeta}{\partial n} = 0. \tag{12.11}$$

PROBLEMS

PROBLEM 1. Determine the deflection of a circular plate (with radius R) with clamped edges, placed horizontally in a gravitational field.

SOLUTION. We take polar coordinates, with the origin at the centre of the plate. The force on unit area of the surface of the plate is $P = \rho hg$. Equation (12.5) becomes $\triangle^2\zeta = 64\beta$, where $\beta = 3\rho g(1-\sigma^2)/16h^2E$; positive values of ζ correspond to displacements downward. Since ζ is a function of r only, we can put $\triangle = r^{-1}\,d(r d/dr)/dr$. The general integral is $\zeta = \beta r^4 + ar^2 + b + cr^2\log(r/R) + d\log(r/R)$. In the case in question we must put $d = 0$, since $\log(r/R)$ becomes infinite at $r = 0$, and $c = 0$, since this term gives a singularity in $\triangle\zeta$ at $r = 0$ (corresponding to a force applied at the centre of the plate; see Problem 3). The constants a and b are determined from the boundary conditions $\zeta = 0$, $d\zeta/dr = 0$ for $r = R$. The result is $\zeta = \beta(R^2 - r^2)^2$.

PROBLEM 2. The same as Problem 1, but for a plate with supported edges.

SOLUTION. The boundary conditions (12.11) for a circular plate are

$$\zeta = 0, \qquad \frac{d^2\zeta}{dr^2} + \frac{\sigma}{r}\frac{d\zeta}{dr} = 0.$$

The solution is similar to that of Problem 1, and the result is

$$\zeta = \beta(R^2 - r^2)\left(\frac{5+\sigma}{1+\sigma}R^2 - r^2\right).$$

PROBLEM 3. Determine the deflection of a circular plate with clamped edges when a force f is applied to its centre.

SOLUTION. We have $\triangle^2\zeta = 0$ everywhere except at the origin. Integration gives

$$\zeta = ar^2 + b + cr^2\log(r/R),$$

the log r term again being omitted. The total force on the plate is equal to the force f at its centre. The integral of $\triangle^2\zeta$ over the surface of the plate must therefore be

$$2\pi\int_0^R r\triangle^2\zeta\,dr = f/D.$$

Hence $c = f/8\pi D$. The constants a and b are determined from the boundary conditions. The result is

$$\zeta = (f/8\pi D)[\tfrac{1}{2}(R^2-r^2)-r^2\log(R/r)].$$

PROBLEM 4. The same as Problem 3, but for a plate with supported edges.

SOLUTION.

$$\zeta = \frac{f}{16\pi D}\left[\frac{3+\sigma}{1+\sigma}(R^2-r^2)-2r^2\log\frac{R}{r}\right].$$

PROBLEM 5. Determine the deflection of a circular plate suspended by its centre and in a gravitational field.

SOLUTION. The equation for ζ and its general solution are the same as in Problem 1. Since the displacement at the centre is $\zeta = 0$, we have $c = 0$. The constants a and b are determined from the boundary conditions (12.6) and (12.7), which are, for circular symmetry,

$$\frac{d\triangle\zeta}{dr} = \frac{d}{dr}\left(\frac{d^2\zeta}{dr^2}+\frac{1}{r}\frac{d\zeta}{dr}\right) = 0, \qquad \frac{d^2\zeta}{dr^2}+\frac{\sigma}{r}\frac{d\zeta}{dr} = 0.$$

The result is

$$\zeta = \beta r^2\left[r^2+8R^2\log\frac{R}{r}+2R^2\frac{3+\sigma}{1+\sigma}\right].$$

PROBLEM 6. A thin layer (of thickness h) is torn off a body by external forces acting against surface tension forces at the surface of separation. With given external forces, equilibrium is established for a definite area of the surface separated and a definite shape of the layer removed (Fig. 5). Derive a formula relating the surface tension to the shape of the layer removed.†

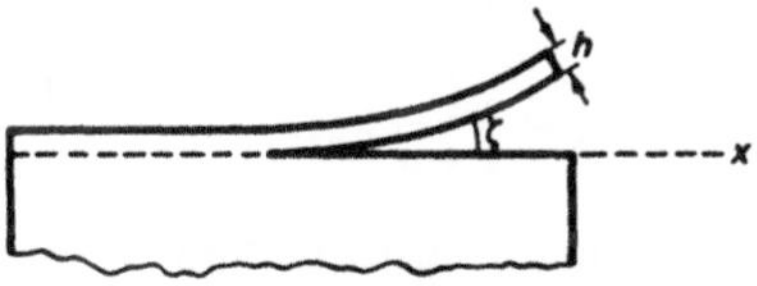

FIG. 5

SOLUTION. The layer removed can be regarded as a plate with one edge (the line of separation) clamped. The bending moment on the layer is given by formula (12.10). The work done by this moment when the length of the separated surface increases by δx is

$$M\partial\delta\zeta/\partial x = M\delta x\partial^2\zeta/\partial x^2 = D(\partial^2\zeta/\partial x^2)^2\,\delta x \tag{1}$$

(the work of the bending force F itself is a second-order quantity). The equilibrium condition is that this work be equal to the change in the energy of the system. The latter is made up of two parts: the change in the surface energy, and the change in the elastic energy of the layer removed owing to the increase in length of its bent part. The first part is $2\alpha\delta x$, where α is the surface-tension coefficient, the factor 2 allowing for the creation of two free surfaces by the separation. The second part is $\frac{1}{2}D(\partial^2\zeta/\partial x^2)^2\,\delta x$, i.e. the energy (11.6) for a length δx of the layer, which is half of the quantity (1). The result is thus

$$\alpha = \tfrac{1}{4}D(\partial^2\zeta/\partial x^2)^2.$$

† This problem was discussed by I. V. Obreimov (1930) in connection with a method which he developed for measuring the surface tension of mica. The measurements which he made by this method were the first direct measurements of the surface tension of solids.

§13. Longitudinal deformations of plates

Longitudinal deformations occurring in the plane of the plate, and not resulting in any bending, form a special case of deformations of thin plates. Let us derive the equations of equilibrium for such deformations.

If the plate is sufficiently thin, the deformation may be regarded as uniform over its thickness. The strain tensor is then a function of x and y only (the xy-plane being that of the plate) and is independent of z. Longitudinal deformations of a plate are usually caused either by forces applied to its edges or by body forces in its plane. The boundary conditions on both surfaces of the plate are then $\sigma_{ik}n_k = 0$, or, since the normal vector is parallel to the z-axis, $\sigma_{iz} = 0$, i.e. $\sigma_{xz} = \sigma_{yz} = \sigma_{zz} = 0$. It should be noticed, however, that in the approximate theory given below these conditions continue to hold even when the external tension forces are applied to the surfaces of the plate, since these forces are still small compared with the resulting longitudinal internal stresses ($\sigma_{xx}, \sigma_{yy}, \sigma_{xy}$) in the plate. Since they are zero at both surfaces, the quantities $\sigma_{xz}, \sigma_{yz}, \sigma_{zz}$ must be small throughout the thickness of the plate, and we can therefore take them as approximately zero everywhere in the plate.

Equating to zero the expressions (11.2), we obtain the relations

$$u_{zz} = -\sigma(u_{xx}+u_{yy})/(1-\sigma), \qquad u_{xz} = u_{yz} = 0. \tag{13.1}$$

Substituting in the general formulae (5.13), we obtain for the non-zero components of the stress tensor

$$\left.\begin{aligned}\sigma_{xx} &= \frac{E}{1-\sigma^2}(u_{xx}+\sigma u_{yy}),\\ \sigma_{yy} &= \frac{E}{1-\sigma^2}(u_{yy}+\sigma u_{xx}),\\ \sigma_{xy} &= \frac{E}{1+\sigma}u_{xy}.\end{aligned}\right\} \tag{13.2}$$

It should be noticed that the formal transformation

$$E \to E/(1-\sigma^2), \qquad \sigma \to \sigma/(1-\sigma) \tag{13.3}$$

converts these expressions into those which give the relation between the stresses $\sigma_{xx}, \sigma_{xy}, \sigma_{yy}$ and the strains u_{xx}, u_{yy}, u_{zz} for a plane deformation (formulae (5.13) with $u_{zz} = 0$).

Having thus eliminated the displacement u_z, we can regard the plate as a two-dimensional medium (an "elastic plane"), of zero thickness, and take the displacement vector $\mathbf{u}$ to be a two-dimensional vector with components u_x and u_y. If P_x and P_y are the components of the external body force per unit area of the plate, the general equations of equilibrium are

$$h\left(\frac{\partial\sigma_{xx}}{\partial x}+\frac{\partial\sigma_{xy}}{\partial y}\right)+P_x = 0,$$

$$h\left(\frac{\partial\sigma_{yx}}{\partial x}+\frac{\partial\sigma_{yy}}{\partial y}\right)+P_y = 0.$$

Substituting the expressions (13.2), we obtain the equations of equilibrium in the form

$$\left.\begin{aligned} Eh\left\{\frac{1}{1-\sigma^2}\frac{\partial^2 u_x}{\partial x^2}+\frac{1}{2(1+\sigma)}\frac{\partial^2 u_x}{\partial y^2}+\frac{1}{2(1-\sigma)}\frac{\partial^2 u_y}{\partial x\partial y}\right\}+P_x=0,\\ Eh\left\{\frac{1}{1-\sigma^2}\frac{\partial^2 u_y}{\partial y^2}+\frac{1}{2(1+\sigma)}\frac{\partial^2 u_y}{\partial x^2}+\frac{1}{2(1-\sigma)}\frac{\partial^2 u_x}{\partial x\partial y}\right\}+P_y=0.\end{aligned}\right\} \quad (13.4)$$

These equations can be written in the two-dimensional vector form

$$\mathbf{grad}\,\mathrm{div}\,\mathbf{u}-\tfrac{1}{2}(1-\sigma)\,\mathbf{curl}\,\mathbf{curl}\,\mathbf{u}=-(1-\sigma^2)\mathbf{P}/Eh, \quad (13.5)$$

where all the vector operators are two-dimensional.

In particular, the equation of equilibrium in the absence of body forces is

$$\mathbf{grad}\,\mathrm{div}\,\mathbf{u}-\tfrac{1}{2}(1-\sigma)\,\mathbf{curl}\,\mathbf{curl}\,\mathbf{u}=0. \quad (13.6)$$

It differs from the equation of equilibrium for a plane deformation of a body infinite in the z-direction (§7) only by the sign of the coefficient (in accordance with (13.3)).† As for a plane deformation, we can introduce the *stress function* defined by

$$\sigma_{xx}=\partial^2\chi/\partial y^2, \qquad \sigma_{xy}=-\partial^2\chi/\partial x\partial y, \qquad \sigma_{yy}=\partial^2\chi/\partial x^2, \quad (13.7)$$

whereby we automatically satisfy the equations of equilibrium in the form

$$\frac{\partial\sigma_{xx}}{\partial x}+\frac{\partial\sigma_{xy}}{\partial y}=0, \qquad \frac{\partial\sigma_{yx}}{\partial x}+\frac{\partial\sigma_{yy}}{\partial y}=0.$$

The stress function, as before, satisfies the biharmonic equation, since for $\triangle\chi$ we have

$$\triangle\chi=\sigma_{xx}+\sigma_{yy}=E(u_{xx}+u_{yy})/(1-\sigma)=\{E/(1-\sigma)\}\,\mathrm{div}\,\mathbf{u};$$

this differs only by a factor from the result for a plane deformation.

It may be pointed out that the stress distribution in a plate deformed by given forces applied to its edges is independent of the elastic constants of the material. For these constants appear neither in the biharmonic equation satisfied by the stress function, nor in the formulae (13.7) which determine the components σ_{ik} from that function (nor, therefore, in the boundary conditions at the edges of the plate).

PROBLEMS

PROBLEM 1. Determine the deformation of a plane disc rotating uniformly about an axis through its centre perpendicular to its plane.

SOLUTION. The required solution differs only in the constant coefficients from the solution obtained in §7, Problem 5, for the plane deformation of a rotating cylinder. The radial displacement $u_r = u(r)$ is given by the formula

$$u=\frac{\rho\Omega^2(1-\sigma^2)}{8E}r\left(\frac{3+\sigma}{1+\sigma}R^2-r^2\right).$$

This is the expression which gives that of §7, Problem 5, if the substitution (13.3) is made.

PROBLEM 2. Determine the deformation of a semi-infinite plate (with a straight edge) under the action of a concentrated force in its plane, applied to a point on the edge.

† A deformation homogeneous in the z-direction for which $\sigma_{zx}=\sigma_{zy}=\sigma_{zz}=0$ everywhere is sometimes called a state of *plane stress*, as distinct from a plane deformation, for which $u_{zx}=u_{zy}=u_{zz}=0$ everywhere.

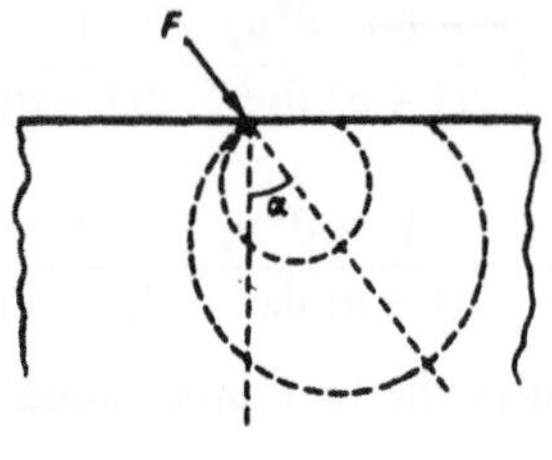

FIG. 6

SOLUTION. We take polar coordinates, with an angle ϕ measured from the direction of the applied force; it takes values from $-(\frac{1}{2}\pi+\alpha)$ to $\frac{1}{2}\pi-\alpha$, where α is the angle between the direction of the force and the normal to the edge of the plate (Fig. 6). At every point of the edge except that where the force is applied (the origin) we must have $\sigma_{\phi\phi}=\sigma_{r\phi}=0$. Using the expressions for $\sigma_{\phi\phi}$ and $\sigma_{r\phi}$ obtained in §7, Problem 11, we find that the stress function must therefore satisfy the conditions

$$\frac{\partial\chi}{\partial r}=\text{constant},\qquad \frac{1}{r}\frac{\partial\chi}{\partial\phi}=\text{constant, for } \phi=-(\tfrac{1}{2}\pi+\alpha),\ (\tfrac{1}{2}\pi-\alpha).$$

Both conditions are satisfied if $\chi=rf(\phi)$. With this substitution, the biharmonic equation

$$\left\{\frac{1}{r}\frac{\partial}{\partial r}\left(r\frac{\partial}{\partial r}\right)+\frac{\partial^2}{\partial\phi^2}\right\}^2\chi=0$$

gives solutions for $f(\phi)$ of the forms $\sin\phi$, $\cos\phi$, $\phi\sin\phi$, $\phi\cos\phi$. The first two of these lead to stresses which are zero identically. The solution which gives the correct value for the force applied at the origin is

$$\chi=-(F/\pi)r\phi\sin\phi,\qquad \sigma_{rr}=-(2F/\pi r)\cos\phi,\qquad \sigma_{r\phi}=\sigma_{\phi\phi}=0,\tag{1}$$

where F is the force per unit thickness of the plate. For, projecting the internal stresses on directions parallel and perpendicular to the force **F**, and integrating over a small semicircle centred at the origin (whose radius then tends to zero), we obtain

$$\int\sigma_{rr}r\cos\phi\,\mathrm{d}\phi=-F,$$

$$\int\sigma_{rr}r\sin\phi\,\mathrm{d}\phi=0,$$

i.e. the values required to balance the external force applied at the origin.

Formulae (1) determine the required stress distribution. It is purely radial: only a radial compression force acts on any area perpendicular to the radius. The lines of equal stress are the circles $r=d\cos\phi$, which pass through the origin and whose centres lie on the line of action of the force **F** (Fig. 6).

The components of the strain tensor are $u_{rr}=\sigma_{rr}/E$, $u_{\phi\phi}=-\sigma\sigma_{rr}/E$, $u_{r\phi}=0$. From these we find by integration (using the expressions (1.8) for the components u_{ik} in polar coordinates) the displacement vector:

$$u_r=-\frac{2F}{\pi E}\log(r/a)\cos\phi-\frac{(1-\sigma)F}{\pi E}\phi\sin\phi,$$

$$u_\phi=\frac{2\sigma F}{\pi E}\sin\phi+\frac{2F}{\pi E}\log(r/a)\sin\phi+\frac{(1-\sigma)F}{\pi E}(\sin\phi-\phi\cos\phi).$$

Here the constants of integration have been chosen so to give zero displacement (translation and rotation) of the plate as a whole: an arbitrarily chosen point at a distance a from the origin on the line of action of the force is assumed to remain fixed.

Using the solution obtained above, we can obtain the solution for any distribution of forces acting on the edge of the plate (cf. §8). It is, of course, inapplicable in the immediate neighbourhood of the origin.

PROBLEM 3. Determine the deformation of an infinite wedge-shaped plate (with angle 2α) due to a force applied at its apex.

SOLUTION. The stress distribution is given by formulae which differ from those of Problem 2 only in their normalization. If the force acts along the mid-line of the wedge (F_1 in Fig. 7), we have $\sigma_{rr}=$

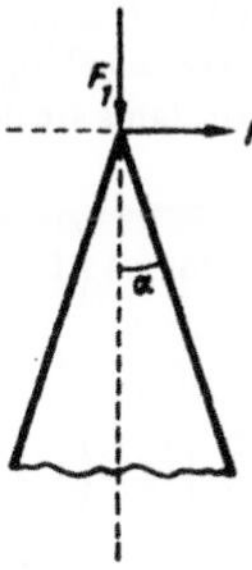

FIG. 7

$-(F_1\cos\phi)/r(\alpha+\frac{1}{2}\sin 2\alpha)$, $\sigma_{r\phi}=\sigma_{\phi\phi}=0$. If, on the other hand, the force acts perpendicular to this direction (F_2 in Fig. 7), then

$$\sigma_{rr} = -(F_2\cos\phi)/r(\alpha-\tfrac{1}{2}\sin 2\alpha).$$

In each case the angle ϕ is measured from the direction of the force.

PROBLEM 4. Determine the deformation of a circular disc (with radius R) compressed by two equal and opposite forces Fh applied at the ends of a diameter (Fig. 8).

SOLUTION. The solution is obtained by superposing three internal stress distributions. Two of these are

$$\sigma^{(1)}{}_{r_1r_1} = -(2F/\pi r_1)\cos\phi_1, \qquad \sigma^{(1)}{}_{r_1\phi_1} = \sigma^{(1)}{}_{\phi_1\phi_1} = 0.$$
$$\sigma^{(1)}{}_{r_2r_2} = -(2F/\pi r_2)\cos\phi_2, \qquad \sigma^{(1)}{}_{r_2\phi_2} = \sigma^{(1)}{}_{\phi_2\phi_2} = 0,$$

where r_1, ϕ_1 and r_2, ϕ_2 are the polar coordinates of an arbitrary point P with origins at A and B respectively. These are the stresses due to a normal force F applied to a point on the edge of a half-plane; see Problem 2. The third distribution, $\sigma^{(3)}{}_{ik} = (F/\pi R)\delta_{ik}$, is a uniform extension of definite intensity. For, if the point P is on the edge of the disc, we have $r_1 = 2R\cos\phi_1$, $r_2 = 2R\cos\phi_2$, so that $\sigma^{(1)}{}_{r_1r_1} = \sigma^{(2)}{}_{r_2r_2} = -F/\pi R$. Since the directions of r_1 and r_2 at this point are perpendicular, we see that the first two stress distributions give a uniform compression on the edge of the disc. These forces can be just balanced by the uniform tension given by the third system, so that the edge of the disc is free from stress, as it should be.

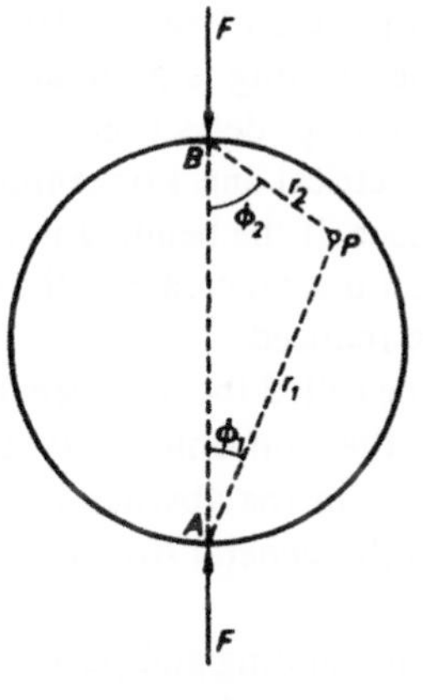

FIG. 8

PROBLEM 5. Determine the stress distribution in an infinite sheet with a circular aperture (radius R) under uniform tension.

SOLUTION. The uniform tension of a continuous sheet corresponds to stresses $\sigma^{(0)}{}_{xx} = T$, $\sigma^{(0)}{}_{yy} = \sigma^{(0)}{}_{xy} = 0$, where T is the tension force. These in turn correspond to the stress function $\chi^{(0)} = \frac{1}{2}Ty^2 = \frac{1}{2}Tr^2\sin^2\phi = \frac{1}{4}Tr^2(1-\cos 2\phi)$. When there is a circular aperture (with the centre as the origin of polar coordinates r, ϕ), we seek the stress function in the form $\chi = \chi^{(0)} + \chi^{(1)}$, $\chi^{(1)} = f(r) + F(r)\cos 2\phi$. The integral of the biharmonic equation which is independent of ϕ is of the form $f(r) = ar^2 \log r + br^2 + c\log r$, and in the integral proportional to $\cos 2\phi$ we have $F(r) = dr^2 + er^4 + g/r^2$. The constants are determined by the conditions $\sigma^{(1)}_{ik} = 0$ for $r = \infty$ and $\sigma_{rr} = \sigma_{r\phi} = 0$ for $r = R$. The result is

$$\chi^{(1)} = \tfrac{1}{4}TR^2\left\{-\log r + \left(1 - \frac{R^2}{2r^2}\right)\cos 2\phi\right\},$$

and the stress distribution is given by

$$\sigma_{rr} = \tfrac{1}{2}T\left(1 - \frac{R^2}{r^2}\right)\left\{1 + \left(1 - \frac{3R^2}{r^2}\right)\cos 2\phi\right\},$$

$$\sigma_{\phi\phi} = \tfrac{1}{2}T\left\{1 + \frac{R^2}{r^2} - \left(1 + \frac{3R^4}{r^4}\right)\cos 2\phi\right\},$$

$$\sigma_{r\phi} = -\tfrac{1}{2}T\left(1 + \frac{2R^2}{r^2} - \frac{3R^4}{r^4}\right)\sin 2\phi.$$

In particular, at the edge of the aperture we have $\sigma_{\phi\phi} = T(1 - 2\cos 2\phi)$, and for $\phi = \pm\frac{1}{2}\pi$, $\sigma_{\phi\phi} = 3T$, i.e. three times the stress at infinity (cf. §7, Problem 12).

§14. Large deflections of plates

The theory of the bending of thin plates given in §§11–13 is applicable only to fairly small deflections. Anticipating the result given below, it may be mentioned here that the condition for that theory to be applicable is that the deflection ζ be small compared with the thickness h of the plate. Let us now derive the equations of equilibrium for a plate undergoing large deflections. The deflection ζ is not now supposed small compared with h. It should be emphasized, however, that the deformation itself must still be small, in the sense that the components of the strain tensor must be small. In practice, this usually implies the condition $\zeta \ll l$, i.e. the deflection must be small compared with the dimension l of the plate.

The bending of a plate in general involves a stretching of it.† For small deflections this stretching can be neglected. For large deflections, however, this is not possible; there is therefore no neutral surface in a plate undergoing large deflections. The existence of a stretching which accompanies the bending is peculiar to plates, and distinguishes them from thin rods, which can undergo large deflections without any general stretching. This property of plates is a purely geometrical one. For example, let a flat circular plate be bent into a segment of a spherical surface. If the bending is such that the circumference of the plate remains constant, its diameter must increase. If the diameter is constant, on the other hand, the circumference must be reduced.

The energy (11.6), which may be called the *pure bending energy*, is only the part of the total energy which arises from the non-uniformity of the tension and compression through the thickness of the plate, in the absence of any general stretching. The total energy includes also a part due to this general stretching; this may be called the *stretching energy*.

Deformations consisting of pure bending and pure stretching have been considered in §§11–13. We can therefore use the results obtained in these sections. It is not necessary to consider the structure of the plate across its thickness, and we can regard it as a two-dimensional surface of negligible thickness.

We first derive an expression for the strain tensor pertaining to the stretching of a plate (regarded as a surface) which is simultaneously bent and stretched in its plane. Let **u** be the

† An exception is, for instance, the bending of a flat plate into a cylindrical surface.

two-dimensional displacement vector (with components u_x, u_y) for pure stretching; ζ, as before, denotes the transverse displacement in bending. Then the element of length $dl = \sqrt{(dx^2+dy^2)}$ of the undeformed plate is transformed by the deformation into an element dl', whose square is given by $dl'^2 = (dx+du_x)^2+(dy+du_y)^2+d\zeta^2$. Putting here $du_x = (\partial u_x/\partial x)\,dx + (\partial u_x/\partial y)\,dy$, and similarly for du_y and $d\zeta$, we obtain to within higher-order terms $dl'^2 = dl^2 + 2u_{\alpha\beta}dx_\alpha dx_\beta$, where the two-dimensional strain tensor is defined as

$$u_{\alpha\beta} = \frac{1}{2}\left(\frac{\partial u_\alpha}{\partial x_\beta}+\frac{\partial u_\beta}{\partial x_\alpha}\right)+\frac{1}{2}\frac{\partial \zeta}{\partial x_\alpha}\frac{\partial \zeta}{\partial x_\beta}. \tag{14.1}$$

(In this and the following sections, Greek suffixes take the two values x and y; as usual, summation over repeated suffixes is understood). The terms quadratic in the derivatives of u_α are here omitted; the same cannot, of course, be done with the derivatives of ζ, since there is no corresponding first-order terms.

The stress tensor $\sigma_{\alpha\beta}$ due to the stretching of the plate is given by formula (13.2), in which $u_{\alpha\beta}$ must be replaced by the total strain tensor given by formula (14.1). The pure bending energy is given by formula (11.6), and can be written $\int \Psi_1(\zeta)\,dx\,dy$, where $\Psi_1(\zeta)$ denotes the integrand in (11.6). The stretching energy per unit volume of the plate is, by the general formulae, $\frac{1}{2}u_{\alpha\beta}\sigma_{\alpha\beta}$. The energy per unit surface area is obtained by multiplying by h, so that the total stretching energy can be written $\int \Psi_2(u_{\alpha\beta})\,df$, where

$$\Psi_2 = \tfrac{1}{2}hu_{\alpha\beta}\sigma_{\alpha\beta}. \tag{14.2}$$

Thus the total free energy of a plate undergoing large deflections is

$$F_{\text{pl}} = \int \{\Psi_1(\zeta)+\Psi_2(u_{\alpha\beta})\}\,df. \tag{14.3}$$

Before deriving the equations of equilibrium, let us estimate the relative magnitude of the two parts of the energy. The first derivatives of ζ are of the order of ζ/l, where l is the dimension of the plate, and the second derivatives are of the order of ζ/l^2. Hence we see from (11.6) that $\Psi_1 \sim Eh^3\zeta^2/l^4$. The order of magnitude of the tensor components $u_{\alpha\beta}$ is ζ^2/l^2, and so $\Psi_2 \sim Eh\zeta^4/l^4$. A comparison shows that the neglect of Ψ_2 in the approximate theory of the bending of plates is valid only if $\zeta^2 \ll h^2$.

The condition of minimum energy is $\delta F + \delta U = 0$, where U is the potential energy in the field of the external forces. We shall suppose that the external stretching forces, if any, can be neglected in comparison with the bending forces. (This is always valid unless the stretching forces are very large, since a thin plate is much more easily bent than stretched.) Then we have for δU the same expression as in §12: $\delta U = -\int P\delta\zeta\,df$, where P is the external force per unit area of the plate. The variation of the integral $\int \Psi_1\,df$ has already been calculated in §12, and is

$$\delta\int \Psi_1\,df = D\int \triangle^2\zeta\,\delta\zeta\,df.$$

The contour integrals in (12.3) are omitted, since they give only the boundary conditions on the equation of equilibrium, and not that equation itself, which is of interest here.

Finally, let us calculate the variation of the integral $\int \Psi_2\,df$. The variation must be taken both with respect to the components of the vector $\mathbf{u}$ and with respect to ζ. We have

$$\delta\int \Psi_2\,df = \int \frac{\partial \Psi_2}{\partial u_{\alpha\beta}}\delta u_{\alpha\beta}\,df.$$

The derivatives of the free energy per unit volume with respect to $u_{\alpha\beta}$ are $\sigma_{\alpha\beta}$; hence $\partial\Psi_2/\partial u_{\alpha\beta} = h\sigma_{\alpha\beta}$. Substituting also for $u_{\alpha\beta}$ the expression (14.1), we obtain

$$\delta\int \Psi_2\,\mathrm{d}f = h\int \sigma_{\alpha\beta}\delta u_{\alpha\beta}\,\mathrm{d}f$$

$$= \tfrac{1}{2}h\int \sigma_{\alpha\beta}\left\{\frac{\partial\delta u_\alpha}{\partial x_\beta} + \frac{\partial\delta u_\beta}{\partial x_\alpha} + \frac{\partial\zeta}{\partial x_\alpha}\frac{\partial\delta\zeta}{\partial x_\beta} + \frac{\partial\delta\zeta}{\partial x_\alpha}\frac{\partial\zeta}{\partial x_\beta}\right\}\mathrm{d}f,$$

or, by the symmetry of $\sigma_{\alpha\beta}$,

$$\delta\int \Psi_2\,\mathrm{d}f = h\int \sigma_{\alpha\beta}\left\{\frac{\partial\delta u_\alpha}{\partial x_\beta} + \frac{\partial\delta\zeta}{\partial x_\beta}\frac{\partial\zeta}{\partial x_\alpha}\right\}\mathrm{d}f.$$

Integrating by parts, we obtain

$$\delta\int \Psi_2\,\mathrm{d}f = -h\int\left\{\frac{\partial\sigma_{\alpha\beta}}{\partial x_\beta}\delta u_\alpha + \frac{\partial}{\partial x_\beta}\left(\sigma_{\alpha\beta}\frac{\partial\zeta}{\partial x_\alpha}\right)\delta\zeta\right\}\mathrm{d}f.$$

The contour integrals along the circumference of the plate are again omitted.

Collecting the above results, we have

$$\delta F_{\mathrm{pl}} + \delta U = \iint\left[\left\{D\triangle^2\zeta - h\frac{\partial}{\partial x_\beta}\left(\sigma_{\alpha\beta}\frac{\partial\zeta}{\partial x_\alpha}\right) - P\right\}\delta\zeta - h\frac{\partial\sigma_{\alpha\beta}}{\partial x_\beta}\delta u_\alpha\right]\mathrm{d}f = 0.$$

In order that this relation should be satisfied identically, the coefficients of $\delta\zeta$ and δu_α must each be zero. Thus we obtain the equations

$$D\triangle^2\zeta - h\frac{\partial}{\partial x_\beta}\left(\sigma_{\alpha\beta}\frac{\partial\zeta}{\partial x_\alpha}\right) = P, \qquad (14.4)$$

$$\partial\sigma_{\alpha\beta}/\partial x_\beta = 0. \qquad (14.5)$$

The unknown functions here are the two components u_x, u_y of the vector $\mathbf{u}$ and the transverse displacement ζ. The solution of the equations gives both the form of the bent plate (i.e. the function $\zeta(x, y)$) and the extension resulting from the bending. Equations (14.4) and (14.5) can be somewhat simplified by introducing the function χ related to $\sigma_{\alpha\beta}$ by (13.7). Equation (14.4) then becomes

$$D\triangle^2\zeta - h\left(\frac{\partial^2\chi}{\partial y^2}\frac{\partial^2\zeta}{\partial x^2} + \frac{\partial^2\chi}{\partial x^2}\frac{\partial^2\zeta}{\partial y^2} - 2\frac{\partial^2\chi}{\partial x\partial y}\frac{\partial^2\zeta}{\partial x\partial y}\right) = P. \qquad (14.6)$$

Equations (14.5) are satisfied automatically by the expressions (13.7). Hence another equation is needed; this can be obtained by eliminating u_α from the relations (13.7) and (13.2).

To do this, we proceed as follows. We express $u_{\alpha\beta}$ in terms of $\sigma_{\alpha\beta}$, obtaining from (13.2)

$$u_{xx} = (\sigma_{xx} - \sigma\sigma_{yy})/E, \qquad u_{yy} = (\sigma_{yy} - \sigma\sigma_{xx})/E, \qquad u_{xy} = (1+\sigma)\sigma_{xy}/E.$$

Substituting here the expression (14.1) for $u_{\alpha\beta}$, and (13.7) for $\sigma_{\alpha\beta}$,we find the equations

$$\frac{\partial u_x}{\partial x}+\frac{1}{2}\left(\frac{\partial \zeta}{\partial x}\right)^2=\frac{1}{E}\left(\frac{\partial^2 \chi}{\partial y^2}-\sigma\frac{\partial^2 \chi}{\partial x^2}\right),$$

$$\frac{\partial u_y}{\partial y}+\frac{1}{2}\left(\frac{\partial \zeta}{\partial y}\right)^2=\frac{1}{E}\left(\frac{\partial^2 \chi}{\partial x^2}-\sigma\frac{\partial^2 \chi}{\partial y^2}\right),$$

$$\frac{\partial u_x}{\partial y}+\frac{\partial u_y}{\partial x}+\frac{\partial \zeta}{\partial x}\frac{\partial \zeta}{\partial y}=-\frac{2(1+\sigma)}{E}\frac{\partial^2 \chi}{\partial x \partial y}.$$

We take $\partial^2/\partial y^2$ of the first, $\partial^2/\partial x^2$ of the second, $-\partial^2/\partial x\partial y$ of the third, and add. The terms in u_x and u_y then cancel, and we have the equation

$$\triangle^2\chi+E\left\{\frac{\partial^2\zeta}{\partial x^2}\frac{\partial^2\zeta}{\partial y^2}-\left(\frac{\partial^2\zeta}{\partial x\partial y}\right)^2\right\}=0. \qquad (14.7)$$

Equations (14.6) and (14.7) form a complete system of equations for large deflections of thin plates (A. Föppl 1907). These equations are very complicated, and cannot be solved exactly, even in very simple cases. It should be noticed that they are non-linear.

We may mention briefly a particular case of deformations of thin plates, that of membranes. A *membrane* is a thin plate subject to large external stretching forces applied at its circumference. In this case we can neglect the additional longitudinal stresses caused by bending of the plate, and therefore suppose that the components of the tensor $\sigma_{\alpha\beta}$ are simply equal to the constant external stretching forces. In equation (14.4) we can then neglect the first term in comparison with the second, and we obtain the equation of equilibrium

$$h\sigma_{\alpha\beta}\frac{\partial^2\zeta}{\partial x_\alpha\partial x_\beta}+P=0, \qquad (14.8)$$

with the boundary condition that $\zeta=0$ at the edge of the membrane. This is a linear equation. The case of isotropic stretching, when the extension of the membrane is the same in all directions, is particularly simple. Let T be the absolute magnitude of the stretching force per unit length of the edge of the membrane. Then $h\sigma_{\alpha\beta}=T\delta_{\alpha\beta}$, and we obtain the equation of equilibrium in the form

$$T\triangle\zeta+P=0. \qquad (14.9)$$

PROBLEMS

Problem 1. Determine the deflection of a plate as a function of the force on it when $\zeta \gg h$.

Solution. An estimate of the terms in equation (14.7) shows that $\chi \sim E\zeta^2$. For $\zeta \gg h$, the first term in (14.6) is small compared with the second, which is of the order of magnitude $h\zeta\chi/l^4 \sim Eh\zeta^3/l^4$ (l being the dimension of the plate). If this is comparable with the external force P, we have $\zeta \sim (l^4P/Eh)^{\frac{1}{3}}$. Hence, in particular, we see that ζ is proportional to the cube root of the force.

Problem 2. Determine the deformation of a circular membrane (with radius R) placed horizontally in a gravitational field.

Solution. We have $P=\rho gh$; in polar coordinates, (14.9) becomes

$$\frac{1}{r}\frac{\mathrm{d}}{\mathrm{d}r}\left(r\frac{\mathrm{d}\zeta}{\mathrm{d}r}\right)=-\frac{\rho gh}{T}.$$

The solution finite for $r=0$ and zero for $r=R$ is $\zeta=\rho gh(R^2-r^2)/4T$.

§15. Deformations of shells

In discussing hitherto the deformations of thin plates, we have always assumed that the plate is flat in its undeformed state. However, deformations of plates which are curved in the undeformed state (called *shells*) have properties which are fundamentally different from those of the deformations of flat plates.

The stretching which accompanies the bending of a flat plate is a second-order effect in comparison with the bending deflection itself. This is seen, for example, from the fact that the strain tensor (14.1), which gives this stretching, is quadratic in ζ. The situation is entirely different in the deformation of shells: here the stretching is a first-order effect, and therefore is important even for small bending deflections. This property is most easily seen from a simple example, that of the uniform stretching of a spherical shell. If every point undergoes the same radial displacement ζ, the length of the equator increases by $2\pi\zeta$. The relative extension is $2\pi\zeta/2\pi R = \zeta/R$, and hence the strain tensor also is proportional to the first power of ζ. This effect tends to zero as $R \to \infty$, i.e. as the curvature tends to zero, and is therefore due to the curvature of the shell.

Let R be the order of magnitude of the radius of curvature of the shell, which is usually of the same order as its dimension. Then the strain tensor for the stretching which accompanies the bending is of the order of ζ/R, the corresponding stress tensor is $\sim E\zeta/R$, and the deformation energy per unit area is, by (14.2), of the order of $Eh(\zeta/R)^2$. The pure bending energy, on the other hand, is of the order of $Eh^3\zeta^2/R^4$, as before. We see that the ratio of the two is of the order of $(R/h)^2$, i.e. it is very large. It should be emphasized that this is true whatever the ratio of the bending deflection ζ to the thickness h, whereas in the bending of flat plates the stretching was important only for $\zeta \gtrsim h$.

In some cases there may be a special type of bending of the shell in which no stretching occurs. For example, a cylindrical shell (open at both ends) can be deformed without stretching if all the generators remain parallel (i.e. if the shell is, as it were, compressed along some generator). Such deformations without stretching are geometrically possible if the shell has free edges (i.e. is not closed) or if it is closed but its curvature has opposite signs at different points. For example, a closed spherical shell cannot be bent without being stretched, but if a hole is cut in it (the edge of the hole not being fixed), then such a deformation becomes possible. Since the pure bending energy is small compared with the stretching energy, it is clear that, if any given shell permits deformation without stretching, then such deformations will, in general, actually occur when arbitrary external forces act on the shell. The requirement that the bending be unaccompanied by stretching places considerable restrictions on the possible displacements u_α. These restrictions are purely geometrical, and can be expressed as differential equations, which must be contained in the complete system of equilibrium equations for such deformations. We shall not pause to discuss this question further.

If, however, the deformation of the shell involves stretching, then the tensile stresses are in general large compared with the bending stresses, which may be neglected. Shells for which this is done are called *membranes*.

The stretching energy of a shell can be calculated as the integral

$$F_{\text{pl}} = \tfrac{1}{2}h \int u_{\alpha\beta}\sigma_{\alpha\beta}\, \mathrm{d}f, \tag{15.1}$$

taken over the surface. Here $u_{\alpha\beta}$ (α, $\beta = 1, 2$) is the two-dimensional strain tensor in the appropriate curvilinear coordinates, and the stress tensor $\sigma_{\alpha\beta}$ is related to $u_{\alpha\beta}$ by formulae

(13.2), which can be written, in two-dimensional tensor notation, as

$$\sigma_{\alpha\beta} = E[(1-\sigma)u_{\alpha\beta} + \sigma\delta_{\alpha\beta}u_{\gamma\gamma}]/(1-\sigma^2). \tag{15.2}$$

A case requiring special consideration is that where the shell is subjected to the action of forces applied to points or lines on the surface and directed through the shell. These may be, in particular, the reaction forces on the shell at points (or lines) where it is fixed. The concentrated forces result in a bending of the shell in small regions near the points where they are applied; let d be the dimension of such a region for a force f applied at a point (so that its area is of the order of d^2). Since the deflection ζ varies considerably over a distance d, the bending energy per unit area is of the order of $Eh^3\zeta^2/d^4$, and the total bending energy (over an area $\sim d^2$) is of the order of $Eh^3\zeta^2/d^2$. The strain tensor for the stretching is again $\sim \zeta/R$, and the total stretching energy due to the concentrated forces is $\sim Eh\zeta^2 d^2/R^2$. Since the bending energy increases and the stretching energy decreases with decreasing d, it is clear that both energies must be taken into account in determining the deformation near the point of application of the forces. The size d of the region of bending is given in order of magnitude by the condition that the sum of these energies be a minimum, whence

$$d \sim \sqrt{(hR)}. \tag{15.3}$$

The energy $\sim Eh^2\zeta^2/R$. Varying this with respect to ζ and equating the result to the work done by the force f, we find the deflection $\zeta \sim fR/Eh^2$.

However, if the forces acting on the shell are sufficiently large, the shape of the shell may be considerably changed by bulges which form in it. The determination of the deformation as a function of the applied loads requires special investigation in this unusual case.†

Let a convex shell (with edges fixed in such a way that it is geometrically rigid) be subjected to the action of a large concentrated force f along the inward normal. For simplicity we shall assume that the shell is part of a sphere with radius R. The region of the bulge will be a spherical cap which is almost a mirror image of its original shape (Fig. 9

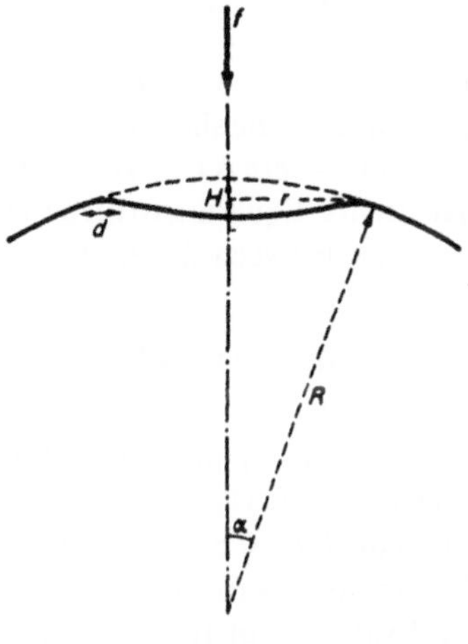

FIG. 9

† The results given below are due to A. V. Pogorelov (1960). A more precise analysis of the problem together with some similar ones is given in his book *Teoriya obolochek pri zakriticheskikh deformatsiyakh* (*Theory of Shells at Supercritical Deformations*), Moscow 1965.

shows a meridional section of the shell). The problem is to determine the size of the bulge as a function of the force.

The major part of the elastic energy is concentrated in a narrow strip near the edge of the bulge, where the bending of the shell is relatively large; we shall call this the *bending strip* and denote its width by d. This energy may be estimated, assuming that the radius r of the bulge region is much less than R, so that the angle $\alpha \ll 1$ (Fig. 9). Then $r = R \sin \alpha \sim R\alpha$, and the depth of the bulge $H = 2R(1 - \cos \alpha) \sim R\alpha^2$. Let ζ denote the displacement of points on the shell in the bending strip. Just as previously, we find that the energies of bending along the meridian and of stretching along the circle of latitude† per unit surface area are respectively, in order of magnitude, $Eh^3\zeta^2/d^4$ and $Eh\zeta^2/R^2$. The order of magnitude of the displacement ζ is in this case determined geometrically: the direction of the meridian changes by an angle $\sim \alpha$ over the width d, and so $\zeta \sim \alpha d \sim rd/R$. Multiplying by the area of the bending strip ($\sim rd$), we obtain the energies Eh^3r^3/R^2d and Ehd^3r^3/R^4. The condition for their sum to be a minimum again gives $d \sim \sqrt{(hR)}$, and the total elastic energy is then $\sim Er^3(h/R)^{5/2}$, or‡

$$\text{constant} \times E h^{5/2} . H^{3/2}/R. \tag{15.4}$$

In this derivation it has been assumed that $d \ll r$; formula (15.4) is therefore valid if the condition

$$Rh/r^2 \ll 1 \tag{15.5}$$

holds. When a bulge is formed, the outer layers of a spherical segment become the inner ones and are therefore compressed, while the inner layers become the outer ones and are stretched. The relative extension (or compression) $\sim h/R$, and so the corresponding total energy in the region of the bulge $\sim E(h/R)^2 hr^2$. With the condition (15.5) it is in fact small in comparison with the energy (15.4) in the bending strip.

The required relation between the depth of the bulge H and the applied force f is obtained by equating f to the derivative of the energy (15.4) with respect to H. Thus we find

$$H \sim f^2 R^2/E^2 h^5. \tag{15.6}$$

It should be noticed that this relation is non-linear.

Finally, let the deformation (bulge) of the shell occur under a uniform external pressure p. In this case the work done is $p\Delta V$, where $\Delta V \sim Hr^2 \sim H^2 R$ is the change in the volume within the shell when the bulge occurs. Equating to zero the derivative with respect to H of the total free energy (the difference between the elastic energy (15.4) and this work), we obtain

$$H \sim h^5 E^2/R^4 p^2. \tag{15.7}$$

The inverse variation (H increasing when p decreases) shows that in this case the bulge is unstable. The value of H given by formula (15.7) corresponds to unstable equilibrium for a given p: bulges with larger values of H grow of their own accord, while smaller ones shrink (it is easy to verify that (15.7) corresponds to a maximum and not a minimum of the total free energy). There is a critical value p_{cr} of the external load beyond which even small

---

† The curvature of the shell does not affect the bending along the meridian in the first approximation, so that this bending occurs without any general stretching along the meridian, as in the cylindrical bending of a flat plate.

‡ A more accurate calculation shows that the constant coefficient is $1.2(1 - \sigma^2)^{-3/4}$.

changes in the shape of the shell increase in size spontaneously. This value may be defined as that which gives $H \sim h$ in (15.7):

$$p_{cr} \sim Eh^2/R^2, \tag{15.8}$$

We shall add to the above brief account of shell theory only a few simple examples in the following Problems.

PROBLEMS

PROBLEM 1. Derive the equations of equilibrium for a spherical shell (with radius R) deformed symmetrically about an axis through its centre.

SOLUTION. We take as two-dimensional coordinates on the surface of the shell the angles θ, ϕ in a system of spherical polar coordinates, whose origin is at the centre of the sphere and polar axis along the axis of symmetry of the deformed shell.

Let P_r be the external radial force per unit surface area. This force must be balanced by a radial resultant of internal stresses acting tangentially on an element of the shell. The condition is

$$h(\sigma_{\phi\phi} + \sigma_{\theta\theta})/R = P_r. \tag{1}$$

This equation is exactly analogous to Laplace's equation for the pressure difference between two media caused by surface tension at the surface of separation.

Next, let $Q_z(\theta)$ be the resultant of all external forces on the part of the shell lying above the co-latitude θ; this resultant is along the polar axis. The force $Q_z(\theta)$ must be balanced by the projection on the polar axis of the stresses $2\pi Rh\sigma_{\theta\theta}\sin\theta$ acting on the cross-section $2\pi Rh\sin\theta$ of the shell at that latitude. Hence

$$2\pi Rh\sigma_{\theta\theta}\sin^2\theta = Q_z(\theta). \tag{2}$$

Equations (1) and (2) determine the stress distribution, and the strain tensor is then given by the formulae

$$u_{\theta\theta} = (\sigma_{\theta\theta} - \sigma\sigma_{\phi\phi})/E, \qquad u_{\phi\phi} = (\sigma_{\phi\phi} - \sigma\sigma_{\theta\theta})/E, \qquad u_{\theta\phi} = 0. \tag{3}$$

Finally, the displacement vector is obtained from the equations

$$u_{\theta\theta} = \frac{1}{R}\left(\frac{du_\theta}{d\theta} + u_r\right), \qquad u_{\theta\phi} = \frac{1}{R}(u_\theta \cot\theta + u_r). \tag{4}$$

PROBLEM 2. Determine the deformation under its own weight of a hemispherical shell convex upwards, the edge of which moves freely on a horizontal support (Fig. 10).

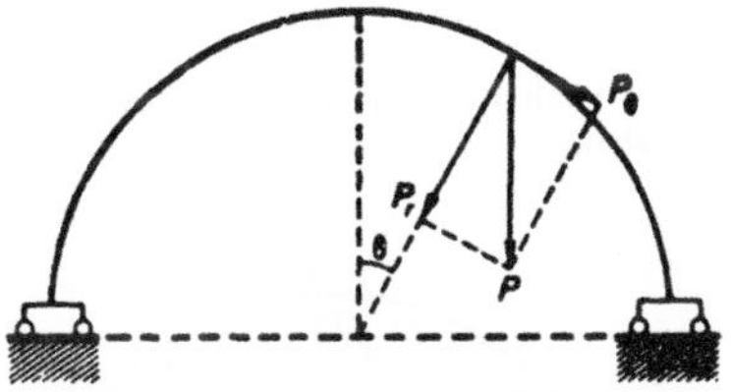

FIG. 10

SOLUTION. We have $P_r = -\rho gh\cos\theta, Q_z = -2\pi R^2\rho gh(1-\cos\theta)$; Q_z is the total weight of the shell above the circle of co-latitude θ. From (1) and (2) of Problem 1 we find

$$\sigma_{\theta\theta} = -\frac{R\rho g}{1+\cos\theta}, \qquad \sigma_{rr} = R\rho g\left(\frac{1}{1+\cos\theta} - \cos\theta\right).$$

From (3) we calculate $u_{\phi\phi}$ and $u_{\theta\theta}$, and then obtain u_θ and u_r from (4); the constant in the integration of the first equation (4) is chosen so that for $\theta = \frac{1}{2}\pi$ we have $u_\theta = 0$. The result is

$$u_\theta = \frac{R^2\rho g(1+\sigma)}{E}\left\{\frac{\cos\theta}{1+\cos\theta} + \log(1+\cos\theta)\right\}\sin\theta,$$

$$u_r = \frac{R^2\rho g(1+\sigma)}{E}\left\{1 - \frac{2+\sigma}{1+\sigma}\cos\theta - \cos\theta\log(1+\cos\theta)\right\}.$$

The value of u_r for $\theta = \frac{1}{2}\pi$ gives the horizontal displacement of the support.

Problem 3. Determine the deformation of a hemispherical shell with clamped edges, convex downwards and filled with liquid (Fig. 11); the weight of the shell itself can be neglected in comparison with that of the liquid.

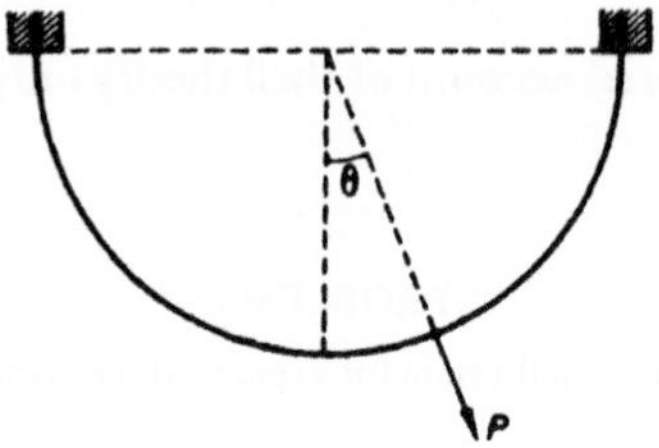

Fig. 11

Solution. We have

$$P_r = \rho_0 g R \cos\theta, \qquad P_\theta = 0,$$

$$Q_z = 2\pi R^2 \int_0^\theta P_r \cos\theta \sin\theta \, d\theta = \frac{2}{3}\pi R^3 \rho_0 g (1 - \cos^3\theta),$$

where ρ_0 is the density of the liquid. We find from (1) and (2) of Problem 1

$$\sigma_{\theta\theta} = \frac{R^2 \rho_0 g}{3h} \cdot \frac{1 - \cos^3\theta}{\sin^2\theta}, \qquad \sigma_{\phi\phi} = \frac{R^2 \rho_0 g}{3h} \cdot \frac{-1 + 3\cos\theta - 2\cos^3\theta}{\sin^2\theta}.$$

The displacements are

$$u_\theta = -\frac{R^3 \rho_0 g (1+\sigma)}{3Eh} \sin\theta \left[\frac{\cos\theta}{1+\cos\theta} + \log(1+\cos\theta)\right],$$

$$u_r = \frac{R^3 \rho_0 g (1+\sigma)}{3Eh}\left[\cos\theta \log(1+\cos\theta) - 1 + \frac{3\cos\theta}{1+\sigma}\right].$$

For $\theta = \frac{1}{2}\pi$, u_r is not zero as it should be. This means that the shell is actually so severely bent near the clamped edge that the above solution is invalid.

Problem 4. A shell in the form of a spherical cap rests on a fixed support (Fig. 12). Determine the bending resulting from the weight Q of the shell.

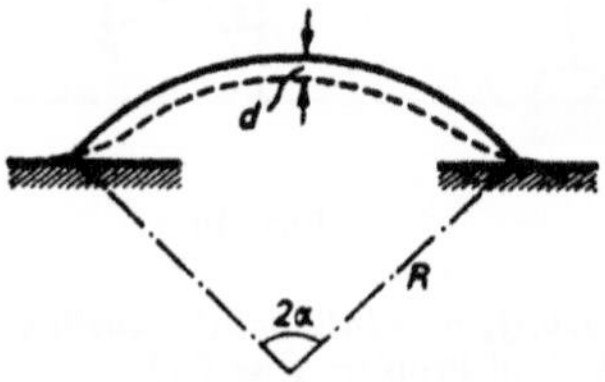

Fig. 12

Solution. The main deformation occurs near the edge, which is bent as shown by the dashed line in Fig. 12. The displacement u_θ is small compared with the radial displacement $u_r \equiv \zeta$. Since ζ decreases rapidly as we move away from the supported edge, the deformation can be regarded as that of a long flat plate (with length $2\pi R \sin\alpha$). This deformation is composed of a bending and a stretching of the plate. The relative extension at each point is ζ/R (R being the radius of the shell), and therefore the stretching energy is $E\zeta^2/2R^2$ per unit volume. Using as the independent variable the distance x from the line of support, we have for the total stretching energy

$$F_{1,\mathrm{pl}} = 2\pi R \sin\alpha \frac{Eh}{2R^2} \int \zeta^2 \, dx.$$

The bending energy is

$$F_{2,\,\mathrm{pl}} = 2\pi R \sin\alpha . \tfrac{1}{2} D \int \left(\frac{d^2\zeta}{dx^2}\right)^2 dx.$$

Varying the sum $F_{\mathrm{pl}} = F_{1,\,\mathrm{pl}} + F_{2,\,\mathrm{pl}}$ with respect to ζ, we obtain

$$\frac{d^4\zeta}{dx^4} + \frac{12(1-\sigma^2)}{h^2 R^2}\zeta = 0.$$

For $x \to \infty$, ζ must tend to zero, and for $x = 0$ we must have the boundary conditions of zero moment of the forces ($\zeta'' = 0$) and equality of the normal force and the corresponding component of the force of gravity:

$$2\pi R \sin\alpha . D\zeta''' = Q \cos\alpha.$$

The solution which satisfies these conditions is $\zeta = Ae^{-\kappa x}\cos\kappa x$, where

$$\kappa = \left[\frac{3(1-\sigma^2)}{h^2 R^2}\right]^{1/4}, \quad A = \frac{Q \cot\alpha}{Eh}\left[\frac{3R^2(1-\sigma^2)}{8\pi h^2}\right]^{1/4}.$$

The bending of the shell is

$$d = \zeta(0)\cos\alpha = A\cos\alpha.$$

§16. Torsion of rods

Let us now consider the deformation of thin rods. This differs from all the cases hitherto considered, in that the displacement vector **u** may be large even for small strains, i.e. when the tensor u_{ik} is small.† For example, when a long thin rod is slightly bent, its ends may move a considerable distance, even though the relative displacements of neighbouring points in the rod are small.

There are two types of deformation of a rod which may be accompanied by a large displacement of certain parts of it. One of these consists in bending the rod, and the other in twisting it. We shall begin by considering the latter case.

A *torsional deformation* is one in which, although the rod remains straight, each transverse section is rotated through some angle relative to those below it. If the rod is long, even a slight torsion causes sufficiently distant cross-sections to turn through large angles. The generators on the sides of the rod, which are parallel to its axis, become helical in form under torsion.

Let us consider a thin straight rod with arbitrary cross-section. We take a coordinate system with the z-axis along the axis of the rod and the origin somewhere inside the rod. We use also the *torsion angle* τ, which is the angle of rotation per unit length of the rod. This means that two neighbouring cross-sections at a distance dz will rotate through a relative angle $d\phi = \tau\, dz$ (so that $\tau = d\phi/dz$). The torsional deformation itself, i.e. the relative displacement of adjoining parts of the rod, is assumed small. The condition for this to be so is that the relative angle turned through by cross-sections of the rod at a distance apart of the order of its transverse dimension R be small, i.e.

$$\tau R \ll 1. \tag{16.1}$$

Let us examine a small portion of the length of the rod near the origin, and determine the displacements **u** of the points of the rod in that portion. As the undisplaced cross-

---

† The only exception is a simple extension of a rod without change of shape, in which case the vector **u** is always small if the tensor u_{ik} is small, i.e. if the extension is small.

section we take that given by the xy-plane. When a radius vector $\mathbf{r}$ turns through a small angle $\delta\phi$, the displacement of its end is given by

$$\delta\mathbf{r} = \delta\boldsymbol{\phi} \times \mathbf{r}, \tag{16.2}$$

where $\delta\boldsymbol{\phi}$ is a vector whose magnitude is the angle of rotation and whose direction is that of the axis of rotation. In the present case, the rotation is about the z-axis, and for points with coordinate z the angle of rotation relative to the xy-plane is τz (since τ can be regarded as a constant in some region near the origin). Then formula (16.2) gives for the components u_x, u_y of the displacement vector

$$u_x = -\tau z y, \qquad u_y = \tau z x. \tag{16.3}$$

When the rod is twisted, the points in it in general undergo a displacement along the z-axis also. Since for $\tau = 0$ this displacement is zero, it may be supposed proportional to τ when τ is small. Thus

$$u_z = \tau\psi(x, y), \tag{16.4}$$

where $\psi(x, y)$ is some function of x and y, called the *torsion function*. As a result of the deformation described by formulae (16.3) and (16.4), each cross-section of the rod rotates about the z-axis, and also becomes curved instead of plane. It should be noted that, by taking the origin at a particular point in the xy-plane, we "fix" a certain point in the cross-section of the rod in such a way that it cannot move in that plane (but it can move in the z-direction). A different choice of origin would not, of course, affect the torsional deformation itself, but would give only an unimportant displacement of the rod as a whole.

Knowing $\mathbf{u}$, we can find the components of the strain tensor. Since $\mathbf{u}$ is small in the region under consideration, we can use the formula

$$u_{ik} = \tfrac{1}{2}(\partial u_i/\partial x_k + \partial u_k/\partial x_i).$$

The result is

$$u_{xx} = u_{yy} = u_{xy} = u_{zz} = 0,$$

$$u_{xz} = \tfrac{1}{2}\tau\left(\frac{\partial\psi}{\partial x} - y\right), \qquad u_{yz} = \tfrac{1}{2}\tau\left(\frac{\partial\psi}{\partial y} + x\right). \tag{16.5}$$

It should be noticed that $u_{ll} = 0$; in other words, torsion does not result in a change in volume, i.e. it is a pure shear deformation.

For the components of the stress tensor we find

$$\sigma_{xx} = \sigma_{yy} = \sigma_{zz} = \sigma_{xy} = 0,$$

$$\sigma_{xz} = 2\mu u_{xz} = \mu\tau\left(\frac{\partial\psi}{\partial x} - y\right), \qquad \sigma_{yz} = 2\mu u_{yz} = \mu\tau\left(\frac{\partial\psi}{\partial y} + x\right). \tag{16.6}$$

Here it is more convenient to use the modulus of rigidity μ in place of E and σ. Since only σ_{xz} and σ_{yz} are different from zero, the general equations of equilibrium $\partial\sigma_{ik}/\partial x_k = 0$ reduce to

$$\frac{\partial\sigma_{zx}}{\partial x} + \frac{\partial\sigma_{zy}}{\partial y} = 0. \tag{16.7}$$

Substituting (16.6), we find that the torsion function must satisfy the equation

$$\triangle\psi = 0, \tag{16.8}$$

where $\triangle$ is the two-dimensional Laplacian.

It is rather more convenient, however, to use a different auxiliary function $\chi(x, y)$, defined by

$$\sigma_{xz} = 2\mu\tau\partial\chi/\partial y, \qquad \sigma_{yz} = -2\mu\tau\partial\chi/\partial x; \tag{16.9}$$

this function satisfies more convenient boundary conditions on the circumference of the rod (see below). Comparing (16.9) and (16.6), we obtain

$$\frac{\partial\psi}{\partial x} = y + 2\frac{\partial\chi}{\partial y}, \qquad \frac{\partial\psi}{\partial y} = -x - 2\frac{\partial\chi}{\partial x}. \tag{16.10}$$

Differentiating the first of these with respect to y, the second with respect to x, and subtracting, we obtain for the function χ the equation

$$\triangle\chi = -1. \tag{16.11}$$

To determine the boundary conditions on the surface of the rod, we note that, since the rod is thin, the external forces on its sides must be small compared with the internal stresses in the rod, and can therefore be put equal to zero in seeking the boundary conditions. This fact is exactly analogous to what we found in discussing the bending of thin plates. Thus we must have $\sigma_{ik}n_k = 0$ on the sides of the rod; since the z-direction is along the axis, $n_z = 0$, and this equation becomes

$$\sigma_{zx}n_x + \sigma_{zy}n_y = 0.$$

Substituting (16.9), we obtain

$$\frac{\partial\chi}{\partial y}n_x - \frac{\partial\chi}{\partial x}n_y = 0.$$

The components of the vector normal to a plane contour (the circumference of the rod) are $n_x = -\mathrm{d}y/\mathrm{d}l$, $n_y = \mathrm{d}x/\mathrm{d}l$, where x and y are coordinates of points on the contour and $\mathrm{d}l$ is an element of arc. Thus we have

$$\frac{\partial\chi}{\partial x}\mathrm{d}x + \frac{\partial\chi}{\partial y}\mathrm{d}y = \mathrm{d}\chi = 0,$$

whence χ = constant, i.e. χ is constant on the circumference. Since only the derivatives of the function χ appear in the definitions (16.9), it is clear that any constant may be added to χ. If the cross-section is singly connected, we can therefore use, without loss of generality, the boundary condition

$$\chi = 0 \tag{16.12}$$

on equation (16.11).†

For a multiply connected cross-section, however, χ will have different constant values on each of the closed curves bounding in the cross-section. Hence we can put $\chi = 0$ on only one of these curves, for instance the outermost (C_0 in Fig. 13). The values of χ on the remaining bounding curves are found from conditions which are a consequence of the

---

† The problem of determining the torsion deformation from equation (16.11) with the boundary condition (16.12) is formally identical with that of determining the bending of a uniformly loaded plane membrane from equation (14.9).

It is useful to note also an analogy with fluid mechanics: an equation of the form (16.11) determines the velocity distribution $v(x, y)$ for a viscous fluid in a pipe, and the boundary condition (16.12) corresponds to the condition $v = 0$ at the fixed walls of the pipe (*see FM*, §17).

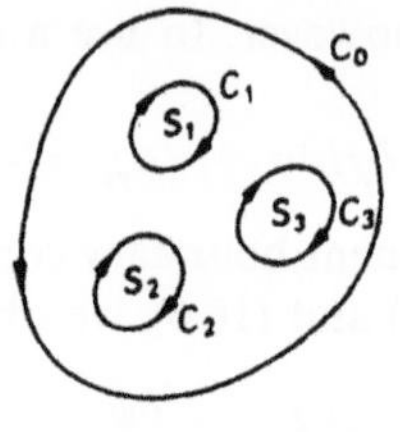

FIG. 13

one-valuedness of the displacement $u_z = \tau\psi(x, y)$ as a function of the coordinates. For, since the torsion function $\psi(x, y)$ is one-valued, the integral of its differential $d\psi$ round a closed contour must be zero. Using the relations (16.10), we therefore have

$$\oint d\psi = \oint\left(\frac{\partial\psi}{\partial x}dx + \frac{\partial\psi}{\partial y}dy\right)$$
$$= -2\oint\left(\frac{\partial\chi}{\partial x}dy - \frac{\partial\chi}{\partial y}dx\right) - \oint(x\,dy - y\,dx)$$
$$= 0,$$

or

$$\oint\frac{\partial\chi}{\partial n}dl = -S, \tag{16.13}$$

where $\partial\chi/\partial n$ is the derivative of the function χ along the outward normal to the curve, and S the area enclosed by the curve. Applying (16.13) to each of the closed curves $C_1, C_2, \ldots,$ we obtain the required conditions.

Let us determine the free energy of a rod under torsion. The energy per unit volume is

$$F = \tfrac{1}{2}\sigma_{ik}u_{ik} = \sigma_{xz}u_{xz} + \sigma_{yz}u_{yz} = (\sigma_{xz}{}^2 + \sigma_{yz}{}^2)/2\mu$$

or, substituting (16.9),

$$F = 2\mu\tau^2\left[\left(\frac{\partial\chi}{\partial x}\right)^2 + \left(\frac{\partial\chi}{\partial y}\right)^2\right] \equiv 2\mu\tau^2(\mathbf{grad}\,\chi)^2,$$

where **grad** denotes the two-dimensional gradient. The torsional energy per unit length of the rod is obtained by integrating over the cross-section of the rod, i.e. it is $\tfrac{1}{2}C\tau^2$, where the constant $C = 4\mu\int(\mathbf{grad}\,\chi)^2\,df$, and is called the *torsional rigidity* of the rod. The total elastic energy of the rod is equal to the integral

$$F_{\text{rod}} = \tfrac{1}{2}\int C\tau^2\,dz, \tag{16.14}$$

taken along its length.

Putting

$$(\mathbf{grad}\,\chi)^2 = \operatorname{div}(\chi\,\mathbf{grad}\,\chi) - \chi\triangle\chi = \operatorname{div}(\chi\,\mathbf{grad}\,\chi) + \chi$$

and transforming the integral of the first term into one along the circumference of the rod, we obtain

$$C = 4\mu\oint\chi\frac{\partial\chi}{\partial n}dl + 4\mu\int\chi\,df. \tag{16.15}$$

If the cross-section is singly connected, the first term vanishes by the boundary condition $\chi = 0$, leaving

$$C = 4\mu \int \chi \, dx \, dy. \tag{16.16}$$

For a multiply connected cross-section (Fig. 13), we put $\chi = 0$ on the outer boundary C_0 and denote by χ_k the constant values of χ on the inner boundaries C_k, obtaining by (16.13)

$$C = 4\mu \sum_k \chi_k S_k + 4\mu \int \chi \, dx \, dy; \tag{16.17}$$

it should be remembered that, in integrating in the first term in (16.15), we go anticlockwise round the contour C_0 and clockwise round all the others.

Let us consider now a more usual case of torsion, where one of the ends of the rod is held fixed and the external forces are applied only to the other end. These forces are such that they cause only a twisting of the rod, and no other deformation such as bending. In other words, they form a couple which twists the rod about its axis. The moment of this couple will be denoted by M.

We should expect that, in such a case, the torsion angle τ is constant along the rod. This can be seen, for example, from the condition that the free energy of the rod be a minimum in equilibrium. The total energy of a deformed rod is equal to the sum $F_{\text{rod}} + U$, where U is the potential energy due to the action of the external forces. Substituting in (16.14) $\tau = d\phi/dz$ and varying with respect to the angle ϕ, we find

$$\delta \tfrac{1}{2} \int C \left(\frac{d\phi}{dz}\right)^2 dz + \delta U = \int C \frac{d\phi}{dz} \frac{d\delta\phi}{dz} dz + \delta U = 0,$$

or, integrating by parts,

$$-\int C \frac{d\tau}{dz} \delta\phi \, dz + \delta U + [C\tau\delta\phi] = 0.$$

The last term on the left is the difference of the values at the limits of integration, i.e. at the ends of the rod. One of these ends, say the lower one, is fixed, so that $\delta\phi = 0$ there. The variation δU of the potential energy is minus the work done by the external forces in rotation through an angle $\delta\phi$. As we know from mechanics, the work done by a couple in such a rotation is equal to the product $M\delta\phi$ of the angle of rotation and the moment of the couple. Since there are no other external forces, $\delta U = -M\delta\phi$, and we have

$$\int C \frac{d\tau}{dz} \delta\phi \, dz + [\delta\phi(-M + C\tau)] = 0. \tag{16.18}$$

The second term on the left has its value at the upper end of the rod. In the integral over z, the variation $\delta\phi$ is arbitrary, and so we must have

$$C \, d\tau/dz = 0,$$

i.e.

$$\tau = \text{constant}. \tag{16.19}$$

Thus the torsion angle is constant along the rod. The total angle of rotation of the upper end of the rod relative to the lower end is τl, where l is the length of the rod.

In equation (16.18), the second term also must be zero, and we obtain the following expression for the constant torsion angle:

$$\tau = M/C. \qquad (16.20)$$

PROBLEMS

PROBLEM 1. Determine the torsional rigidity of a rod whose cross-section is a circle with radius R.

SOLUTION. The solutions of Problems 1–4 are formally identical with those of problems of the motion of a viscous fluid in a pipe of corresponding cross-section (see the last footnote to this section). The discharge Q is here represented by C.

For a rod of circular cross-section we have, taking the origin at the centre of the circle, $\chi = \frac{1}{4}(R^2 - x^2 - y^2)$, and the torsional rigidity is $C = \frac{1}{2}\mu\pi R^4$. For the function ψ we have, from (16.10), ψ = constant. A constant ψ, however, corresponds by (16.4) to a simple displacement of the whole rod along the z-axis, and so we can suppose that $\psi = 0$. Thus the transverse sections of a circular rod undergoing torsion remain plane.

PROBLEM 2. The same as Problem 1, but for an elliptical cross-section with semi-axes a and b.

SOLUTION. The torsional rigidity is $C = \pi\mu a^3b^3/(a^2 + b^2)$. The distribution of longitudinal displacements is given by the torsion function $\psi = (b^2 - a^2)xy/(b^2 + a^2)$, where the coordinate axes coincide with those of the ellipse.

PROBLEM 3. The same as Problem 1, but for an equilateral triangular cross-section with side a.

SOLUTION. The torsional rigidity is $C = \sqrt{3}\mu a^4/80$. The torsion function is

$$\psi = y(x\sqrt{3} + y)(x\sqrt{3} - y)/6a$$

the origin being at the centre of the triangle and the x-axis along an altitude.

PROBLEM 4. The same as Problem 1, but for a rod in the form of a long thin plate (with width d and thickness $h \ll d$).

SOLUTION. The problem is equivalent to that of viscous fluid flow between plane parallel walls. The result is that $C = \frac{1}{3}\mu dh^3$.

PROBLEM 5. The same as Problem 1, but for a cylindrical pipe with internal and external radii R_1 and R_2 respectively.

SOLUTION. The function $\chi = \frac{1}{4}(R_2{}^2 - r^2)$ (in polar coordinates) satisfies the condition (16.13) at both boundaries of the annular cross-section of the pipe. From formula (16.17) we then find $C = \frac{1}{2}\mu\pi(R_2{}^4 - R_1{}^4)$.

PROBLEM 6. The same as Problem 1, but for a thin-walled pipe with arbitrary cross-section.

SOLUTION. Since the walls are thin, we can assume that χ varies through the wall thickness h, from zero on one side to χ_1 on the other, according to the linear law $\chi = \chi_1 y/h$ (y being a coordinate measured through the wall). Then the condition (16.13) gives $\chi_1 L/h = S$, where L is the perimeter of the pipe cross-section and S the area which it encloses. The second term in the expression (16.17) is small compared with the first, and we obtain $C = 4hS^2\mu/L$. If the pipe is cut longitudinally along a generator, the torsional rigidity falls sharply, becoming (by the result of Problem 4) $C = \frac{1}{3}\mu Lh^3$.

§17. Bending of rods

A bent rod is stretched at some points and compressed at others. Lines on the convex side of the bent rod are extended, and those on the concave side are compressed. As with plates, there is a neutral surface in the rod, which undergoes neither extension nor compression. It separates the region of compression from the region of extension.

Let us begin by investigating a bending deformation in a small portion of the length of the rod, where the bending may be supposed slight; by this we here mean that not only the strain tensor but also the magnitudes of the displacements of points in the rod are small. We take a coordinate system with the origin on the neutral surface in the portion considered, and the z-axis parallel to the axis of the undeformed rod. Let the bending occur in the zx-plane. In a rod undergoing only small deflections we can suppose that the bending occurs in a single plane. This follows from the result of differential geometry that the deviation of a slightly bent curve from a plane (its torsion) is of a higher order of smallness than its curvature.

As in the bending of plates and the twisting of rods, the external forces on the sides of a thin bent rod are small compared with the internal stresses, and can be taken as zero in determining the boundary conditions at the sides of the rod. Thus we have everywhere on the sides of the rod $\sigma_{ik}n_k = 0$, or, since $n_z = 0$, $\sigma_{xx}n_x + \sigma_{xy}n_y = 0$, and similarly for $i = y, z$. We take a point on the circumference of a cross-section for which the normal $\mathbf{n}$ is parallel to the x-axis. There will be another such point somewhere on the opposite side of the rod. At both these points $n_y = 0$, and the above equation gives $\sigma_{xx} = 0$. Since the rod is thin, however, σ_{xx} must be small everywhere in the cross-section if it vanishes on either side. We can therefore put $\sigma_{xx} = 0$ everywhere in the rod. In a similar manner, it can be seen that all the components of the stress tensor except σ_{zz} must be zero. That is, in the bending of a thin rod only the extension (or compression) component of the internal stress tensor is large. A deformation in which only the component σ_{zz} of the stress tensor is non-zero is just a simple extension or compression (§5). Thus there is a simple extension or compression in every volume element of a bent rod. The amount of this varies, of course, from point to point in every cross-section, and so the whole rod is bent.

It is easy to determine the relative extension at any point in the rod. Let us consider an element of length $\mathrm{d}z$ parallel to the axis of the rod and near the origin. On bending, the length of this element becomes $\mathrm{d}z'$. The only elements which remain unchanged are those which lie in the neutral surface. Let R be the radius of curvature of the neutral surface near the origin. The lengths $\mathrm{d}z$ and $\mathrm{d}z'$ can be regarded as elements of arcs of circles whose radii are respectively R and $R + x$, x being the coordinate of the point where $\mathrm{d}z'$ lies. Hence

$$\mathrm{d}z' = \frac{R+x}{R}\,\mathrm{d}z = \left(1 + \frac{x}{R}\right)\mathrm{d}z.$$

The relative extension is therefore $(\mathrm{d}z' - \mathrm{d}z)/\mathrm{d}z = x/R$.

The relative extension of the element $\mathrm{d}z$, however, is equal to the component u_{zz} of the strain tensor. Thus

$$u_{zz} = x/R. \tag{17.1}$$

We can now find σ_{zz} by using the relation $\sigma_{zz} = Eu_{zz}$ which holds for a simple extension. This gives

$$\sigma_{zz} = Ex/R. \tag{17.2}$$

The position of the neutral surface in a bent rod has now to be determined. This can be done from the condition that the deformation considered must be pure bending, with no general extension or compression of the rod. The total internal stress force on a cross-section of the rod must therefore be zero, i.e. the integral $\int \sigma_{zz}\,\mathrm{d}f$, taken over a cross-section, must vanish. Using the expression (17.2) for σ_{zz}, we obtain the condition

$$\int x\,\mathrm{d}f = 0. \tag{17.3}$$

We can now bring in the centre of mass of the cross-section, which is that of a uniform flat disc of the same shape. The coordinates of the centre of mass are, as we know, given by the integrals $\int x\,\mathrm{d}f/\int \mathrm{d}f$, $\int y\,\mathrm{d}f/\int \mathrm{d}f$. Thus the condition (17.3) signifies that, in a coordinate system with the origin in the neutral surface, the x coordinate of the centre of mass of any cross-section is zero. The neutral surface therefore passes through the centres of mass of the cross-sections of the rod.

Two components of the strain tensor besides u_{zz} are non-zero, since for a simple extension we have $u_{xx} = u_{yy} - \sigma u_{zz}$. Knowing the strain tensor, we can easily find the displacement also:

$$u_{zz} = \partial u_z/\partial z = x/R, \qquad \partial u_x/\partial x = \partial u_y/\partial y = -\sigma x/R,$$

$$\frac{\partial u_z}{\partial x} + \frac{\partial u_x}{\partial z} = 0, \qquad \frac{\partial u_x}{\partial y} + \frac{\partial u_y}{\partial x} = 0, \qquad \frac{\partial u_y}{\partial z} + \frac{\partial u_z}{\partial y} = 0.$$

Integration of these equations gives the following expressions for the components of the displacement:

$$\left.\begin{aligned} u_x &= -\frac{1}{2R}\{z^2 + \sigma(x^2 - y^2)\}, \\ u_y &= -\sigma xy/R, \quad u_z = xz/R. \end{aligned}\right\} \tag{17.4}$$

The constants of integration have been put equal to zero; this means that we "fix" the origin.

It is seen from formulae (17.4) that the points initially on a cross-section z = constant $\equiv z_0$ will be found, after the deformation, on the surface $z = z_0 + u_z = z_0(1 + x/R)$. We see that, in the approximation used, the cross-sections remain plane but are turned through an angle relative to their initial positions. The shape of the cross-section changes, however; for example, when a rod of rectangular cross-section (sides a, b) is bent, the sides $y = \pm\frac{1}{2}b$ of the cross-section become $y = \pm\frac{1}{2}b + u_y = \pm\frac{1}{2}b(1 - \sigma x/R)$, i.e. no longer parallel but still straight. The sides $x = \pm\frac{1}{2}a$, however, are bent into the parabolic curves

$$x = \pm\tfrac{1}{2}a + u_x = \pm\tfrac{1}{2}a - \frac{1}{2R}\left[z_0{}^2 + \sigma(\tfrac{1}{4}a^2 - y^2)\right]$$

(Fig. 14).

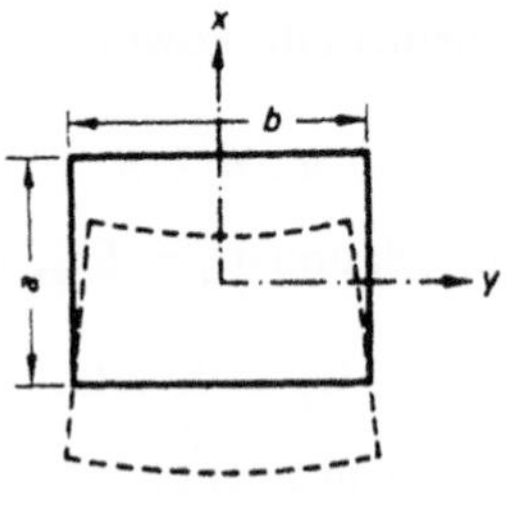

FIG. 14

The free energy per unit volume of the rod is

$$\tfrac{1}{2}\sigma_{ik}u_{ik} = \tfrac{1}{2}\sigma_{zz}u_{zz} = \tfrac{1}{2}Ex^2/R^2.$$

Integrating over the cross-section of the rod, we have

$$\tfrac{1}{2}(E/R^2)\int x^2\,\mathrm{d}f. \tag{17.5}$$

This is the free energy per unit length of a bent rod. The radius of curvature R is that of the neutral surface. However, since the rod is thin, R can here be regarded, to the same approximation, as the radius of curvature of the bent rod itself, regarded as a line (often called an "elastic line").

In the expression (17.5) it is convenient to introduce the moment of inertia of the cross-section. The moment of inertia about the y-axis in its plane is defined as

$$I_y = \int x^2 \, \mathrm{d}f, \tag{17.6}$$

analogously to the ordinary moment of inertia, but with the surface element $\mathrm{d}f$ instead of the mass element. Then the free energy per unit length of the rod can be written

$$\tfrac{1}{2} E I_y / R^2. \tag{17.7}$$

We can also determine the moment of the internal stress forces on a given cross-section of the rod (the *bending moment*). A force $\sigma_{zz} \, \mathrm{d}f = (xE/R) \, \mathrm{d}f$ acts in the z-direction on the surface element $\mathrm{d}f$ of the cross-section. Its moment about the y-axis is $x\sigma_{zz} \, \mathrm{d}f$. Hence the total moment of the forces about this axis is

$$M_y = (E/R) \int x^2 \, \mathrm{d}f = E I_y / R. \tag{17.8}$$

Thus the curvature $1/R$ of the elastic line is proportional to the bending moment on the cross-section concerned.

The magnitude of I_y depends on the direction of the y-axis in the cross-sectional plane. It is convenient to express I_y in terms of the principal moments of inertia. If θ is the angle between the y-axis and one of the principal axes of inertia in the cross-section, we know from mechanics that

$$I_y = I_1 \cos^2\theta + I_2 \sin^2\theta, \tag{17.9}$$

where I_1 and I_2 are the principal moments of inertia. The planes through the z-axis and the principal axes of inertia are called the *principal planes of bending*.

If, for example, the cross-section is rectangular (with sides a, b), its centre of mass is at the centre of the rectangle, and the principal axes of inertia are parallel to the sides. The principal moments of inertia are

$$I_1 = a^3 b/12, \qquad I_2 = ab^3/12. \tag{17.10}$$

For a circular cross-section with radius R, the centre of mass is at the centre of the circle, and the principal axes are arbitrary. The moment of inertia about any axis lying in the cross-section and passing through the centre is

$$I = \tfrac{1}{4} \pi R^4. \tag{17.11}$$

§18. The energy of a deformed rod

In §17 we have discussed only a small portion of the length of a bent rod. In going on to investigate the deformation throughout the rod, we must begin by finding a suitable method of describing this deformation. It is important to note that, when a rod undergoes large bending deflections,† there is in general a twisting of it as well, so that the resulting deformation is a combination of pure bending and torsion.

† By this, it should be remembered, we mean that the vector **u** is not small, but the strain tensor is still small.

To describe the deformation, it is convenient to proceed as follows. We divide the rod into infinitesimal elements, each of which is bounded by two adjacent cross-sections. For each such element we use a coordinate system ξ, η, ζ, so chosen that all the systems are parallel in the undeformed state, and their ζ-axes are parallel to the axis of the rod. When the rod is bent, the coordinate system in each element is rotated, and in general differently in different elements. Any two adjacent systems are rotated through an infinitesimal relative angle.

Let $d\boldsymbol{\phi}$ be the vector of the angle of relative rotation of two systems at a distance dl apart along the rod (we know that an infinitesimal angle of rotation can be regarded as a vector parallel to the axis of rotation; its components are the angles of rotation about each of the three axes of coordinates).

To describe the deformation, we use the vector

$$\boldsymbol{\Omega} = d\boldsymbol{\phi}/dl, \tag{18.1}$$

which gives the "rate" of rotation of the coordinate axes along the rod. If the deformation is a pure torsion, the coordinate system rotates only about the axis of the rod, i.e. about the ζ-axis. In this case, therefore, the vector $\boldsymbol{\Omega}$ is parallel to the axis of the rod, and is just the torsion angle τ used in §16. Correspondingly, in the general case of an arbitrary deformation we can call the component Ω_ζ of the vector $\boldsymbol{\Omega}$ the *torsion angle*. For a pure bending of the rod in a single plane, on the other hand, the vector $\boldsymbol{\Omega}$ has no component Ω_ζ, i.e. it lies in the $\xi\eta$-plane at each point. If we take the plane of bending as the $\xi\zeta$-plane, then the rotation is about the η-axis at every point, i.e. $\boldsymbol{\Omega}$ is parallel to the η-axis.

We take a unit vector $\mathbf{t}$ tangential to the rod (regarded as an elastic line). The derivative $d\mathbf{t}/dl$ is the curvature vector of the line; its magnitude is $1/R$, where R is the radius of curvature,† and its direction is that of the principal normal to the curve. The change in a vector due to an infinitesimal rotation is equal to the vector product of the rotation vector and the vector itself. Hence the change in the vector $\mathbf{t}$ between two neighbouring points of the elastic line is given by $d\mathbf{t} = d\boldsymbol{\phi}\times\mathbf{t}$, or, dividing by dl,

$$d\mathbf{t}/dl = \boldsymbol{\Omega}\times\mathbf{t}. \tag{18.2}$$

Multiplying this equation vectorially by $\mathbf{t}$, we have

$$\boldsymbol{\Omega} = \mathbf{t}\times d\mathbf{t}/dl + \mathbf{t}(\mathbf{t}\cdot\boldsymbol{\Omega}). \tag{18.3}$$

The direction of the tangent vector at any point is the same as that of the ζ-axis at that point. Hence $\mathbf{t}\cdot\boldsymbol{\Omega} = \Omega_\zeta$. Using the unit vector $\mathbf{n}$ along the principal normal ($\mathbf{n} = R\, d\mathbf{t}/dl$), we can therefore put

$$\boldsymbol{\Omega} = \mathbf{t}\times\mathbf{n}/R + \mathbf{t}\Omega_\zeta. \tag{18.4}$$

The first term on the right is a vector with two components Ω_ξ, Ω_η. The unit vector $\mathbf{t}\times\mathbf{n}$ is the binormal unit vector. Thus the components Ω_ξ, Ω_η form a vector along the binormal to the rod, whose magnitude equals the curvature $1/R$.

By using the vector $\boldsymbol{\Omega}$ to characterize the deformation and ascertaining its properties, we can derive an expression for the elastic free energy of a bent rod. The elastic energy per unit length of the rod is a quadratic function of the deformation, i.e., in this case, a quadratic

† It may be recalled that any curve in space is characterized at each point by a curvature and a torsion. This torsion (which we shall not use) should not be confused with the torsional deformation, which is a twisting of a rod about its axis.

function of the components of the vector $\mathbf{\Omega}$. It is easy to see that there can be no terms in this quadratic form proportional to $\Omega_\xi\Omega_\zeta$ and $\Omega_\eta\Omega_\zeta$. For, since the rod is uniform along its length, all quantities, and in particular the energy, must remain constant when the direction of the positive ζ-axis is reversed, i.e. when ζ is replaced by $-\zeta$, whereas the products mentioned change sign.

For $\Omega_\xi = \Omega_\eta = 0$ we have a pure torsion, and the expression for the energy must be that obtained in §16. Thus the term in Ω_ζ^2 in the free energy is $\frac{1}{2}C\Omega_\zeta^2$.

Finally, the terms quadratic in Ω_ξ and Ω_η can be obtained by starting from the expression (17.7) for the energy of a slightly bent short section of the rod. Let us suppose that the rod is only slightly bent. We take the $\xi\zeta$-plane as the plane of bending, so that the component Ω_ξ is zero; there is also no torsion in a slight bending. The expression for the energy must then be that given by (17.7), i.e. $\frac{1}{2}EI_\eta/R^2$. We have seen, however, that $1/R^2$ is the square of the two-dimensional vector $(\Omega_\xi, \Omega_\eta)$. Hence the energy must be of the form $\frac{1}{2}EI_\eta\Omega_\eta^2$. For an arbitrary choice of the ξ and η axes this expression becomes, as we know from mechanics,

$$\tfrac{1}{2}E\,(I_{\eta\eta}\Omega_\eta^2 + 2I_{\eta\xi}\Omega_\eta\Omega_\xi + I_{\xi\xi}\Omega_\xi^2),$$

where $I_{\eta\eta}, I_{\eta\xi}, I_{\xi\xi}$ are the components of the inertia tensor for the cross-section of the rod. It is convenient to take the ξ and η axes to coincide with the principal axes of inertia. We then have simply $\frac{1}{2}E\,(I_1\Omega_\xi^2 + I_2\Omega_\eta^2)$, where I_1, I_2 are the principal moments of inertia. Since the coefficients of Ω_ξ^2 and Ω_η^2 are constants, the resulting expression must be valid for large deflections also.

Finally, integrating over the length of the rod, we obtain the following expression for the elastic free energy of a bent rod:

$$F_{\text{rod}} = \int \{\tfrac{1}{2}I_1 E\Omega_\xi^2 + \tfrac{1}{2}I_2 E\Omega_\eta^2 + \tfrac{1}{2}C\Omega_\zeta^2\}\,\mathrm{d}l. \tag{18.5}$$

Next, we can express in terms of $\mathbf{\Omega}$ the moment of the forces acting on a cross-section of the rod. This is easily done by again using the results previously obtained for pure torsion and pure bending. In pure torsion, the moment of the forces about the axis of the rod is $C\tau$. Hence we conclude that, in the general case, the moment M_ζ about the ζ-axis must be $C\Omega_\zeta$. Next, in a slight deflection in the $\xi\zeta$-plane, the moment about the η-axis is EI_2/R. In such a bending, however, the vector $\mathbf{\Omega}$ is along the η-axis, so that $1/R$ is just the magnitude of $\mathbf{\Omega}$, and $EI_2/R = EI_2\Omega$. Hence we conclude that, in the general case, we must have $M_\xi = EI_1\Omega_\xi$, $M_\eta = EI_2\Omega_\eta$ (the ξ and η axes being along the principal axes of inertia in the cross-section). Thus the components of the moment vector $\mathbf{M}$ are

$$M_\xi = EI_1\Omega_\xi, \qquad M_\eta = EI_2\Omega_\eta, \qquad M_\zeta = C\Omega_\zeta. \tag{18.6}$$

The elastic energy (18.5), expressed in terms of the moment of the forces, is

$$F_{\text{rod}} = \int \left\{ \frac{M_\xi^2}{2I_1E} + \frac{M_\eta^2}{2I_2E} + \frac{M_\zeta^2}{2C} \right\} \mathrm{d}l. \tag{18.7}$$

An important case of the bending of rods is that of a slight bending, in which the deviation from the initial position is everywhere small compared with the length of the rod. In this case torsion can be supposed absent, and we can put $\Omega_\zeta = 0$, so that (18.4) gives simply

$$\mathbf{\Omega} = \mathbf{t}\times\mathbf{n}/R \equiv \mathbf{t}\times\mathrm{d}\mathbf{t}/\mathrm{d}l. \tag{18.8}$$

We take a coordinate system x, y, z fixed in space, with the z-axis along the axis of the undeformed rod (instead of the system ξ, η, ζ for each point in the rod), and denote by X, Y the coordinates x, y for points on the elastic line; X and Y give the displacement of points on the line from their positions before the deformation.

Since the bending is only slight, the tangent vector **t** is almost parallel to the z-axis, and the difference in direction can be approximately neglected. The unit tangent vector is the derivative $\mathbf{t} = d\mathbf{r}/dl$ of the position vector **r** of a point on the curve with respect to its length. Hence

$$d\mathbf{t}/dl = d^2\mathbf{r}/dl^2 \cong d^2\mathbf{r}/dz^2;$$

the derivative with respect to the length can be approximately replaced by the derivative with respect to z. In particular, the x and y components of this vector are respectively d^2X/dz^2 and d^2Y/dz^2. The components Ω_ξ, Ω_η are, to the same accuracy, equal to Ω_x, Ω_y, and we have from (18.8)

$$\Omega_\xi = -d^2Y/dz^2, \qquad \Omega_\eta = d^2X/dz^2. \tag{18.9}$$

Substituting these expressions in (18.5), we obtain the elastic energy of a slightly bent rod in the form

$$F_{\text{rod}} = \tfrac{1}{2}E\int\left\{I_1\left(\frac{d^2Y}{dz^2}\right)^2 + I_2\left(\frac{d^2X}{dz^2}\right)^2\right\}dz. \tag{18.10}$$

Here I_1 and I_2 are the moments of inertia about the axes of x and y respectively, which are the principal axes of inertia.

In particular, for a rod with circular cross-section, $I_1 = I_2 = I$, and the integrand is just the sum of the squared second derivatives, which in the approximation considered is the square of the curvature:

$$\left(\frac{d^2X}{dz^2}\right)^2 + \left(\frac{d^2Y}{dz^2}\right)^2 \cong \frac{1}{R^2}.$$

Hence formula (18.10) can be plausibly generalized to the case of slight bending of a circular rod having any shape (not necessarily straight) in its undeformed state. To do so, we must write the bending energy as

$$F_{\text{rod}} = \tfrac{1}{2}EI\int\left(\frac{1}{R} - \frac{1}{R_0}\right)^2 dz, \tag{18.11}$$

where R_0 is the radius of curvature at any point of the undeformed rod. This expression has a minimum, as it should, in the undeformed state ($R = R_0$), and for $R_0 \to \infty$ it becomes formula (18.10).

§19. The equations of equilibrium of rods

We can now derive the equations of equilibrium for a bent rod. We again consider an infinitesimal element bounded by two adjoining cross-sections of the rod, and calculate the total force acting on it. We denote by **F** the resultant internal stress on a cross-section.†

† This notation will not lead to any confusion with the free energy, which does not appear in §§19–21.

The components of this vector are the integrals of $\sigma_{i\zeta}$ over the cross-section:

$$F_i = \int \sigma_{i\zeta}\, df. \tag{19.1}$$

If we regard the two adjoining cross-sections as the ends of the element, a force $\mathbf{F}+d\mathbf{F}$ acts on the upper end, and $-\mathbf{F}$ on the lower end; the sum of these is the differential $d\mathbf{F}$. Next, let $\mathbf{K}$ be the external force on the rod per unit length. Then an external force $\mathbf{K}\,dl$ acts on the element of length dl. The resultant of the forces on the element is therefore $d\mathbf{F}+\mathbf{K}\,dl$. This must be zero in equilibrium. Thus we have

$$d\mathbf{F}/dl = -\mathbf{K}. \tag{19.2}$$

A second equation is obtained from the condition that the total moment of the forces on the element be zero. Let $\mathbf{M}$ be the moment of the internal stresses on the cross-section. This is the moment about a point (the origin) which lies in the plane of the cross-section; its components are given by formulae (18.6). We shall calculate the total moment, on the element considered, about a point O lying in the plane of its upper end. Then the internal stresses on this end give a moment $\mathbf{M}+d\mathbf{M}$. The moment about O of the internal stresses on the lower end of the element is composed of the moment $-\mathbf{M}$ of those forces about the origin O' in the plane of the lower end and the moment about O of the total force $-\mathbf{F}$ on that end. This latter moment is $-d\mathbf{l}\times-\mathbf{F}$, where $d\mathbf{l}$ is the vector of the element of length of the rod between O' and O. The moment due to the external forces $\mathbf{K}$ is of a higher order of smallness. Thus the total moment acting on the element considered is $d\mathbf{M}+d\mathbf{l}\times\mathbf{F}$. In equilibrium, this must be zero:

$$d\mathbf{M}+d\mathbf{l}\times\mathbf{F} = 0.$$

Dividing this equation by dl and using the fact that $d\mathbf{l}/dl = \mathbf{t}$ is the unit vector tangential to the rod (regarded as a line), we have

$$d\mathbf{M}/dl = \mathbf{F}\times\mathbf{t}. \tag{19.3}$$

Equations (19.2) and (19.3) form a complete set of equilibrium equations for a rod bent in any manner.

If the external forces on the rod are concentrated, i.e. applied only at isolated points of the rod, the equilibrium equations at all other points are much simplified. For $\mathbf{K} = 0$ we have from (19.2)

$$\mathbf{F} = \text{constant}, \tag{19.4}$$

i.e. the stress resultant is constant along any portion of the rod between points where forces are applied. The values of the constant are found from the fact that the difference $\mathbf{F}_2-\mathbf{F}_1$ of the forces at two points 1 and 2 is

$$\mathbf{F}_2-\mathbf{F}_1 = -\Sigma\mathbf{K}, \tag{19.5}$$

where the sum is over all forces applied to the segment of the rod between the two points. It should be noticed that, in the difference $\mathbf{F}_2-\mathbf{F}_1$, the point 2 is further from the point from which l is measured than is the point 1; this is important in determining the signs in equation (19.5). In particular, if only one concentrated force $\mathbf{f}$ acts on the rod, and is applied at its free end, then $\mathbf{F} = \text{constant} = \mathbf{f}$ at all points of the rod.

The second equilibrium equation (19.3) is also simplified. Putting $\mathbf{t} = d\mathbf{l}/dl = d\mathbf{r}/dl$

(where **r** is the radius vector from any fixed point to the point considered) and integrating, we obtain

$$\mathbf{M} = \mathbf{F}\times\mathbf{r} + \text{constant}, \tag{19.6}$$

since **F** is constant.

If concentrated forces also are absent, and the rod is bent by the application of concentrated moments, i.e. of concentrated couples, then $\mathbf{F}$ = constant at all points of the rod, while **M** is discontinuous at points where couples are applied, the discontinuity being equal to the moment of the couple.

Let us consider also the boundary conditions at the ends of a bent rod. Various cases are possible.

The end of the rod is said to be *clamped* (Fig. 4a, §12) if it cannot move either longitudinally or transversely, and moreover its direction (i.e. the direction of the tangent to the rod) cannot change. In this case the boundary conditions are that the coordinates of the end of the rod and the unit tangential vector **t** there are given. The reaction force and moment exerted on the rod by the clamp are determined by solving the equations.

The opposite case is that of a free end, whose position and direction are arbitrary. In this case the boundary conditions are that the force **F** and moment **M** must be zero at the end of the rod.†

If the end of the rod is fixed to a hinge, it cannot be displaced, but its direction can vary. In this case the moment of the forces on the freely turning end must be zero.

Finally, if the rod is supported (Fig. 4b), it can slide at the point of support but cannot undergo transverse displacements. In this case the direction **t** of the rod at the support and the point on the rod at which it is supported are unknown. The moment of the forces at the point of support must be zero, since the rod can turn freely, and the force **F** at that point must be perpendicular to the rod; a longitudinal force would cause a further sliding of the rod at this point.

The boundary conditions for other modes of fixing the rod can easily be established in a similar manner. We shall not pause to add to the typical examples already given.

It was mentioned at the beginning of §18 that a rod with arbitrary cross-section undergoing large deflections is in general twisted also, even if no external twisting moment is applied to the rod. An exception occurs when a rod is bent in one of its principal planes, in which case there is no torsion. For a rod with circular cross-section no torsion results for any bending (if there is no external twisting moment, of course). This can be seen as follows. The twisting is given by the component $\Omega_\zeta = \mathbf{\Omega}\cdot\mathbf{t}$ of the vector $\mathbf{\Omega}$. Let us calculate the derivative of this along the rod. To do so, we use the fact that $\Omega_\zeta = M_\zeta/C$:

$$\frac{\mathrm{d}}{\mathrm{d}l}(\mathbf{M}\cdot\mathbf{t}) = C\frac{\mathrm{d}\Omega_\zeta}{\mathrm{d}l} = \frac{\mathrm{d}\mathbf{M}}{\mathrm{d}l}\cdot\mathbf{t} + \mathbf{M}\cdot\frac{\mathrm{d}\mathbf{t}}{\mathrm{d}l}.$$

Substituting (19.3), we see that the first term is zero, so that

$$C\mathrm{d}\Omega_\zeta/\mathrm{d}l = \mathbf{M}\cdot\mathrm{d}\mathbf{t}/\mathrm{d}l.$$

For a rod with circular cross-section, $I_1 = I_2 \equiv I$; by (18.3) and (18.6), we can therefore write **M** in the form

$$\mathbf{M} = EI\mathbf{t}\times\mathrm{d}\mathbf{t}/\mathrm{d}l + \mathbf{t}C\Omega_\zeta. \tag{19.7}$$

† If a concentrated force **f** is applied to the free end of the rod, the boundary condition is $\mathbf{F} = \mathbf{f}$, not $\mathbf{F} = 0$.

Multiplying by $\mathbf{dt}/\mathbf{d}l$, we have zero on the right-hand side, so that

$$d\Omega_\zeta/dl = 0,$$

whence

$$\Omega_\zeta = \text{constant}, \tag{19.8}$$

i.e. the torsion angle is constant along the rod. If no twisting moments are applied to the ends of the rod, then Ω_ζ is zero at the ends, and there is no torsion anywhere in the rod.

For a rod with circular cross-section, we can therefore put for pure bending

$$M = EI\mathbf{t}\times d\mathbf{t}/dl = EI\frac{d\mathbf{r}}{dl}\times\frac{d^2\mathbf{r}}{dl^2}. \tag{19.9}$$

Substituting this in (19.3), we obtain the equation for pure bending of a circular rod:

$$EI\frac{d\mathbf{r}}{dl}\times\frac{d^3\mathbf{r}}{dl^3} = \mathbf{F}\times\frac{d\mathbf{r}}{dl}. \tag{19.10}$$

PROBLEMS

PROBLEM 1. Reduce to quadratures the problem of determining the shape of a rod with circular cross-section bent in one plane by concentrated forces.

SOLUTION. Let us consider a portion of the rod lying between points where the forces are applied; on such a portion $\mathbf{F}$ is constant. We take the plane of the bent rod as the xy-plane, with the y-axis parallel to the force $\mathbf{F}$, and introduce the angle θ between the tangent to the rod and the y-axis. Then $dx/dl = \sin\theta$, $dy/dl = \cos\theta$, where x, y are the coordinates of a point on the rod. Expanding the vector products in (19.10), we obtain the following equation for θ as a function of the arc length l: $EId^2\theta/dl^2 - F\sin\theta = 0$. A first integration gives $\frac{1}{2}EI(d\theta/dl)^2 + F\cos\theta = c_1$, and

$$l = \pm\sqrt{(\tfrac{1}{2}EI)}\int\frac{d\theta}{\sqrt{(c_1 - F\cos\theta)}} + c_2. \tag{1}$$

The function $\theta(l)$ can be obtained in terms of elliptic functions. The coordinates

$$x = \int\sin\theta\,dl, \qquad y = \int\cos\theta\,dl$$

are

$$x = \pm\sqrt{[2EI(c_1 - F\cos\theta)/F^2]} + \text{constant},$$

$$y = \pm\sqrt{(\tfrac{1}{2}EI)}\int\frac{\cos\theta\,d\theta}{\sqrt{(c_1 - F\cos\theta)}} + \text{constant}. \tag{2}$$

The moment $\mathbf{M}$ (19.9) is parallel to the z-axis, and its magnitude is $M = EId\theta/dl$.

PROBLEM 2. Determine the shape of a bent rod with one end clamped and the other under a force $\mathbf{f}$ perpendicular to the original direction of the rod (Fig. 15).

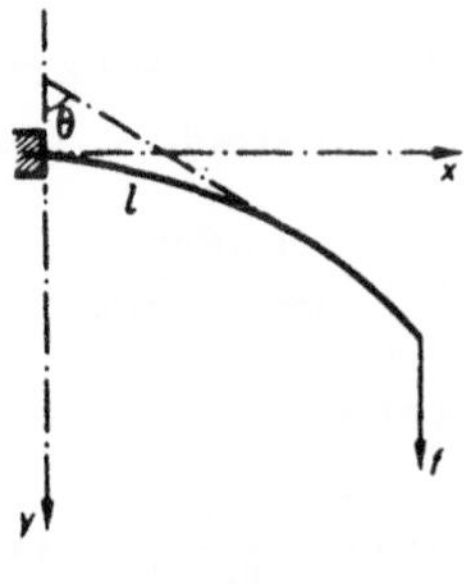

FIG. 15

SOLUTION. We have $\mathbf{F}$ = constant = $\mathbf{f}$ everywhere on the rod. At the clamped end ($l = 0$), $\theta = \frac{1}{2}\pi$, and at the free end ($l = L$, the length of the rod) $M = 0$, i.e. $\theta' = 0$. Putting $\theta(L) \equiv \theta_0$, we have in (1), Problem 1, $c_1 = f\cos\theta_0$ and

$$l = \sqrt{(\tfrac{1}{2}EI/f)} \int_{\theta}^{\frac{1}{2}\pi} \frac{d\theta}{\sqrt{(\cos\theta_0 - \cos\theta)}}.$$

Hence we obtain the equation for θ_0:

$$L = \sqrt{(\tfrac{1}{2}EI/f)} \int_{\theta_0}^{\frac{1}{2}\pi} \frac{d\theta}{\sqrt{(\cos\theta_0 - \cos\theta)}}.$$

The shape of the rod is given by

$$x = \sqrt{(2EI/f)}\,[\sqrt{(\cos\theta_0)} - \sqrt{(\cos\theta_0 - \cos\theta)}],$$

$$y = \sqrt{(EI/2f)} \int_{\theta}^{\frac{1}{2}\pi} \frac{\cos\theta\, d\theta}{\sqrt{(\cos\theta_0 - \cos\theta)}}.$$

PROBLEM 3. The same as Problem 2, but for a force $\mathbf{f}$ parallel to the original direction of the rod.

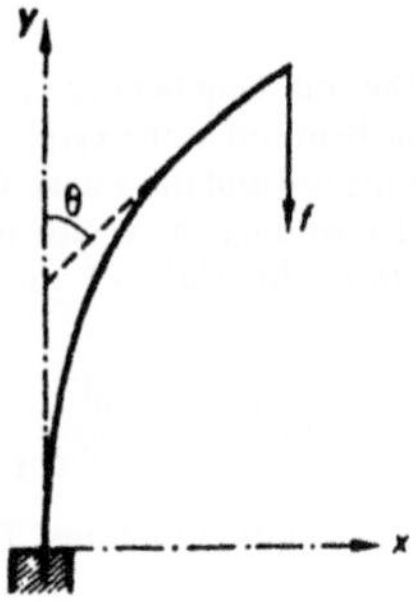

FIG. 16

SOLUTION. We have $\mathbf{F} = -\mathbf{f}$; the coordinate axes are taken as shown in Fig. 16. The boundary conditions are $\theta = 0$ for $l = 0$, $\theta' = 0$ for $l = L$. Then

$$l = \sqrt{(\tfrac{1}{2}EI/f)} \int_{0}^{\theta} \frac{d\theta}{\sqrt{(\cos\theta - \cos\theta_0)}},$$

where θ_0 is determined by the equation $l(\theta_0) = L$. For x and y we obtain

$$x = \sqrt{(2EI/f)}\,[\sqrt{(1 - \cos\theta_0)} - \sqrt{(\cos\theta - \cos\theta_0)}],$$

$$y = \sqrt{(EI/2f)} \int_{0}^{\theta} \frac{\cos\theta\, d\theta}{\sqrt{(\cos\theta - \cos\theta_0)}}.$$

For a small deflection, $\theta_0 \ll 1$, and we can write

$$L \cong \sqrt{(EI/f)} \int_{0}^{\theta_0} \frac{d\theta}{\sqrt{(\theta_0^2 - \theta^2)}} = \tfrac{1}{2}\pi\sqrt{(EI/f)},$$

i.e. θ_0 does not appear. This shows that, in accordance with the result of §21, Problem 3, the solution in question exists only for $f \geqslant \pi^2 EI/4L^2$, i.e. when the rectilinear shape ceases to be stable.

PROBLEM 4. The same as Problem 2, but for the case where both ends of the rod are supported and a force **f** is applied at its centre. The distance between the supports is L_0.

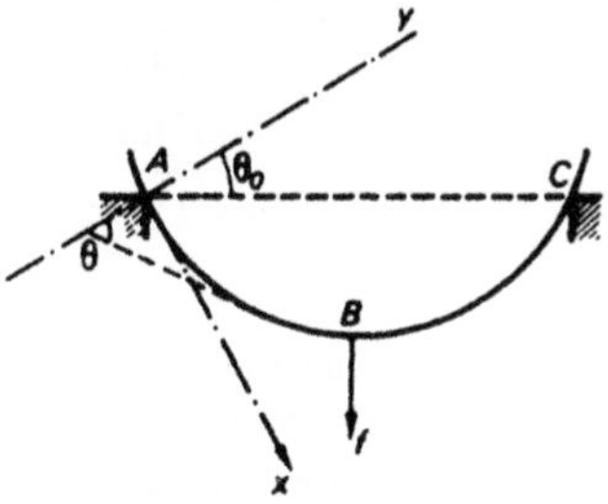

FIG. 17

SOLUTION. We take the coordinate axes as shown in Fig. 17. The force **F** is constant on each of the segments AB and BC, and on each is perpendicular to the direction of the rod at the point of support A or C. The difference between the values of **F** on AB and BC is **f**, and so we conclude that, on AB, $F \sin\theta_0 = -\frac{1}{2}f$, where θ_0 is the angle between the y-axis and the line AC. At the point A ($l = 0$) we have the conditions $\theta = \frac{1}{2}\pi$ and $M = 0$, i.e. $\theta' = 0$, so that on AB

$$l = \sqrt{\frac{EI\sin\theta_0}{f}}\int_{\theta}^{\frac{1}{2}\pi}\frac{d\theta}{\sqrt{\cos\theta}}, \qquad x = 2\sqrt{\frac{EI\sin\theta_0\cos\theta}{f}},$$

$$y = \sqrt{\frac{EI\sin\theta_0}{f}}\int_{\theta}^{\frac{1}{2}\pi}\sqrt{\cos\theta}\, d\theta.$$

The angle θ_0 is determined from the condition that the projection of AB on the straight line AC must be $\frac{1}{2}L_0$, whence

$$\tfrac{1}{2}L_0 = \sqrt{\frac{EI\sin\theta_0}{f}}\int_{\theta_0}^{\frac{1}{2}\pi}\frac{\cos(\theta-\theta_0)}{\sqrt{\sin\theta}}\, d\theta.$$

For some value θ_0 lying between 0 and $\frac{1}{2}\pi$ the derivative $df/d\theta_0$ (f being regarded as a function of θ_0) passes through zero to positive values. A further decrease in θ_0, i.e. increase in the deflection, would mean a decrease in f. This means that the solution found here becomes unstable, the rod collapsing between the supports.

PROBLEM 5. Reduce to quadratures the problem of three-dimensional bending of a rod under the action of concentrated forces.

SOLUTION. Let us consider a segment of the rod between points where forces are applied, on which $\mathbf{F}$ = constant. Integrating (19.10), we obtain

$$EI\frac{d\mathbf{r}}{dl}\times\frac{d^2\mathbf{r}}{dl^2} = \mathbf{F}\times\mathbf{r} + c\mathbf{F}; \tag{1}$$

the constant of integration has been written as a vector $c\mathbf{F}$ parallel to $\mathbf{F}$, since, by appropriately choosing the origin, i.e. by adding a constant vector to $\mathbf{r}$, we can eliminate any vector perpendicular to $\mathbf{F}$. Multiplying (1) scalarly and vectorially by $\mathbf{r}'$ (the prime denoting differentiation with respect to l), and using the fact that $\mathbf{r}'\cdot\mathbf{r}'' = 0$ (since $\mathbf{r}'^2 = 1$), we obtain $\mathbf{F}\cdot\mathbf{r}\times\mathbf{r}' + c\mathbf{F}\cdot\mathbf{r}' = 0$, $EI\mathbf{r}'' = (\mathbf{F}\times\mathbf{r})\times\mathbf{r}' + c\mathbf{F}\times\mathbf{r}'$. In components (with the z-axis parallel to **F**) we obtain $(xy' - yx') + cz' = 0$, $EIz'' = -F(xx' + yy')$. Using cylindrical polar coordinates r, ϕ, z, we have

$$r^2\phi' + cz' = 0, \qquad EIz'' = -Frr'. \tag{2}$$

The second of these gives

$$z' = F(A - r^2)/2EI, \tag{3}$$

where A is a constant. Combining (2) and (3) with the identity $r'^2+r^2\phi'^2+z'^2 = 1$, we find

$$dl = \frac{r\,dr}{\sqrt{[r^2-(r^2+c^2)(A-r^2)^2F^2/4E^2I^2]}},$$

and then (2) and (3) give

$$z = \frac{F}{2EI}\int\frac{(A-r^2)r\,dr}{\sqrt{[r^2-F^2(r^2+c^2)(A-r^2)^2/4E^2I^2]}},$$

$$\phi = -\frac{cF}{2EI}\int\frac{(A-r^2)dr}{r\sqrt{[r^2-F^2(r^2+c^2)(A-r^2)^2/4E^2I^2]}},$$

which gives the shape of the bent rod.

PROBLEM 6. A rod with circular cross-section is subjected to torsion (with torsion angle τ) and twisted into a spiral. Determine the force and moment which must be applied to the ends of the rod to keep it in this state.

SOLUTION. Let R be the radius of the cylinder on whose surface the spiral lies (and along whose axis we take the z-direction) and α the angle between the tangent to the spiral and a plane perpendicular to the z-axis; the pitch h of the spiral is related to α and R by $h = 2\pi R\tan\alpha$. The equation of the spiral is $x = R\cos\phi$, $y = R\sin\phi$, $z = \phi R\tan\alpha$, where ϕ is the angle of rotation about the z-axis. The element of length is $dl = (R/\cos\alpha)d\phi$. Substituting these expressions in (19.7), we calculate the components of the vector $\mathbf{M}$, and then the force $\mathbf{F}$ from formula (19.3); $\mathbf{F}$ is constant everywhere on the rod. The result is that the force $\mathbf{F}$ is parallel to the z-axis and its magnitude is $F = F_z = (C\tau/R)\sin\alpha-(EI/R^2)\cos^2\alpha\sin\alpha$. The moment $\mathbf{M}$ has a z-component $M_z = C\tau\sin\alpha + (EI/R)\cos^3\alpha$ and a ϕ-component, along the tangent to the cross-section of the cylinder, $M_\phi = FR$.

PROBLEM 7. Determine the form of a flexible wire (whose resistance to bending can be neglected in comparison with its resistance to stretching) suspended at two points and in a gravitation field.

SOLUTION. We take the plane of the wire as the xy-plane, with the y-axis vertically downwards. In equation (19.3) we can neglect the term $d\mathbf{M}/dl$, since $\mathbf{M}$ is proportional to EI. Then $\mathbf{F}\times\mathbf{t} = 0$, i.e. $\mathbf{F}$ is parallel to $\mathbf{t}$ at every point, and we can put $\mathbf{F} = F\mathbf{t}$. Equation (19.2) then gives

$$\frac{d}{dl}\left(F\frac{dx}{dl}\right) = 0, \qquad \frac{d}{dl}\left(F\frac{dy}{dl}\right) = q,$$

where q is the weight of the wire per unit length; hence $F\,dx/dl = c$, $F\,dy/dl = ql$, and so $F = \sqrt{(c^2+q^2l^2)}$, so that $dx/dl = A/\sqrt{(A^2+l^2)}$, $dy/dl = l/\sqrt{(A^2+l^2)}$, where $A = c/q$. Integration gives $x = A\sinh^{-1}(l/A)$, $y = \sqrt{(A^2+l^2)}$, whence $y = A\cosh(x/A)$, i.e. the wire takes the form of a catenary. The choice of origin and the constant A are determined by the fact that the curve must pass through the two given points and have a given length.

§20. Small deflections of rods

The equations of equilibrium are considerably simplified in the important case of small deflections of rods. This case holds if the direction of the vector $\mathbf{t}$ tangential to the rod varies only slowly along its length, i.e. the derivative $d\mathbf{t}/dl$ is small. In other words, the radius of curvature of the bent rod is everywhere large compared with the length of the rod. In practice, this condition amounts to requiring that the transverse deflection of the rod be small compared with its length. It should be emphasized that the deflection need not be small compared with the thickness of the rod, as it had to be in the approximate theory of small deflections of plates given in §§11–12.†

Differentiating (19.3) with respect to the length, we have

$$\frac{d^2\mathbf{M}}{dl^2} = \frac{d\mathbf{F}}{dl}\times\mathbf{t}+\mathbf{F}\times\frac{d\mathbf{t}}{dl}. \tag{20.1}$$

† We shall not give the complex theory of the bending of rods which are not straight when undeformed, but only consider one simple example (see Problems 8 and 9).

The second term contains the small quantity $d\mathbf{t}/dl$, and so can usually be neglected (some exceptional cases are discussed below). Substituting in the first term $d\mathbf{F}/dl = -\mathbf{K}$, we obtain the equation of equilibrium in the form

$$d^2\mathbf{M}/dl^2 = \mathbf{t}\times\mathbf{K}. \tag{20.2}$$

We write this equation in components, substituting in it from (18.6) and (18.9)

$$M_x = -EI_1 Y'', \qquad M_y = EI_2 X'', \qquad M_z = 0, \tag{20.3}$$

where the prime denotes differentiation with respect to z. The unit vector $\mathbf{t}$ may be supposed to be parallel to the z-axis. Then (20.2) gives

$$EI_2 X^{(iv)} - K_x = 0, \qquad EI_1 Y^{(iv)} - K_y = 0. \tag{20.4}$$

These equations give the deflections X and Y as functions of z, i.e. the shape of a slightly bent rod.

The stress resultant $\mathbf{F}$ on a cross-section of the rod can also be expressed in terms of the derivatives of X and Y. Substituting (20.3) in (19.3), we obtain

$$F_x = -EI_2 X''', \qquad F_y = -EI_1 Y'''. \tag{20.5}$$

We see that the second derivatives give the moment of the internal stresses, while the third derivatives give the stress resultant. The force (20.5) is called the *shearing force.* If the bending is due to concentrated forces, the shearing force is constant along each segment of the rod between points where forces are applied, and has a discontinuity at each of these points equal to the force applied there.

The quantities EI_2 and EI_1 are called the *flexural rigidities* of the rod in the xz and yz planes respectively.†

If the external forces applied to the rod act in one plane, the bending takes place in one plane, though not in general the same plane. The angle between the two planes is easily found. If α is the angle between the plane of action of the forces and the first principal plane of bending (the xz-plane), the equations of equilibrium become $X^{(iv)} = (K/I_2E)\cos\alpha$, $Y^{(iv)} = (K/I_1E)\sin\alpha$. The two equations differ only in the coefficient of K. Hence X and Y are proportional, and $Y = (XI_2/I_1)\tan\alpha$. The angle θ between the plane of bending and the xz-plane is given by

$$\tan\theta = (I_2/I_1)\tan\alpha. \tag{20.6}$$

For a rod with circular cross-section $I_1 = I_2$ and $\alpha = \theta$, i.e. the bending occurs in the plane of action of the forces. The same is true for a rod with any cross-section when $\alpha = 0$, i.e.

---

† An equation of the form

$$DX^{(iv)} - K_x = 0 \tag{20.4a}$$

also describes the bending of a thin plate in certain limiting cases. Let a rectangular plate (with sides a, b and thickness h) be fixed along its sides a (parallel to the y-axis) and bent along its sides b (parallel to the z-axis) by a load uniform in the y-direction. In the general case of arbitrary a and b, the two-dimensional equation (12.5), with the appropriate boundary conditions at the fixed and free edges, must be used to determine the bending. In the limiting case $a \gg b$, however, the deformation may be regarded as uniform in the y-direction, and then the two-dimensional equilibrium equation becomes of the form (20.4a), with the flexural rigidity replaced by $D = Eh^3a/12(1-\sigma^2)$. Equation (20.4a) is also applicable to the opposite limiting case $a \ll b$, when the plate can be regarded as a rod of length b with a narrow rectangular cross-section (a rectangle with sides a and h); in this case, however, the flexural rigidity is $D = EI_2 = Eh^3a/12$.

when the forces act in a principal plane. The magnitude of the deflection $\zeta = \sqrt{(X^2+Y^2)}$ satisfies the equation

$$EI\zeta^{(iv)} = K, \qquad I = I_1 I_2/\sqrt{(I_1{}^2\cos^2\alpha + I_2{}^2\sin^2\alpha)}. \tag{20.7}$$

The shearing force **F** is in the same plane as **K**, and its magnitude is

$$F = -EI\zeta'''. \tag{20.8}$$

Here I is the "effective" moment of inertia of the cross-section of the rod.

We can write down explicitly the boundary conditions on the equations of equilibrium for a slightly bent rod. If the end of the rod is clamped, we must have $X = Y = 0$ there, and also $X' = Y' = 0$, since its direction cannot change. Thus the conditions at a clamped end are

$$X = Y = 0, \qquad X' = Y' = 0. \tag{20.9}$$

The reaction force and moment at the point of support are determined from the known solution by formulae (20.3) and (20.5).

When the bending is sufficiently slight, the hinging and supporting of a point on the rod are equivalent as regards the boundary conditions. The reason is that, in the latter case, the longitudinal displacement of the rod at its point of support is of the second order of smallness compared with the transverse deflection, and can therefore be neglected. The boundary conditions of zero transverse displacement and moment give

$$X = Y = 0, \qquad X'' = Y'' = 0. \tag{20.10}$$

The direction of the end of the rod and the reaction force at the point of support are obtained by solving the equations.

Finally, at a free end, the force **F** and moment **M** must be zero. According to (20.3) and (20.5), this gives the conditions

$$X'' = Y'' = 0, \qquad X''' = Y''' = 0. \tag{20.11}$$

If a concentrated force is applied at the free end, then **F** must be equal to this force, and not to zero.

It is not difficult to generalize equations (20.4) to the case of a rod with variable cross-section. For such a rod the moments of inertia I_1 and I_2 are functions of z. Formulae (20.3), which determine the moment at any cross-section, are still valid. Substitution in (20.2) now gives

$$E\frac{d^2}{dz^2}\left(I_1\frac{d^2Y}{dz^2}\right) = K_y, \qquad E\frac{d^2}{dz^2}\left(I_2\frac{d^2X}{dz^2}\right) = K_x, \tag{20.12}$$

in which I_1 and I_2 must be differentiated. The shearing force is

$$F_x = -E\frac{d}{dz}\left(I_2\frac{d^2X}{dz^2}\right), \qquad F_y = -E\frac{d}{dz}\left(I_1\frac{d^2Y}{dz^2}\right). \tag{20.13}$$

Let us return to equations (20.1). Our neglect of the second term on the right-hand side may in some cases be illegitimate, even if the bending is slight. The cases involved are those in which a large internal stress resultant acts along the rod, i.e. F_z is very large. Such a force is usually caused by a strong tension of the rod by external stretching forces applied to its ends. We denote by T the constant lengthwise stress F_z. If the rod is strongly compressed instead of being extended, T will be negative. In expanding the vector product $\mathbf{F}\times d\mathbf{t}/dl$ we must now retain the terms in T, but those in F_x and F_y can again be neglected. Substituting

X'', Y'', 1 for the components of the vector $\mathrm{d}\mathbf{t}/\mathrm{d}l$, we obtain the equations of equilibrium in the form

$$\left.\begin{aligned} I_2EX^{(iv)} - TX'' - K_x = 0, \\ I_1EY^{(iv)} - TY'' - K_y = 0. \end{aligned}\right\} \tag{20.14}$$

The expressions (20.5) for the shearing force will now contain additional terms giving the projections of the force T (along the vector $\mathbf{t}$) on the x and y axes:

$$F_x = -EI_2X''' + TX', \qquad F_y = -EI_1Y''' + TY'. \tag{20.15}$$

These formulae can also, of course, be obtained directly from (19.3).

In some cases a large force T can result from the bending itself, even if no stretching forces are applied. Let us consider a rod with both ends clamped or hinged to fixed supports, so that no longitudinal displacement is possible. Then the bending of the rod must result in an extension of it, which leads to a force T in the rod. It is easy to estimate the magnitude of the deflection for which this force becomes important. The length $L + \Delta L$ of the bent rod is given by

$$L + \Delta L = \int_0^L \sqrt{(1 + X'^2 + Y'^2)}\,\mathrm{d}z,$$

taken along the straight line joining the points of support. For slight bending the square root can be expanded in series, and we find

$$\Delta L = \tfrac{1}{2}\int_0^L (X'^2 + Y'^2)\,\mathrm{d}z.$$

The stress force in simple stretching is equal to the relative extension multiplied by Young's modulus and by the area S of the cross-section of the rod. Thus the force T is

$$T = \frac{ES}{2L}\int_0^L (X'^2 + Y'^2)\,\mathrm{d}z. \tag{20.16}$$

If δ is the order of magnitude of the transverse bending, the derivatives X' and Y' are of the order of δ/L, so that the integral in (20.16) is of the order of δ^2/L, and $T \sim ES(\delta/L)^2$. The orders of magnitude of the first and second terms in (20.14) are respectively $EI\delta/L^4$ and $T\delta/L^2 \sim ES\delta^3/L^4$. The moment of inertia I is of the order of h^4, and $S \sim h^2$, where h is the thickness of the rod. Substituting, we easily find that the first and second terms in (20.14) are comparable in magnitude if $\delta \sim h$. Thus, when a rod with fixed ends is bent, the equations of equilibrium can be used in the form (20.4) only if the deflection is small in comparison with the thickness of the rod. If δ is not small compared with h (but still, of course, small compared with L), equations (20.14) must be used. The force T in these equations is not known *a priori*. It must first be regarded as a parameter in the solution, and then determined by formula (20.16) from the solution obtained; this gives the relation between T and the bending forces applied to the rod.

The opposite limiting case is that where the resistance of the rod to bending is small compared with its resistance to stretching, so that the first terms in equations (20.14) can be neglected in comparison with the second terms. Physically this case can be realized either by a very strong tension force T or by a small value of EI, which can result from a small thickness h. Rods under strong tension are called *strings*. In such cases the equations of equilibrium are

$$TX'' + K_x = 0, \qquad TY'' + K_y = 0. \tag{20.17}$$

The ends of the string are fixed, in the sense that their coordinates are given, i.e.

$$X = Y = 0. \tag{20.18}$$

The direction of the ends cannot be decided arbitrarily, but is given by the solution of the equations.

In conclusion, we may show how the equations of equilibrium of a slightly bent rod may be obtained from the variational principle, using the expression (18.10) for the elastic energy:

$$F_{\text{rod}} = \tfrac{1}{2}E\int\{I_1 Y''^2 + I_2 X''^2\}\,\mathrm{d}z.$$

In equilibrium the sum of this energy and the potential energy due to the external forces **K** acting on the rod must be a minimum, i.e. we must have $\delta F_{\text{rod}} - \int(K_x\delta X + K_y\delta Y)\,\mathrm{d}z = 0$, where the second term is the work done by the external forces in an infinitesimal displacement of the rod. In varying F_{rod}, we effect a repeated integration by parts:

$$\begin{aligned}\tfrac{1}{2}\delta\int X''^2\,\mathrm{d}z &= \int X''\delta X''\,\mathrm{d}z\\ &= [X''\delta X'] - \int X'''\delta X'\,\mathrm{d}z\\ &= [X''\delta X'] - [X'''\delta X] + \int X^{(\mathrm{iv})}\delta X\,\mathrm{d}z,\end{aligned}$$

and similarly for the integral of Y''^2. Collecting terms, we obtain

$$\begin{aligned}&\int[(EI_1 Y^{(\mathrm{iv})} - K_y)\delta Y + (EI_2 X^{(\mathrm{iv})} - K_x)\delta X]\,\mathrm{d}z +\\ &+ EI_1[(Y''\delta Y' - Y'''\delta Y)] + EI_2[(X''\delta X' - X'''\delta X)] = 0.\end{aligned}$$

The integral gives the equilibrium equations (20.4), since the variations δX and δY are arbitrary. The integrated terms give the boundary conditions on these equations; for example, at a free end the variations $\delta X, \delta Y, \delta X', \delta Y'$ are arbitrary, and the corresponding conditions (20.11) are obtained. Also, the coefficients of δX and δY in these terms give the expressions (20.5) for the components of the shearing force, and those of $\delta X'$ and $\delta Y'$ give the expressions (20.3) for the components of the bending moment.

Finally, the equations of equilibrium (20.14) in the presence of a tension force T can be obtained by the same method if we include in the energy a term $T\Delta L = \tfrac{1}{2}T\int(X'^2 + Y'^2)\,\mathrm{d}z$, which is the work done by the force T over a distance ΔL equal to the extension of the rod.

PROBLEMS

PROBLEM 1. Determine the shape of a rod (with length l) bent by its own weight, for various modes of support at the ends.

SOLUTION. The required shape is given by a solution of the equation $\zeta^{(\mathrm{iv})} = q/EI$, where q is the weight per unit length, with the appropriate boundary conditions at its ends, as shown in the text. The following shapes and maximum displacements are obtained for various modes of support at the ends of the rod. The origin is at one end of the rod in each case.

(a) Both ends clamped:

$$\zeta = qz^2(z-l)^2/24EI, \qquad \zeta(\tfrac{1}{2}l) = ql^4/384EI.$$

(b) Both ends supported:

$$\zeta = qz(z^3 - 2lz^2 + l^3)/24EI, \qquad \zeta(\tfrac{1}{2}l) = 5ql^4/384EI.$$

(c) One end ($z = l$) clamped, the other supported:

$$\zeta = qz(2z^3 - 3lz^2 + l^3)/48EI, \qquad \zeta(0{\cdot}42l) = 0{\cdot}0054ql^4/EI.$$

(d) One end ($z = 0$) clamped, the other free:

$$\zeta = qz^2(z^2 - 4lz + 6l^2)/24EI, \qquad \zeta(l) = ql^4/8EI.$$

PROBLEM 2. Determine the shape of a rod bent by a force f applied to its mid-point.

SOLUTION. We have $\zeta^{(iv)} = 0$ everywhere except at $z = \frac{1}{2}l$. The boundary conditions at the ends of the rod ($z = 0$ and $z = l$) are determined by the mode of support; at $z = \frac{1}{2}l$, ζ, ζ' and ζ'' must be continuous, and the discontinuity in the shearing force $F = -EI\zeta'''$ must be equal to f.

The shape of the rod (for $0 \leqslant z \leqslant \frac{1}{2}l$) and the maximum displacement are given by the following formulae:

(a) Both ends clamped:

$$\zeta = fz^2(3l-4z)48EI, \qquad \zeta(\tfrac{1}{2}l) = fl^3/192EI.$$

(b) Both ends supported:

$$\zeta = fz(3l^2 - 4z^2)/48EI, \qquad \zeta(\tfrac{1}{2}l) = fl^3/48EI.$$

The rod is symmetrical about its mid-point, so that the functions $\zeta(z)$ in $\frac{1}{2}l \leqslant z \leqslant l$ are obtained simply by replacing z by $l - z$.

PROBLEM 3. The same as Problem 2, but for a rod clamped at one end ($z = 0$) and free at the other end ($z = l$), to which a force f is applied.

SOLUTION. At all points of the rod F = constant $= f$, so that $\zeta''' = -f/EI$. Using the conditions $\zeta = 0$, $\zeta' = 0$ for $z = 0$, $\zeta'' = 0$ for $z = l$, we obtain

$$\zeta = fz^2(3l-z)/6EI, \qquad \zeta(l) = fl^3/3EI.$$

PROBLEM 4. Determine the shape of a rod with fixed ends, bent by a couple at its mid-point.

SOLUTION. At all points of the rod $\zeta^{(iv)} = 0$, and at $z = \frac{1}{2}l$ the moment $M = EI\zeta''$ has a discontinuity equal to the moment m of the applied couple. The results are:

(a) Both ends clamped:

$$\zeta = mz^2(l-2z)/8EIl \quad \text{for} \quad 0 \leqslant z \leqslant \tfrac{1}{2}l,$$

$$\zeta = -m(l-z)^2[l - 2(l-z)]/8EIl \quad \text{for} \quad \tfrac{1}{2}l \leqslant z \leqslant l.$$

(b) Both ends hinged:

$$\zeta = mz(l^2 - 4z^2)/24EIl \quad \text{for} \quad 0 \leqslant z \leqslant \tfrac{1}{2}l,$$

$$\zeta = -m(l-z)[l^2 - 4(l-z)^2]/24EIl \quad \text{for} \quad \tfrac{1}{2}l \leqslant z \leqslant l.$$

The rod is bent in opposite directions on the two sides of $z = \frac{1}{2}l$.

PROBLEM 5. The same as Problem 4, but for the case where one end is clamped and the other end free, the couple being applied at the latter end.

SOLUTION. At all points of the rod $M = EI\zeta'' = m$, and at $z = 0$ we have $\zeta = 0$, $\zeta' = 0$. The shape is given by $\zeta = mz^2/2EI$.

PROBLEM 6. Determine the shape of a circular rod with hinged ends stretched by a force T and bent by a force f applied at its mid-point.

SOLUTION. On the segment $0 \leqslant z \leqslant \frac{1}{2}l$ the shearing force is $\frac{1}{2}f$, so that (20.15) gives the equation

$$\zeta''' - T\zeta'/EI = -f/2EI.$$

The boundary conditions are $\zeta = \zeta'' = 0$ for $z = 0$ and l; $\zeta' = 0$ for $z = \frac{1}{2}l$ (since ζ' is continuous). The shape of the rod (in the segment $0 \leqslant z \leqslant \frac{1}{2}l$) is given by

$$\zeta = \frac{f}{2T}\left(z - \frac{\sinh kz}{k\cosh\frac{1}{2}kl}\right), \quad k = \sqrt{(T/EI)}.$$

For small k this gives the result obtained in Problem 2 (b). For large k it becomes $\zeta = fz/2T$, i.e., in accordance with equations (20.17), a flexible wire under a force f takes the form of two straight pieces intersecting at $z = \frac{1}{2}l$.

If the force T is due to the stretching of the rod by the transverse force, it must be determined by formula (20.16). Substituting the above result, we obtain the equation

$$\frac{1}{k^6}\left[\frac{3}{2}+\frac{1}{2}\tanh^2\frac{1}{2}kl-\frac{3}{kl}\tanh\frac{1}{2}kl\right]=\frac{8E^2I^3}{f^2S},$$

which determines T as an implicit function of f.

PROBLEM 7. A circular rod with infinite length lies in an elastic substance, i.e. when it is bent a force $K = -\alpha\zeta$ proportional to the deflection acts on it. Determine the shape of the rod when a concentrated force f acts on it.

SOLUTION. We take the origin at the point where the force f is applied. The equation $EI\zeta^{(iv)} = -\alpha\zeta$ holds everywhere except at $z = 0$. The solution must satisfy the condition $\zeta = 0$ at $z = \pm-\infty$, and at $z = 0$ ζ' and ζ'' must be continuous; the difference between the shearing forces $F = -EI\zeta'''$ for $z \to 0+$ and $z \to 0-$ must be f. The required solution is

$$\zeta = \frac{f}{8\beta^3 EI}e^{-\beta|z|}[\cos\beta|z| + \sin\beta|z|],\quad \beta = \left(\frac{\alpha}{4EI}\right)^{1/4}.$$

PROBLEM 8. Derive the equation of equilibrium for a slightly bent thin circular rod which, in its undeformed state, is an arc of a circle and is bent in its plane by radial forces.

SOLUTION. Taking the origin of polar coordinates r, ϕ at the centre of the circle, we write the equation of the deformed rod as $r = a+\zeta(\phi)$, where a is the radius of the arc and ζ a small radial displacement. Using the expression for the radius of curvature in polar coordinates, we find as far as the first order in ζ

$$\frac{1}{R}=\frac{r^2-rr''+2r'^2}{(r^2+r'^2)^{3/2}}\cong\frac{1}{a}-\frac{\zeta+\zeta''}{a^2},$$

where the prime denotes differentiation with respect to ϕ. According to (18.11), the elastic bending energy is

$$F_{\text{rod}} = \tfrac{1}{2}EI\int_0^{\phi_0}\left(\frac{1}{R}-\frac{1}{a}\right)^2 a\,d\phi = \frac{EI}{2a^3}\int_0^{\phi_0}(\zeta+\zeta'')^2\,d\phi,$$

ϕ_0 being the angle subtended by the arc at its centre. The equation of equilibrium is obtained from the variational principle

$$\delta F_{\text{rod}} - \int_0^{\phi_0}\delta\zeta K_r a\,d\phi = 0,$$

where K_r is the external radial force per unit length, with the auxiliary condition

$$\int_0^{\phi_0}\zeta\,d\phi = 0,$$

which is, in this approximation, the statement of the fact that the total length of the rod is unchanged, i.e. it undergoes no general extension. Using Lagrange's method, we put

$$\delta F_{\text{rod}} - \int_0^{\phi_0} aK_r\delta\zeta\,d\phi + a\alpha\int_0^{\phi_0}\delta\zeta\,d\phi = 0,$$

where α is a constant. Varying the integrand in F_{rod} and integrating the $\delta\zeta''$ term twice by parts, we obtain

$$\int\left\{\frac{EI}{a^3}(\zeta+2\zeta''+\zeta^{(iv)})-aK_r+\alpha a\right\}\delta\zeta\,d\phi+\frac{EI}{a^3}[(\zeta+\zeta'')\delta\zeta']-\frac{EI}{a^3}[(\zeta'+\zeta''')\delta\zeta]=0.$$

Hence we find the equation of equilibrium†

$$EI(\zeta^{(iv)}+2\zeta''+\zeta)/a^4 - K_r + \alpha = 0, \tag{1}$$

† In the absence of external forces, $K_r = 0$ and $\alpha = 0$; the non-zero solutions of the resulting homogeneous equation correspond to a simple rotation or translation of the whole rod.

the shearing force $F = -EI(\zeta' + \zeta''')/a^3$, and the bending moment $M = EI(\zeta + \zeta'')/a^2$; cf. the end of §20. The constant α is determined from the condition that the rod as a whole be not stretched.

PROBLEM 9. Determine the deformation of a circular ring bent by two forces f applied along a diameter (Fig. 18).

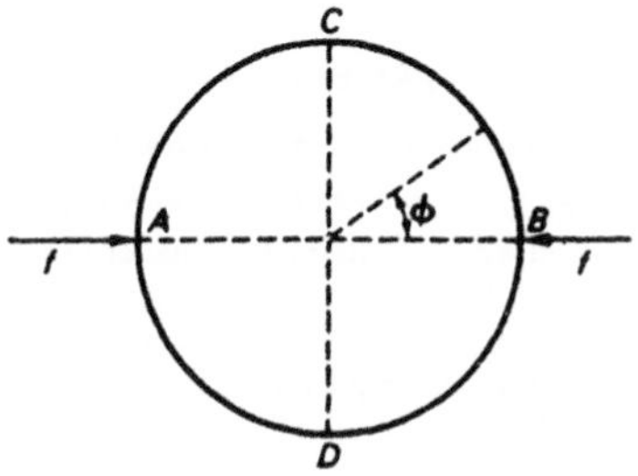

FIG. 18

SOLUTION. Integrating equation (1), Problem 8, along the circumference of the ring, we have $2\pi\alpha a = \int K_r a \, d\phi = 2f$. We have equation (1) with $K_r = 0$ everywhere except at $\phi = 0$ and $\phi = \pi$:

$$\zeta^{(iv)} + 2\zeta'' + \zeta + fa^3/\pi EI = 0.$$

The required deformation of the ring is symmetrical about the diameters AB and CD, and so we must have $\zeta' = 0$ at A, B, C and D. The difference in the shearing forces for $\phi \to 0\pm$ must be f. The solution of the equation of equilibrium which satisfies these conditions is

$$\zeta = \frac{fa^3}{EI}\left(\frac{1}{\pi} + \frac{1}{4}\phi\cos\phi - \frac{1}{8}\pi\cos\phi - \frac{1}{4}\sin\phi\right), \qquad 0 \leqslant \phi \leqslant \pi.$$

In particular, the points A and B approach through a distance

$$|\zeta(0) + \zeta(\pi)| = \frac{fa^3}{EI}\left(\frac{\pi}{4} - \frac{2}{\pi}\right).$$

§21. The stability of elastic systems

The behaviour of a rod subject to longitudinal compressing forces is the simplest example of the important phenomenon of *elastic instability*, first discovered by L. Euler.

In the absence of transverse bending forces K_x, K_y, the equations of equilibrium (20.14) for a compressed rod have the evident solution $X = Y = 0$, which corresponds to the rod's remaining straight under a longitudinal force $|T|$. This solution, however, gives a stable equilibrium of the rod only if the compressing force $|T|$ is less than a certain critical value T_{cr}. For $|T| < T_{cr}$, the straight rod is stable with respect to any small perturbation. In other words, if the rod is slightly bent by some small force, it will tend to return to its original position when that force ceases to act.

If, on the other hand, $|T| > T_{cr}$, the straight rod is in unstable equilibrium. An infinitesimal bending suffices to destroy the equilibrium, and a large bending of the rod results. It is clear that, if this is so, the compressed rod cannot actually remain straight.

The behaviour of the rod after it ceases to be stable must satisfy the equations for bending with large deflections. The value T_{cr} of the critical load, however, can be obtained from the equations for small deflections. For $|T| = T_{cr}$, the straight rod is in neutral equilibrium. This means that, besides the solution $X = Y = 0$, there must also be states where the rod is slightly bent but still in equilibrium. Hence the critical value of T_{cr} is the value of $|T|$ for which the equations

$$EI_2 X^{(iv)} + |T|X'' = 0, \qquad EI_1 Y^{(iv)} + |T|Y'' = 0 \qquad (21.1)$$

have a non-zero solution. This solution gives also the nature of the deformation of the rod immediately after it ceases to be stable.

The following Problems give some typical cases of the loss of stability in various elastic systems.

PROBLEMS

PROBLEM 1. Determine the critical compression force for a rod with hinged ends.

SOLUTION. Since we are seeking the smallest value of $|T|$ for which equations (21.1) have a non-zero solution, it is sufficient to consider only the equation which contains the smaller of I_1 and I_2. Let $I_2 < I_1$. Then we seek a solution of the equation $EI_2X^{(iv)}+|T|X''=0$ in the form $X = A+Bz+C\sin kz+D\cos kz$, where $k=\sqrt{(|T|/EI_2)}$. The non-zero solution which satisfies the conditions $X = X''=0$ for $z=0$ and $z=l$ is $X = C\sin kz$, with $\sin kl = 0$. Hence we find the required critical force to be $T_{cr}=\pi^2EI_2/l^2$. On ceasing to be stable, the rod takes the form shown in Fig. 19a.

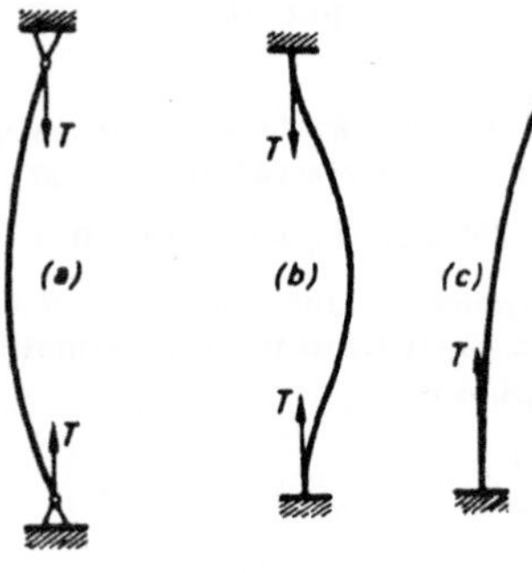

FIG. 19

PROBLEM 2. The same as Problem 1, but for a rod with clamped ends (Fig. 19b).

SOLUTION. $T_{cr}=4\pi^2EI_2/l^2$.

PROBLEM 3. The same as Problem 1, but for a rod with one end clamped and the other free (Fig. 19c).

SOLUTION. $T_{cr}=\pi^2EI_2/4l^2$.

PROBLEM 4. Determine the critical compression force for a circular rod with hinged ends in an elastic medium (see §20, Problem 7).

SOLUTION. The equations (21.1) must now be replaced by $EIX^{(iv)}+|T|X''+\alpha X=0$. A similar treatment gives the solution $X = A\sin n\pi z/l$,

$$T_{cr}=\frac{\pi^2EI}{l^2}\left(n^2+\frac{\alpha l^4}{n^2\pi^4EI}\right),$$

where n is the integer for which T_{cr} is least. Where α is large, $n > 1$, i.e. the rod exhibits several undulations as soon as it ceases to be stable.

PROBLEM 5. A circular rod is subjected to torsion, its ends being clamped. Determine the critical torsion beyond which the straight rod becomes unstable.

SOLUTION. The critical value of the torsion angle is determined by the appearance of non-zero solutions of the equations for slight bending of a twisted rod. To derive these equations, we substitute the expression (19.7) $\mathbf{M} = EI\mathbf{t}\times d\mathbf{t}/dl+C\tau\mathbf{t}$, where τ is the constant torsion angle, in equation (19.3). This gives

$$EI\mathbf{t}\times\frac{d^2\mathbf{t}}{dl^2}+C\tau\frac{d\mathbf{t}}{dl}-\mathbf{F}\times\mathbf{t}=0.$$

We differentiate; since the bending is not large, $\mathbf{t}$ may be regarded as a constant vector $\mathbf{t}_0$ along the axis of the rod (the z-axis) in differentiating the first and third terms. Since also $d\mathbf{F}/dl = 0$ (there being no external forces except at the ends of the rod), we obtain

$$EI\mathbf{t}_0 \times \frac{d^3\mathbf{t}}{dl^3} + C\tau \frac{d^2\mathbf{t}}{dl^2} = 0,$$

or, in components,

$$Y^{(iv)} - \kappa X''' = 0,$$
$$X^{(iv)} + \kappa Y''' = 0,$$

where $\kappa = C\tau/EI$. Taking as the unknown function $\xi = X + iY$, we obtain $\xi^{(iv)} - i\kappa\xi''' = 0$. We seek a solution which satisfies the conditions $\xi = 0$, $\xi' = 0$ for $z = 0$ and $z = l$, in the form $\xi = a(1 + i\kappa z - e^{i\kappa z}) + bz^2$, and obtain as the compatibility condition of the equations for a and b the relation $e^{i\kappa l} = (2 + i\kappa l)/(2 - i\kappa l)$, whence $\frac{1}{2}\kappa l = \tan\frac{1}{2}\kappa l$. The smallest root of this equation is $\frac{1}{2}\kappa l = 4{\cdot}49$, so that $\tau_{cr} = 8{\cdot}98EI/Cl$.

PROBLEM 6. The same as Problem 5, but for a rod with hinged ends.

SOLUTION. In this case we have $\xi = a(1 - e^{i\kappa z} - \frac{1}{2}\kappa^2 z^2) + bz$, where κ is given by

$$e^{i\kappa l} = 1, \text{ i.e. } \kappa l = 2\pi.$$

Hence the required critical torsion angle is $\tau_{cr} = 2\pi EI/Cl$.

PROBLEM 7. Determine the limit of stability of a vertical rod under its own weight, the lower end being clamped.

SOLUTION. If the longitudinal stress $F_z \equiv T$ varies along the rod, $dF_z/dl \neq 0$ in the first term of (20.1), and equations (20.14) are replaced by

$$I_2 E X^{(iv)} - (TX')' - K_x = 0,$$
$$I_1 E Y^{(iv)} - (TY')' - K_y = 0.$$

In the case considered, there are no transverse bending forces anywhere in the rod, and $T = -q(l - z)$, where q is the weight of the rod per unit length and z is measured from the lower end. Assuming that $I_2 < I_1$, we consider the equation

$$I_2 E X''' = TX' = -q(l - z)X';$$

for $z = l$, $X''' = 0$ automatically. The general integral of this equation for the function $u = X'$ is

$$u = \eta^{\frac{1}{3}}[aJ_{-\frac{1}{3}}(\eta) + bJ_{\frac{1}{3}}(\eta)],$$

where

$$\eta = \tfrac{2}{3}\sqrt{[q(l - z)^3/EI_2]}.$$

The boundary conditions $X' = 0$ for $z = 0$ and $X'' = 0$ for $z = l$ give for the function $u(\eta)$ the conditions $u = 0$ for $\eta = \eta_0 \equiv \frac{2}{3}\sqrt{(ql^3/EI_2)}$, $u'\eta^{1/3} = 0$ for $\eta = 0$. In order to satisfy these conditions we must put $b = 0$ and $J_{-\frac{1}{3}}(\eta_0) = 0$. The smallest root of this equation is $\eta_0 = 1{\cdot}87$, and so the critical length is $l_{cr} = 1{\cdot}98(EI_2/q)^{1/3}$.

PROBLEM 8. A rod has an elongated cross-section, so that $I_2 \gg I_1$. One end is clamped and a force f is applied to the other end, which is free, so as to bend it in the principal xz-plane (in which the flexural rigidity is EI_2). Determine the critical force f_{cr} at which the rod bent in a plane becomes unstable and the rod is bent sideways (in the yz-plane), at the same time undergoing torsion.

SOLUTION. Since the rigidity EI_2 is large compared with EI_1 (and with the torsional rigidity C),† the instability as regards sideways bending occurs while the deflection in the xz-plane is still small. To determine the point where instability sets in, we must form the equations for slight sideways bending of the rod, retaining the terms proportional to the products of the force f in the xz-plane and the small displacements. Since there is a concentrated force only at the free end of the rod, we have $\mathbf{F} = \mathbf{f}$ at all points, and at the free end ($z = l$) the moment $\mathbf{M} = 0$; from formula (19.6) we find the components of the moment relative to a fixed system of coordinates x, y, z: $M_x = 0$, $M_y = (l - z)f$, $M_z = (Y - Y_0)f$, where $Y_0 = Y(l)$. Taking the components along coordinate axes ξ, η, ζ fixed at each point to the rod, we obtain as far as the first-order terms in the displacements $M_\xi = \phi(l - z)f$, $M_\eta = (l - z)f$, $M_\zeta = (l - z)f dY/dz + f(Y - Y_0)$, where ϕ is the total angle of rotation of a cross-section of the rod under torsion; the torsion angle $\tau = d\phi/dz$ is not constant along the rod. According to (18.6) and (18.9), however, we have for a small deflection

$$M_\xi = -EI_1 Y'', \qquad M_\eta = EI_2 X'', \qquad M_\zeta = C\phi';$$

---

† For example, for a narrow rectangular cross-section with sides b and h ($b \gg h$), we have $EI_1 = bh^3E/12$, $EI_2 = b^3hE/12$, $C = bh^3\mu/3$.

comparing, we obtain the equations of equilibrium

$$EI_2X'' = (l-z)f, \qquad EI_1Y'' = -\phi(l-z)f,$$
$$C\phi' = (l-z)fY' + (Y-Y_0)f.$$

The first of these equations gives the main bending of the rod, in the xz-plane; we require the value of f for which non-zero solutions of the second and third equations appear. Eliminating Y, we find

$$\phi'' + k^2(l-z)^2\phi = 0, \qquad k^2 = f^2/EI_1C.$$

The general integral of this equation is

$$\phi = a\sqrt{(l-z)}\,J_{\frac{1}{4}}[\tfrac{1}{2}k(l-z)^2] + b\sqrt{(l-z)}\,J_{-\frac{1}{4}}[\tfrac{1}{2}k(l-z)^2].$$

At the clamped end ($z = 0$) we must have $\phi = 0$, and at the free end the twisting moment $C\phi' = 0$. From the second condition we have $a = 0$, and then the first gives $J_{-\frac{1}{4}}(\tfrac{1}{2}kl^2) = 0$. The smallest root of this equation is $\tfrac{1}{2}kl^2 = 2{\cdot}006$, whence $f_{cr} = 4{\cdot}01\sqrt{(EI_1C)}/l^2$.

CHAPTER III

ELASTIC WAVES

§22. Elastic waves in an isotropic medium

IF motion occurs in a deformed body, its temperature is not in general constant, but varies in both time and space. This considerably complicates the exact equations of motion in the general case of arbitrary motions.

Usually, however, matters are simplified in that the transfer of heat from one part of the body to another (by simple thermal conduction) occurs very slowly. If the heat exchange during times of the order of the period of oscillatory motions in the body is negligible, we can regard any part of the body as thermally insulated, i.e. the motion is adiabatic. In adiabatic deformations, however, σ_{ik} is given in terms of u_{ik} by the usual formulae, the only difference being that the ordinary (isothermal) values of E and σ must be replaced by their adiabatic values (see §6). We shall assume in what follows that this condition is fulfilled, and accordingly E and σ in this chapter will be understood to have their adiabatic values.

In order to obtain the equations of motion for an elastic medium, we must equate the internal stress force $\partial\sigma_{ik}/\partial x_k$ to the product of the acceleration $\ddot{u}_i$ and the mass per unit volume of the body, i.e. its density ρ:

$$\rho\ddot{u}_i = \partial\sigma_{ik}/\partial x_k. \tag{22.1}$$

This is the general equation of motion.†

In particular, the equations of motion for an isotropic elastic medium can be written down at once by analogy with the equation of equilibrium (7.2). We have

$$\rho\ddot{\mathbf{u}} = \frac{E}{2(1+\sigma)}\triangle\mathbf{u} + \frac{E}{2(1+\sigma)(1-2\sigma)}\,\mathbf{grad}\operatorname{div}\mathbf{u}. \tag{22.2}$$

Since all deformations are supposed small, the motions considered in the theory of elasticity are small *elastic oscillations* or *elastic waves*. We shall begin by discussing a plane elastic wave in an infinite isotropic medium, i.e. a wave in which the deformation $\mathbf{u}$ is a function only of one coordinate (x, say) and of the time. All derivatives with respect to y and z in equations (22.2) are then zero, and we obtain for the components of the vector $\mathbf{u}$ the equations

$$\frac{\partial^2 u_x}{\partial x^2} - \frac{1}{c_l^2}\frac{\partial^2 u_x}{\partial t^2} = 0, \qquad \frac{\partial^2 u_y}{\partial x^2} - \frac{1}{c_t^2}\frac{\partial^2 u_y}{\partial t^2} = 0 \tag{22.3}$$

† It is assumed that the velocity $\mathbf{v}$ of a point in the medium is equal to the derivative $\dot{\mathbf{u}}$ of its displacement. We must emphasize, however, that the identity of these two quantities is by no means self-evident. In crystals, $\mathbf{u}$ is the displacement of a lattice site, but $\mathbf{v}$ is defined in continuum mechanics as the momentum of unit mass of the substance. The equation $\mathbf{v} = \dot{\mathbf{u}}$ is, strictly speaking, valid only for perfect crystals, with one atom at every lattice site and none elsewhere. If the crystal contains defects (vacancies or interstitial atoms), mass transport relative to the lattice (i.e. a non-zero momentum) can occur even without deformation of the lattice if there is diffusion of defects through it. The identity of $\mathbf{v}$ and $\dot{\mathbf{u}}$ implies that such effects are neglected, on the grounds that diffusion is slow or that the defect concentration is low.

(the equation for u_z is the same as that for u_y); here†

$$c_l = \sqrt{\frac{E(1-\sigma)}{\rho(1+\sigma)(1-2\sigma)}}, \qquad c_t = \sqrt{\frac{E}{2\rho(1+\sigma)}}. \tag{22.4}$$

Equations (22.3) are ordinary wave equations in one dimension, and the quantities c_l and c_t which appear in them are the velocities of propagation of the wave. We see that the velocity of propagation for the component u_x is different from that for u_y and u_z.

Thus an elastic wave is essentially two waves propagated independently. In one (u_x) the displacement is in the direction of propagation; this is called the *longitudinal wave*, and is propagated with velocity c_l. In the other wave (u_y, u_z) the displacement is in a plane perpendicular to the direction of propagation; this is called the *transverse wave*, and is propagated with velocity c_t. It is seen from (22.4) that the velocity of longitudinal waves is always greater than that of transverse waves: we always have‡

$$c_l > \sqrt{(4/3)}c_t. \tag{22.5}$$

The velocities c_l and c_t are often called the *longitudinal and transverse velocities of sound.*

We know that the volume change in a deformation is given by the sum of the diagonal terms in the strain tensor, i.e. by $u_{ii} \equiv \operatorname{div} \mathbf{u}$. In the transverse wave there is no component u_x, and, since the other components do not depend on y or z, $\operatorname{div} \mathbf{u} = 0$ for such a wave. Thus transverse waves do not involve any change in volume of the parts of the body. For longitudinal waves, however, $\operatorname{div} \mathbf{u} \neq 0$, and these waves involve compressions and expansions in the body.

The separation of the wave into two parts propagated independently with different velocities can also be effected in the general case of an arbitrary (not plane) elastic wave in an infinite medium. We rewrite equation (22.2) in terms of the velocities c_l and c_t:

$$\ddot{\mathbf{u}} = c_t^2 \triangle \mathbf{u} + (c_l^2 - c_t^2)\,\mathbf{grad}\operatorname{div}\mathbf{u}. \tag{22.6}$$

We then represent the vector $\mathbf{u}$ as the sum of two parts:

$$\mathbf{u} = \mathbf{u}_l + \mathbf{u}_t, \tag{22.7}$$

of which one satisfies

$$\operatorname{div}\mathbf{u}_t = 0 \tag{22.8}$$

and the other satisfies

$$\mathbf{curl}\,\mathbf{u}_l = 0. \tag{22.9}$$

We know from vector analysis that this representation (i.e. the expression of a vector as the sum of the curl of a vector and the gradient of a scalar) is always possible.

Substituting $\mathbf{u} = \mathbf{u}_l + \mathbf{u}_t$ in (22.6), we obtain

$$\ddot{\mathbf{u}}_l + \ddot{\mathbf{u}}_t = c_t^2 \triangle (\mathbf{u}_l + \mathbf{u}_t) + (c_l^2 - c_t^2)\,\mathbf{grad}\operatorname{div}\mathbf{u}_l. \tag{22.10}$$

We take the divergence of both sides. Since $\operatorname{div}\mathbf{u}_t = 0$, the result is

$$\operatorname{div}\ddot{\mathbf{u}}_l = c_t^2 \triangle \operatorname{div}\mathbf{u}_l + (c_l^2 - c_t^2)\triangle \operatorname{div}\mathbf{u}_l,$$

or $\operatorname{div}(\ddot{\mathbf{u}}_l - c_l^2 \triangle \mathbf{u}_l) = 0$. The curl of the expression in parentheses is also zero, by (22.9). If

† We may give also expressions for c_l and c_t in terms of the moduli of compression and rigidity and the Lamé coefficients: $c_l = \sqrt{\{(3K+4\mu)/3\rho\}} = \sqrt{\{(\lambda+2\mu)/\rho\}}$, $c_t = \sqrt{(\mu/\rho)}$.

‡ Since σ actually varies only between 0 and $\frac{1}{2}$ (see the second footnote to §5), we always have $c_l > \sqrt{2}c_t$.

the curl and divergence of a vector both vanish in all space, that vector must be zero identically. Thus

$$\frac{\partial^2 \mathbf{u}_l}{\partial t^2} - c_l^2 \triangle \mathbf{u}_l = 0. \tag{22.11}$$

Similarly, taking the curl of equation (22.10) we have, since the curls of $\mathbf{u}_l$ and of any gradient are zero, $\mathbf{curl}\,(\ddot{\mathbf{u}}_t - c_t^2 \triangle \mathbf{u}_t) = 0$. Since the divergence of the expression in parentheses is also zero, we obtain an equation of the same form as (22.11):

$$\frac{\partial^2 \mathbf{u}_t}{\partial t^2} - c_t^2 \triangle \mathbf{u}_t = 0. \tag{22.12}$$

Equations (22.11) and (22.12) are ordinary wave equations in three dimensions. Each of them represents the propagation of an elastic wave, with velocity c_l and c_t respectively. One wave ($\mathbf{u}_t$) does not involve a change in volume (since $\operatorname{div} \mathbf{u}_t = 0$), while the other ($\mathbf{u}_l$) is accompanied by volume compressions and expansions.

In a monochromatic elastic wave, the displacement vector is

$$\mathbf{u} = \operatorname{re}\{\mathbf{u}_0(\mathbf{r}) e^{-i\omega t}\}, \tag{22.13}$$

where $\mathbf{u}_0$ is a function of the coordinates which satisfies the equation

$$c_t^2 \triangle \mathbf{u}_0 + (c_l^2 - c_t^2)\,\mathbf{grad}\,\operatorname{div} \mathbf{u}_0 + \omega^2 \mathbf{u}_0 = 0, \tag{22.14}$$

obtained by substituting (22.13) in (22.6). The longitudinal and transverse parts of a monochromatic wave satisfy the equations

$$\triangle \mathbf{u}_l + k_l^2 \mathbf{u}_l = 0, \qquad \triangle \mathbf{u}_t + k_t^2 \mathbf{u}_t = 0, \tag{22.15}$$

where $k_l = \omega/c_l$, $k_t = \omega/c_t$ are the wave numbers of the longitudinal and transverse waves.

Finally, let us consider the reflection and refraction of a plane monochromatic elastic wave at the boundary between two different elastic media. It must be borne in mind that the nature of the wave is in general changed when it is reflected or refracted. If a purely transverse or purely longitudinal wave is incident on a surface of separation, the result is a mixed wave containing both transverse and longitudinal parts. The nature of the wave remains unchanged (as we see from symmetry) only when it is incident normally on the surface of separation, or when a transverse wave whose oscillations are parallel to the plane of separation is incident (at any angle).

The relations giving the directions of the reflected and refracted waves can be obtained immediately from the constancy of the frequency and of the tangential components of the wave vector.† Let θ and θ' be the angles of incidence and reflection (or refraction) and c, c' the velocities of the two waves. Then

$$\frac{\sin\theta}{\sin\theta'} = \frac{c}{c'}. \tag{22.16}$$

For example, let the incident wave be transverse. Then $c = c_{t1}$ is the velocity of transverse waves in medium 1. For the transverse reflected wave we have $c' = c_{t1}$ also, so

† See *FM*, §66. The arguments given there are applicable in their entirety.

that (22.16) gives $\theta = \theta'$, i.e. the angle of incidence is equal to the angle of reflection. For the longitudinal reflected wave, however, $c' = c_{l1}$, and so

$$\frac{\sin\theta}{\sin\theta'} = \frac{c_{t1}}{c_{l1}}.$$

For the transverse part of the refracted wave $c' = c_{t2}$, and for a transverse incident wave

$$\frac{\sin\theta}{\sin\theta'} = \frac{c_{t1}}{c_{t2}}.$$

Similarly, for the longitudinal refracted wave

$$\frac{\sin\theta}{\sin\theta'} = \frac{c_{t1}}{c_{l2}}.$$

PROBLEMS

PROBLEM 1. Determine the reflection coefficient for a longitudinal monochromatic wave incident at any angle on the surface of a body (with a vacuum outside).†

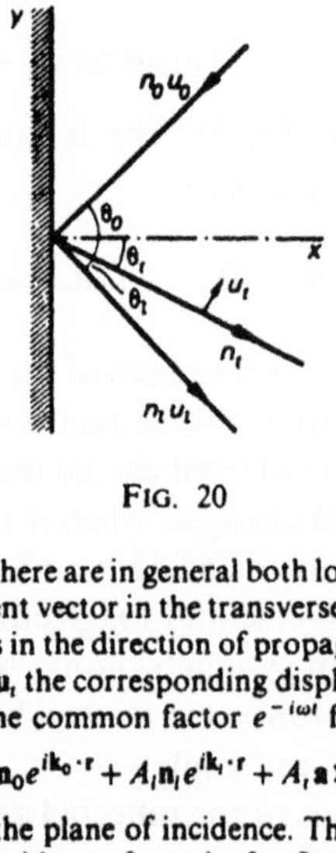

FIG. 20

SOLUTION. When the wave is reflected, there are in general both longitudinal and transverse reflected waves. It is clear from symmetry that the displacement vector in the transverse reflected wave lies in the plane of incidence (Fig. 20, where $\mathbf{n}_0$, $\mathbf{n}_l$ and $\mathbf{n}_t$ are unit vectors in the direction of propagation of the incident, longitudinal reflected and transverse reflected waves, and $\mathbf{u}_0$, $\mathbf{u}_l$, $\mathbf{u}_t$ the corresponding displacement vectors). The total displacement in the body is given by the sum (omitting the common factor $e^{-i\omega t}$ for brevity)

$$\mathbf{u} = A_0\mathbf{n}_0 e^{i\mathbf{k}_0\cdot\mathbf{r}} + A_l\mathbf{n}_l e^{i\mathbf{k}_l\cdot\mathbf{r}} + A_t\mathbf{a}\times\mathbf{n}_t e^{i\mathbf{k}_t\cdot\mathbf{r}},$$

where $\mathbf{a}$ is a unit vector perpendicular to the plane of incidence. The magnitudes of the wave vectors are $k_0 = k_l = \omega/c_l$, $k_t = \omega/c_t$, and the angles of incidence θ_0 and of reflection θ_l, θ_t are related by $\theta_l = \theta_0$, $\sin\theta_t = (c_t/c_l)\sin\theta_0$. For the components of the strain tensor at the boundary we obtain

$$u_{xx} = ik_0(A_0 + A_l)\cos^2\theta_0 + iA_t k_t \cos\theta_t \sin\theta_t, \quad u_{ll} = ik_0(A_0 + A_l),$$
$$u_{xy} = ik_0(A_0 - A_l)\sin\theta_0\cos\theta_0 + \tfrac{1}{2}iA_t k_t(\cos^2\theta_t - \sin^2\theta_t).$$

again omitting the common exponential factor. The components of the stress tensor can be calculated from the general formula (5.11), which can here be conveniently written

$$\sigma_{ik} = 2\rho c_t^2 u_{ik} + \rho(c_l^2 - 2c_t^2)u_{ll}\delta_{ik}.$$

---

† The more general case of the reflection of sound waves from a solid-liquid interface, and the similar problem of the reflection of a wave incident from a liquid on to a solid, are discussed by L. M. Brekhovskikh, *Waves in Layered Media*, 2nd edition, §7, Academic Press, New York 1980.

The boundary conditions at the free surface of the medium are $\sigma_{ik}n_k = 0$, whence

$$\sigma_{xx} = \sigma_{yx} = 0,$$

giving two equations which express A_l and A_t in terms of A_0. The result is

$$A_l = A_0 \frac{c_t^2 \sin 2\theta_t \sin 2\theta_0 - c_l^2 \cos^2 2\theta_t}{c_t^2 \sin 2\theta_t \sin 2\theta_0 + c_l^2 \cos^2 2\theta_t},$$

$$A_t = -A_0 \frac{2c_l c_t \sin 2\theta_0 \cos 2\theta_t}{c_t^2 \sin 2\theta_t \sin 2\theta_0 + c_l^2 \cos^2 2\theta_t}.$$

For $\theta_0 = 0$ we have $A_l = -A_0$, $A_t = 0$, i.e. the wave is reflected as a purely longitudinal wave. The ratio of the energy flux density components normal to the surface in the reflected and incident longitudinal waves is $R_l = |A_l/A_0|^2$. The corresponding ratio for the reflected transverse wave is

$$R_t = \frac{c_t \cos\theta_t}{c_l \cos\theta_0}\left|\frac{A_t}{A_0}\right|^2.$$

The sum of R_l and R_t is, of course, 1.

Problem 2. The same as Problem 1, but for a transverse incident wave (with the oscillations in the plane of incidence).†

Solution. The wave is reflected as a transverse and a longitudinal wave, with $\theta_t = \theta_0$, $c_t \sin\theta_l = c_l \sin\theta_0$. The total displacement vector is

$$\mathbf{u} = \mathbf{a}\times\mathbf{n}_0 A_0 e^{i\mathbf{k}_0\cdot\mathbf{r}} + \mathbf{n}_l A_l e^{i\mathbf{k}_l\cdot\mathbf{r}} + \mathbf{a}\times\mathbf{n}_t A_t e^{i\mathbf{k}_t\cdot\mathbf{r}}.$$

The expressions for the amplitudes of the reflected waves are

$$\frac{A_t}{A_0} = \frac{c_t^2 \sin 2\theta_l \sin 2\theta_0 - c_l^2 \cos^2 2\theta_0}{c_t^2 \sin 2\theta_l \sin 2\theta_0 + c_l^2 \cos^2 2\theta_0},$$

$$\frac{A_l}{A_0} = \frac{2c_l c_t \sin 2\theta_0 \cos 2\theta_0}{c_t^2 \sin 2\theta_l \sin 2\theta_0 + c_l^2 \cos^2 2\theta_0}.$$

Problem 3. Determine the characteristic frequencies of radial vibrations of an elastic sphere with radius R.

Solution. We take spherical polar coordinates, with the origin at the centre of the sphere. For radial vibrations, $\mathbf{u}$ is along the radius, and is a function of r and t only. Hence $\mathbf{curl}\,\mathbf{u} = 0$. We define the displacement "potential" ϕ by $u_r = u = \partial\phi/\partial r$. The equation of motion, expressed in terms of ϕ, is just the wave equation $c_l^2 \triangle\phi = \ddot{\phi}$, or, for oscillations periodic in time ($\propto e^{-i\omega t}$),

$$\triangle\phi \equiv \frac{1}{r^2}\frac{\partial}{\partial r}\left(r^2 \frac{\partial\phi}{\partial r}\right) = -k^2\phi,\ k = \omega/c_l. \tag{1}$$

The solution which is finite at the origin is $\phi = (A/r)\sin kr$ (the time factor is omitted). The radial stress is

$$\sigma_{rr} = \rho\{(c_l^2 - 2c_t^2)u_{ii} + 2c_t^2 u_{rr}\}$$

$$= \rho\{(c_l^2 - 2c_t^2)\triangle\phi + 2c_t^2\phi''\}$$

or, using (1),

$$\sigma_{rr}/\rho = -\omega^2\phi - 4c_t^2\phi'/r. \tag{2}$$

The boundary condition $\sigma_{rr}(R) = 0$ leads to the equation

$$\frac{\tan kR}{kR} = \frac{1}{1 - (kRc_l/2c_t)^2}, \tag{3}$$

whose roots determine the characteristic frequencies $\omega = kc_l$ of the vibrations.

Problem 4. Determine the frequency of radial vibrations of a spherical cavity in an infinite elastic medium for which $c_l \gg c_t$ (M. A. Isakovich 1949).

Solution. In an infinite medium, radial oscillations of the cavity are accompanied by the emission of longitudinal sound waves, leading to loss of energy and hence to damping of the oscillations. When $c_l \gg c_t$ (i.e.

---

† If the oscillations are perpendicular to the plane of incidence, the wave is entirely reflected as a wave of the same kind, and so $R_t = 1$.

$K \gg \mu$), this emission is weak, and we can speak of the characteristic frequencies of oscillations with a small coefficient of damping.

We seek a solution of equation (1), Problem 3, in the form of an outgoing spherical wave $\phi = Ae^{ikr}/r$, $k = \omega/c_l$ and, using (2), obtain from the boundary condition $\sigma_{rr}(R) = 0$ the result $(kRc_l/c_t)^2 = 4(1 - ikR)$. Hence, when $c_l \gg c_t$,

$$\omega = \frac{2c_t}{R}\left(1 - i\frac{c_t}{c_l}\right).$$

The real part of ω gives the characteristic frequency of oscillation; the imaginary part gives the damping coefficient. In an incompressible medium ($c_l \to \infty$) there would of course be no damping. These vibrations are specifically due to the shear resistance of the medium ($\mu \neq 0$). It should be noticed that they have $kR = 2c_t/c_l \ll 1$, i.e. the corresponding wavelength is large compared with R; it is interesting to compare this with the result for vibrations of an elastic sphere, where with $c_l \gg c_t$ the first characteristic frequency is given by (3): $kR = \pi$.

§23. Elastic waves in crystals

The propagation of elastic waves in anisotropic media, i.e. in crystals, is more complicated than for the case of isotropic media. To investigate such waves, we must return to the general equations of motion $\rho\ddot{u}_i = \partial\sigma_{ik}/\partial x_k$ and use for σ_{ik} the general expression (10.3) $\sigma_{ik} = \lambda_{iklm}u_{lm}$. According to what was said at the beginning of §22, λ_{iklm} always denotes the adiabatic moduli of elasticity.

Substituting for σ_{ik} in the equations of motion, we obtain

$$\rho\ddot{u}_i = \lambda_{iklm}\frac{\partial u_{lm}}{\partial x_k} = \tfrac{1}{2}\lambda_{iklm}\frac{\partial}{\partial x_k}\left(\frac{\partial u_l}{\partial x_m} + \frac{\partial u_m}{\partial x_l}\right)$$

$$= \tfrac{1}{2}\lambda_{iklm}\frac{\partial^2 u_l}{\partial x_k \partial x_m} + \tfrac{1}{2}\lambda_{iklm}\frac{\partial^2 u_m}{\partial x_k \partial x_l}.$$

Since the tensor λ_{iklm} is symmetrical with respect to the suffixes l and m we can interchange these in the first term, which then becomes identical with the second term. Thus the equations of motion are

$$\rho\ddot{u}_i = \lambda_{iklm}\frac{\partial^2 u_m}{\partial x_k \partial x_l}. \tag{23.1}$$

Let us consider a monochromatic elastic wave in a crystal. We can seek a solution of the equations of motion in the form $u_i = u_{0i}e^{i(\mathbf{k}\cdot\mathbf{r} - \omega t)}$, where the u_{0i} are constants, the relation between the wave vector $\mathbf{k}$ and the frequency ω being such that this function actually satisfies equation (23.1). Differentiation of u_i with respect to time results in multiplication by $-i\omega$, and differentiation with respect to x_k leads to multiplication by ik_k. Hence the above substitution converts equation (23.1) into $\rho\omega^2 u_i = \lambda_{iklm}k_k k_l u_m$. Putting $u_i = \delta_{im}u_m$, we can write this as

$$(\rho\omega^2\delta_{im} - \lambda_{iklm}k_k k_l)u_m = 0. \tag{23.2}$$

This is a set of three homogeneous equations of the first degree for the unknowns u_x, u_y, u_z. Such equations have non-zero solutions only if the determinant of the coefficients is zero. Thus we must have

$$|\lambda_{iklm}k_k k_l - \rho\omega^2\delta_{im}| = 0. \tag{23.3}$$

This equation (the *dispersion equation*) determines the relation between the wave frequency and the wave vector, called the *dispersion relation*. The equation (23.3) is cubic in ω^2, and has three roots $\omega^2 = \omega_j^2(\mathbf{k})$, which are in general different; the dispersion relation is said to have three branches. Substituting each root in turn back into (23.2) and solving, we find the directions of the displacement vector **u** in these waves—the directions of *polarization* of the waves; since the equations (23.2) are homogeneous, they of course do not determine the magnitude of **u**, which remains arbitrary.† The directions of polarization of the three waves with the same wave vector **k** are mutually perpendicular. This important result follows directly from the fact that (23.3) may be regarded as an equation for the principal values of the symmetrical tensor of rank two‡ $\lambda_{iklm}k_k k_l$; the equations (23.2) determine the principal directions of this tensor, which are known to be mutually perpendicular. None of these directions is, however, in general either purely longitudinal or purely transverse with respect to the direction of **k**.

The velocity with which the wave is propagated (its *group velocity*) is given by the derivative

$$\mathbf{U} = \partial\omega/\partial\mathbf{k}; \tag{23.4}$$

see *FM*, §67. In an isotropic medium, the dependence of ω on **k** reduces to a direct proportionality to the magnitude k, and the group velocity is parallel to the wave vector. In crystals, this is not so, and the direction of propagation of the wave is in general different from that of **k**. Only certain exceptional directions (the symmetry axes of the crystal) can be those of both **k** and **U**.

It is seen from the dispersion equation (23.3) that, in a crystal, ω is a first-order homogeneous function of the components of **k**. (If the ratio ω/k is treated as the unknown, the coefficients in the equation are independent of k.) Thus **U** is a zero-order homogeneous function of k_x, k_y, k_z. In other words, the velocity of propagation of the wave depends on its direction but not on the frequency.

If we construct in **k**-space (i.e. in the coordinates k_x, k_y, k_z) a surface of constant frequency, $\omega(\mathbf{k})$ = constant, for any branch of the dispersion relation, then the vector (23.4) is along the normal to the surface. Evidently, if this surface is everywhere convex, there is a one-to-one relation between the directions of **U** and **k**: a definite direction of **U** corresponds to each direction of **k**, and vice versa. If, however, the constant-frequency surface is not everywhere convex, the relation is no longer one-to-one: there is again one direction of **U** for each direction of **k** (in a given branch of the dispersion relation), but a particular direction of **U** may occur for various directions of **k**.

PROBLEMS

PROBLEM 1. Determine the dispersion relation of elastic waves in a cubic crystal which are propagated (a) in the (001) crystal plane, that of a cube face, (b) in the [111] crystal direction, that of a cube diagonal.

SOLUTION. In a cubic crystal, the non-zero elastic moduli are $\lambda_{xxxx} \equiv \lambda_1$, $\lambda_{xxyy} \equiv \lambda_2$, $\lambda_{xyxy} \equiv \lambda_3$ (and the equal components with x and y replaced by other pairs from x, y, z; see §10); the x, y, and z axes are along the edges of the cube.

---

† In an isotropic body the branches are $\omega = c_l k$ (longitudinally polarized waves) and two coincident roots $\omega = c_t k$ corresponding to waves with two independent transverse directions of polarization.

‡ From the symmetry of λ_{iklm},

$$\lambda_{iklm}k_k k_l = \lambda_{kiml}k_k k_l = \lambda_{mlki}k_k k_l.$$

The last expression differs from the first only in the naming of the dummy suffixes k and l, so that $\lambda_{iklm}k_k k_l$ is in fact symmetrical in the suffixes i and m.

(a) We take the (001) plane as the xy-plane. Let θ be the angle between the wave vector $\mathbf{k}$ in this plane and the x-axis. By constructing and solving the dispersion equation (23.3), we find three branches of the dispersion relation:

$$\rho\omega_{1,2}^2 = \tfrac{1}{2}k^2\{\lambda_1 + \lambda_3 \pm [(\lambda_1 - \lambda_3)^2 - 4(\lambda_1 + \lambda_2)(\lambda_1 - \lambda_2 - 2\lambda_3)\sin^2\theta\cos^2\theta]^{1/2}\},$$

$$\rho\omega_3^2 = \lambda_3 k^2.$$

The third-branch wave is transverse and is polarized along the z-axis. The waves of the first two branches are polarized in the xy-plane. It is evident from symmetry that the propagation velocity $\mathbf{U} = \partial\omega/\partial\mathbf{k}$ of all these waves is also in the xy-plane; the expressions obtained are therefore sufficient to calculate it.

When $\theta = 0$ ($\mathbf{k}$ is along the x-axis),

$$\rho\omega_1^2 = \lambda_1 k^2, \qquad \rho\omega_2^2 = \lambda_3 k^2,$$

wave 1 being longitudinal (polarized along the x-axis) and wave 2 transverse (polarized along the y-axis).

When $\theta = \tfrac{1}{4}\pi$ ($\mathbf{k}$ along the diagonal of the cube face),

$$\rho\omega_1^2 = \tfrac{1}{2}(\lambda_1 + \lambda_2 + 2\lambda_3)k^2,$$

$$\rho\omega_2^2 = \tfrac{1}{2}(\lambda_1 - \lambda_2)k^2.$$

Wave 1 is longitudinal; wave 2 is transverse and is polarized in the xy-plane.

(b) Here the wave vector components are $k_x = k_y = k_z = k/\sqrt{3}$. The solutions of the dispersion equation are

$$\rho\omega_1^2 = \tfrac{1}{3}k^2(\lambda_1 + 2\lambda_2 + 4\lambda_3),$$

$$\rho\omega_{2,3}^2 = \tfrac{1}{3}k^2(\lambda_1 + \lambda_2 + \lambda_3).$$

Wave 1 is longitudinal; waves 2 and 3 are transverse.

PROBLEM 2. Determine the dispersion relation for elastic waves in a crystal of the hexagonal system.

SOLUTION. A hexagonal crystal has five independent elastic moduli (§10, Problem 1), for which we use the notation

$$\lambda_{xxxx} = \lambda_{yyyy} = a, \qquad \lambda_{xyxy} = b, \qquad \lambda_{xxyy} = a - 2b,$$

$$\lambda_{xxzz} = \lambda_{yyzz} = c, \qquad \lambda_{xzxz} = \lambda_{yzyz} = d, \qquad \lambda_{zzzz} = f.$$

The z-axis is along the sixth-order axis of symmetry; the directions of the x and y axes may be chosen arbitrarily. We shall take the xz-plane so as to contain the wave vector $\mathbf{k}$. Then $k_x = k\sin\theta$, $k_y = 0$, $k_z = k\cos\theta$, where θ is the angle between $\mathbf{k}$ and the z-axis. By constructing and solving the equation (23.3), we find

$$\rho\omega_1^2 = k^2(b\sin^2\theta + d\cos^2\theta),$$

$$\rho\omega_{2,3}^2 = \tfrac{1}{2}k^2\{a\sin^2\theta + f\cos^2\theta + d \pm [\{(a-d)\sin^2\theta + (d-f)\cos^2\theta\}^2 + 4(c+d)^2\sin^2\theta\cos^2\theta]^{1/2}\}.$$

When $\theta = 0$,

$$\rho\omega_{1,2}^2 = k^2 d, \qquad \rho\omega_3^2 = k^2 f;$$

wave 3 is longitudinal, waves 1 and 2 transverse.

§24. Surface waves

A particular kind of elastic waves are those propagated near the surface of a body without penetrating into it (*Rayleigh waves*). We write the equation of motion in the form (22.11) and (22.12):

$$\frac{\partial^2 u}{\partial t^2} - c^2 \triangle u = 0, \tag{24.1}$$

where u is any component of the vectors $\mathbf{u}_l$, $\mathbf{u}_t$, and c is the corresponding velocity c_l or c_t, and seek solutions corresponding to these surface waves. The surface of the elastic medium is supposed plane and of infinite extent. We take this plane as the xy-plane; let the medium be in $z < 0$.

Let us consider a plane monochromatic surface wave propagated along the x-axis. Accordingly $u = e^{i(kx - \omega t)}f(z)$. Substituting this expression in (24.1), we obtain for the function $f(z)$ the equation

$$\frac{d^2 f}{dz^2} = \left(k^2 - \frac{\omega^2}{c^2}\right)f.$$

If $k^2 - \omega^2/c^2 < 0$, this equation gives a periodic function f, i.e. we obtain an ordinary plane wave which is not damped inside the body. We must therefore suppose that $k^2 - \omega^2/c^2 > 0$. Then the solutions for f are

$$f(z) = \text{constant} \times \exp\left(\pm\sqrt{\left[k^2 - \frac{\omega^2}{c^2}\right]}z\right).$$

The solution with the minus sign would correspond to an unlimited increase in the deformation for $z \to -\infty$. This solution is clearly impossible, and so the plus sign must be taken.

Thus we have the following solution of the equations of motion:

$$u = \text{constant} \times e^{i(kx - \omega t)}e^{\kappa z}, \tag{24.2}$$

where

$$\kappa = \sqrt{(k^2 - \omega^2/c^2)}. \tag{24.3}$$

It corresponds to a wave which is exponentially damped towards the interior of the medium, i.e. is propagated only near the surface. The quantity κ determines the rapidity of the damping.

The true displacement vector $\mathbf{u}$ in the wave is the sum of the vectors $\mathbf{u}_l$ and $\mathbf{u}_t$, the components of each of which satisfy the equation (24.1) with $c = c_l$ for $\mathbf{u}_l$ and c_t for $\mathbf{u}_t$. For volume waves in an infinite medium, the two parts are independently propagated waves. For surface waves, however, this division into two independent parts is not possible, on account of the boundary conditions. The displacement vector $\mathbf{u}$ must be a definite linear combination of the vectors $\mathbf{u}_l$ and $\mathbf{u}_t$. It should also be mentioned that these latter vectors have no longer the simple significance of the displacement components parallel and perpendicular to the direction of propagation.

To determine the linear combination of the vectors $\mathbf{u}_l$ and $\mathbf{u}_t$ which gives the true displacement $\mathbf{u}$, we must use the conditions at the boundary of the body. These give a relation between the wave vector $\mathbf{k}$ and the frequency ω, and therefore the velocity of propagation of the wave. At the free surface we must have $\sigma_{ik}n_k = 0$. Since the normal vector $\mathbf{n}$ is parallel to the z-axis, it follows that $\sigma_{xz} = \sigma_{yz} = \sigma_{zz} = 0$, whence

$$u_{xz} = 0, \qquad u_{yz} = 0, \qquad \sigma(u_{xx} + u_{yy}) + (1 - \sigma)u_{zz} = 0. \tag{24.4}$$

Since all quantities are independent of the coordinate y, the second of these conditions gives

$$u_{yz} = \frac{1}{2}\left(\frac{\partial u_y}{\partial z} + \frac{\partial u_z}{\partial y}\right) = \tfrac{1}{2}\partial u_y/\partial z = 0.$$

Using (24.2), we therefore have

$$u_y = 0. \tag{24.5}$$

Thus the displacement vector $\mathbf{u}$ in a surface wave is in a plane through the direction of propagation perpendicular to the surface.

The transverse part $\mathbf{u}_t$ of the wave must satisfy the condition (22.8) div $\mathbf{u}_t = 0$, or

$$\frac{\partial u_{tx}}{\partial x} + \frac{\partial u_{tz}}{\partial z} = 0.$$

The dependence of u_{tx} and u_{tz} on x and z is determined by the factor $e^{ikx + \kappa_t z}$, where κ_t is given by the expression (24.3) with $c = c_t$, i.e.

$$\kappa = \sqrt{(k^2 - \omega^2/c_t^2)}.$$

Hence the above condition leads to the equation

$$iku_{tx} + \kappa_t u_{tz} = 0, \quad \text{or} \quad u_{tx}/u_{tz} = -\kappa_t/ik.$$

Thus we can write

$$u_{tx} = \kappa_t a e^{ikx+\kappa_t z - i\omega t}, \qquad u_{tz} = -ika e^{ikx+\kappa_t z - i\omega t}, \tag{24.6}$$

where a is some constant.

The longitudinal part $\mathbf{u}_l$ satisfies the condition (22.9) **curl** $\mathbf{u}_l = 0$, or

$$\frac{\partial u_{lx}}{\partial z} - \frac{\partial u_{lz}}{\partial x} = 0,$$

whence

$$iku_{lz} - \kappa_l u_{lx} = 0 \quad (\kappa_l = \sqrt{[k^2 - \omega^2/c_l^2]}).$$

Thus we must have

$$u_{lx} = kbe^{ikx+\kappa_l z - i\omega t}, \qquad u_{lz} = -i\kappa_l b e^{ikx+\kappa_l z - i\omega t}, \tag{24.7}$$

where b is a constant.

We now use the first and third conditions (24.4). Expressing u_{ik} in terms of the derivatives of u_i, and using the velocities c_l, c_t, we can write these conditions as

$$\left.\begin{aligned} &\frac{\partial u_x}{\partial z} + \frac{\partial u_z}{\partial x} = 0, \\ &c_l^2 \frac{\partial u_z}{\partial z} + (c_l^2 - 2c_t^2)\frac{\partial u_x}{\partial x} = 0. \end{aligned}\right\} \tag{24.8}$$

Here we must substitute $u_x = u_{lx} + u_{tx}$, $u_z = u_{lz} + u_{tz}$. The result is that the first condition (24.8) gives

$$a(k^2 + \kappa_t^2) + 2bk\kappa_l = 0. \tag{24.9}$$

The second condition leads to the equation

$$2ac_t^2\kappa_t k + b[c_l^2(\kappa_l^2 - k^2) + 2c_t^2 k^2] = 0.$$

Dividing this equation by c_t^2 and substituting

$$\kappa_l^2 - k^2 = -\omega^2/c_l^2 = -(k^2 - \kappa_t^2)c_t^2/c_l^2,$$

we can write it as

$$2a\kappa_t k + b(k^2 + \kappa_t^2) = 0. \tag{24.10}$$

The condition for the two homogeneous equations (24.9) and (24.10) to be compatible is $(k^2 + \kappa_t^2)^2 = 4k^2\kappa_t\kappa_l$ or, squaring and substituting the values of κ_t^2 and κ_l^2,

$$\left(2k^2 - \frac{\omega^2}{c_t^2}\right)^4 = 16k^4\left(k^2 - \frac{\omega^2}{c_t^2}\right)\left(k^2 - \frac{\omega^2}{c_l^2}\right). \tag{24.11}$$

From this equation we obtain the relation between ω and k. It is convenient to put

$$\omega = c_t k\xi; \tag{24.12}$$

k^8 then cancels from both sides of the equation, and, expanding, we obtain for ξ the equation

$$\xi^6 - 8\xi^4 + 8\xi^2\left(3 - 2\frac{c_t^2}{c_l^2}\right) - 16\left(1 - \frac{c_t^2}{c_l^2}\right) = 0. \tag{24.13}$$

Hence we see that ξ depends only on the ratio c_t/c_l, which is a constant characteristic of any given substance and in turn depends only on Poisson's ratio:

$$c_t/c_l = \sqrt{\{(1-2\sigma)/2(1-\sigma)\}}.$$

The quantity ξ must, of course, be real and positive, and $\xi < 1$ (so that κ_t and κ_l are real). Equation (24.13) has only one root satisfying these conditions, and so a single value of ξ is obtained for any given value of c_t/c_l.†

Thus, for both surface waves and volume waves, the frequency is proportional to the wave number. The proportionality coefficient is the velocity of propagation of the wave,

$$U = c_t \xi. \tag{24.14}$$

This gives the velocity of propagation of surface waves in terms of the velocities c_t and c_l of the transverse and longitudinal volume waves. The ratio of the amplitudes of the transverse and longitudinal parts of the wave is given in terms of ξ by the formula

$$\frac{a}{b} = -\frac{2-\xi^2}{2\sqrt{(1-\xi^2)}}. \tag{24.15}$$

The ratio c_t/c_l actually varies from $1/\sqrt{2}$ to 0 for various substances, corresponding to the variation of σ from 0 to $\frac{1}{2}$; ξ then varies from 0·874 to 0·955. Fig. 21 shows a graph of ξ as a function of σ.

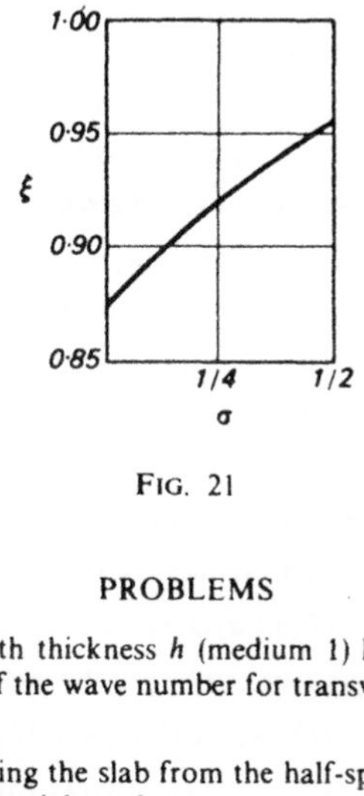

FIG. 21

PROBLEMS

PROBLEM 1. A plane-parallel slab with thickness h (medium 1) lies on an elastic half-space (medium 2). Determine the frequency as a function of the wave number for transverse waves in the slab whose direction of oscillation is parallel to its boundaries.

SOLUTION. We take the plane separating the slab from the half-space as the xy-plane, the half-space being in $z < 0$ and the slab in $0 \leqslant z \leqslant h$. In the slab we have

$$u_{x1} = u_{z1} = 0, \qquad u_{y1} = f(z)e^{i(kx-\omega t)},$$

and in medium 2 a damped wave:

$$u_{x2} = u_{z2} = 0, \qquad u_{y2} = Ae^{\kappa_2 z}e^{i(kx-\omega t)}, \qquad \kappa_2 = \sqrt{(k^2-\omega^2/c_{t2}{}^2)}.$$

For the function $f(z)$ we have the equation

$$f'' + \kappa_1{}^2 f = 0, \qquad \kappa_1 = \sqrt{(\omega^2/c_{t1}{}^2 - k^2)}$$

† In going from (24.11) to (24.13), the root $\omega^2 = 0$ ($\kappa_t = \kappa_l = k$) is lost; it corresponds to $\xi = 0$, which also is less than unity. However, it can be seen from (24.9) and (24.10) that this root gives $a = -b$ and hence a total displacement $\mathbf{u} = \mathbf{u}_l + \mathbf{u}_t = 0$, so that there is no motion at all.

(we shall see below that $\kappa_1^2 > 0$), whence $f(z) = B\sin\kappa_1 z + C\cos\kappa_1 z$. At the free surface of the slab ($z = h$) we must have $\sigma_{zy} = 0$, i.e. $\partial u_{y1}/\partial z = 0$. At the boundary between the two media ($z = 0$) the conditions are $u_{y1} = u_{y2}$, $\mu_1 \partial u_{y1}/\partial z = \mu_2 \partial u_{y2}/\partial z$, μ_1 and μ_2 being the moduli of rigidity for the two media. From these conditions we find three equations for A, B, C, and the compatibility condition is $\tan\kappa_1 h = \mu_2\kappa_2/\mu_1\kappa_1$. This equation gives ω as an implicit function of k; it has solutions only for real κ_1 and κ_2, and so $c_{t2} > \omega/k > c_{t1}$. Hence we see that such waves can be propagated only if $c_{t2} > c_{t1}$.

PROBLEM 2. Determine the depth of penetration and the velocity of surface waves propagated in the crystal direction [100] (cube edge) on the plane surface (001) (cube face) of a cubic crystal. The crystal is assumed to have highly anisotropic elastic properties, in the sense that, if $\eta = (\lambda_1 - \lambda_2)/2\lambda_3$, either (a) $\eta \gg 1$ or (b) $\eta \ll 1$. The elastic moduli are denoted as in §23, Problem 1.†

SOLUTION. We take the crystal surface as the xy-plane, with the medium in the half-space $z < 0$, and the wave propagated in the x-direction. As in the text, it can be shown that $u_y = 0$, i.e. the displacement vector $\mathbf{u}$ is in the xz-plane. We seek it in the form

$$u_z = ae^{\kappa z}e^{i(kx-\omega t)}, \quad u_x = ibe^{\kappa z}e^{i(kx-\omega t)} \tag{1}$$

and use the ratios $\gamma = \kappa/k$ and $\Gamma = b/a = -i(u_x/u_z)_{z=0}$; γ measures the depth of penetration of the surface wave in units of the wavelength $1/k$, and Γ gives the axis ratio of the wave polarization ellipse at the surface. The stress tensor components are

$$\sigma_{xx} = \lambda_1 u_{xx} + \lambda_2 u_{zz},$$
$$\sigma_{zz} = \lambda_1 u_{zz} + \lambda_2 u_{xx},$$
$$\sigma_{xz} = 2\lambda_3 u_{xz}.$$

The volume equations of motion (22.1) with $\mathbf{u}$ from (1) give the two algebraic equations

$$\left.\begin{aligned} a(\lambda_2+\lambda_3)\gamma + b(\lambda_3\gamma^2 - \lambda_1 + \rho U^2) &= 0,\\ a(\lambda_1\gamma^2 - \lambda_3 + \rho U^2) - b(\lambda_2+\lambda_3)\gamma &= 0,\end{aligned}\right\} \tag{2}$$

where $U = \omega/k$ is the velocity of propagation.‡ Hence

$$\Gamma = \frac{(\lambda_1+\lambda_3)\gamma}{\lambda_1 - \lambda_3\gamma^2 - \rho U^2}, \tag{3}$$

and the compatibility condition for the two equations (2) gives

$$\lambda_1\lambda_3\gamma^4 - \gamma^2[\lambda_1(\lambda_1-\rho U^2) + \lambda_3(\lambda_3-\rho U^2) - (\lambda_2+\lambda_3)^2] + (\lambda_1-\rho U^2)(\lambda_3-\rho U^2) = 0. \tag{4}$$

This equation determines two values γ_1 and γ_2 for given values of ω and k.

Accordingly, we now seek the displacement vector in the form

$$\left.\begin{aligned} u_z &= (a_1e^{\gamma_1 kz} + a_2e^{\gamma_2 kz})e^{i(kx-\omega t)},\\ u_x &= i(a_1\Gamma_1 e^{\gamma_1 kz} + a_2\Gamma_2 e^{\gamma_2 kz})e^{i(kx-\omega t)}.\end{aligned}\right\} \tag{5}$$

Substituting these expressions in the boundary conditions $\sigma_{xz} = \sigma_{zz} = 0$ for $z = 0$, we find the two equations

$$\left.\begin{aligned} a_1(1+\Gamma_1\gamma_1) + a(1+\Gamma_2\gamma_2) &= 0,\\ a_1(\lambda_1\gamma_1 - \lambda_2\Gamma_1) + a_2(\lambda_1\gamma_2 - \lambda_2\Gamma_2) &= 0.\end{aligned}\right\} \tag{6}$$

The compatibility condition for these equations can be put, by means of (3), in the form

$$\{(\lambda_2+\rho U^2)(\lambda_3-\rho U^2)\lambda_2 + \lambda_1^2\lambda_3\gamma_1^2\gamma_2^2 + \\ + \lambda_1\lambda_2(\gamma_1^2+\gamma_2^2+\gamma_1\gamma_2)(\lambda_3-\rho U^2) - \lambda_1\lambda_3(\lambda_2+\rho U^2)\gamma_1\gamma_2\}(\gamma_1-\gamma_2) = 0. \tag{7}$$

When $\gamma_1 \neq \gamma_2$, the factor $\gamma_1 - \gamma_2$ may be omitted. The sum $\gamma_1^2 + \gamma_2^2$ and the product $\gamma_1^2\gamma_2^2$ are given by the coefficients in the quadratic (in γ^2) equation (4), and (7) becomes

$$\rho U^2[\lambda_1\lambda_3(\lambda_1-\rho U^2)]^{\frac{1}{2}} = (\lambda_3-\rho U^2)^{\frac{1}{2}}(\lambda_1^2 - \lambda_2^2 - \lambda_1\rho U^2). \tag{8}$$

Let us now consider the two cases mentioned in stating the problem.

---

† When $\eta = 1$, the crystal behaves as an isotropic body regarding its elastic properties; see the third footnote to §10.

‡ Because of the symmetric position of the x-axis and the (first-order) homogeneity of $\omega(k_x, k_y)$, the velocity $U = \partial\omega/\partial k$ is also in the x-direction, and its magnitude is ω/k.

(a) When $\eta \gg 1$, λ_3 may be regarded as a small quantity. We then find from (7)

$$\rho U^2 = \lambda_3\left[1 - \frac{\lambda_1^2}{4\eta^2(\lambda_1+\lambda_2)^2}\right].$$

This velocity U is much less than the velocity $\sqrt{(\lambda_1/\rho)}$ (see §23, Problem 1(a)) of a longitudinal volume wave in the same direction, and in that sense the surface wave is slow, like the transverse volume waves. From (4), we then find the two values of γ:

$$\gamma_1 = \left[\frac{\lambda_1}{2\eta(\lambda_1+\lambda_3)}\right]^{3/2} \ll 1, \qquad \gamma_2 = \left[\frac{2\eta(\lambda_1+\lambda_3)}{\lambda_1}\right]^{1/2} \gg 1,$$

and from (3)

$$\Gamma_1 = \lambda_2\gamma_1/\lambda_1 \sim \eta^{-3/2} \ll 1, \qquad \Gamma_2 = \lambda_1\gamma_2/\lambda_2 \sim \eta^{1/2} \gg 1.$$

Lastly, from (6),

$$a_1/a_2 = \Gamma_2\gamma_2 = \lambda_1\gamma_2^2/\lambda_2 \sim \eta \gg 1.$$

Thus, since $\gamma_1 \gg 1$ and $a_1 \gg a_2$, the depth of penetration of the surface wave is (on account of the first terms in (5)) much greater than the wavelength.† Its polarization ellipse in the xz-plane is elongated in the direction of the z-axis normal to the surface ($\Gamma_1 \ll 1$).

(b) When $\eta \ll 1$, $\lambda_1 - \lambda_2$ is small. Then, from (4),

$$\rho U^2 = (\lambda_1 - \lambda_2)\left[1 - \frac{\eta(\lambda_1+\lambda_3)}{2\lambda_1}\right].$$

The values of γ from (4) are now complex: $\gamma_{1,2} = \gamma' \pm i\gamma''$, where

$$\gamma' = \tfrac{1}{2}\eta(1+\lambda_3/\lambda_1), \qquad \gamma'' = 1 - \tfrac{1}{4}\eta(1+\lambda_3/\lambda_1).$$

In consequence, $\Gamma_{1,2} = {}_{1,2} = \pm i$, $a_1/a_2 = i$. In this case too, therefore, the wave is slow and deeply penetrating ($\gamma' \ll 1$). Since γ is complex, the wave damping into the medium is here not monotonic but oscillatory; the oscillation period (in the z-direction) is $\sim 1/\gamma'' k$, about the same as the wavelength and therefore much less than the penetration depth.

§25. Vibration of rods and plates

Waves propagated in thin rods and plates are fundamentally different from those propagated in a medium infinite in all directions. Here we are speaking of waves of length large compared with the thickness of the rod or plate. If the wavelength is small compared with this thickness, the rod or plate is effectively infinite in all directions as regards the propagation of the wave, and we return to the results obtained for infinite media.

Waves in which the oscillations are parallel to the axis of the rod or the plane of the plate must be distinguished from those in which they are perpendicular to it. We shall begin by studying longitudinal waves in rods.

A longitudinal deformation of the rod (uniform over any cross-section), with no external force on the sides of the rod, is a simple extension or compression. Thus longitudinal waves in a rod are simple extensions or compressions propagated along its length. In a simple extension, however, only the component σ_{zz} of the stress tensor (the z-axis being along the rod) is different from zero; it is related to the strain tensor by $\sigma_{zz} = Eu_{zz} = E\partial u_z/\partial z$ (see §5). Substituting this in the general equation of motion $\rho\ddot{u}_z = \partial\sigma_{zk}/\partial x_k$, we find

$$\frac{\partial^2 u_z}{\partial z^2} - \frac{\rho}{E}\frac{\partial^2 u_z}{\partial t^2} = 0. \tag{25.1}$$

† The possibility of deeply penetrating slow surface waves in a crystal was first noted by S. V. Gerus and V. V. Tarasenko (1975).

This is the equation of longitudinal vibrations in rods. We see that it is an ordinary wave equation. The velocity of propagation of longitudinal waves in rods is

$$\sqrt{(E/\rho)}. \tag{25.2}$$

Comparing this with the expression (22.4) for c_l, we see that it is less than the velocity of propagation of longitudinal waves in an infinite medium.

Let us now consider longitudinal waves in thin plates. The equations of motion for such vibrations can be written down at once by substituting $-\rho h\partial^2 u_x/\partial t^2$ and $-\rho h\partial^2 u_y/\partial t^2$ for P_x and P_y in the equilibrium equations (13.4):

$$\left.\begin{aligned}\frac{\rho}{E}\frac{\partial^2 u_x}{\partial t^2} &= \frac{1}{1-\sigma^2}\frac{\partial^2 u_x}{\partial x^2}+\frac{1}{2(1+\sigma)}\frac{\partial^2 u_x}{\partial y^2}+\frac{1}{2(1-\sigma)}\frac{\partial^2 u_y}{\partial x\partial y},\\ \frac{\rho}{E}\frac{\partial^2 u_y}{\partial t^2} &= \frac{1}{1-\sigma^2}\frac{\partial^2 u_y}{\partial y^2}+\frac{1}{2(1+\sigma)}\frac{\partial^2 u_y}{\partial x^2}+\frac{1}{2(1-\sigma)}\frac{\partial^2 u_x}{\partial x\partial y}.\end{aligned}\right\} \tag{25.3}$$

We take the case of a plane wave propagated along the x-axis, i.e. a wave in which the deformation depends only on the coordinate x, and not on y. Then equations (25.3) are much simplified, becoming

$$\frac{\partial^2 u_x}{\partial t^2}-\frac{E}{\rho(1-\sigma^2)}\frac{\partial^2 u_x}{\partial x^2}=0, \quad \frac{\partial^2 u_y}{\partial t^2}-\frac{E}{2\rho(1+\sigma)}\frac{\partial^2 u_y}{\partial x^2}=0. \tag{25.4}$$

We thus again obtain wave equations. The coefficients are different for u_x and u_y. The velocity of propagation of a wave with oscillations parallel to the direction of propagation (u_x) is

$$\sqrt{[E/\rho(1-\sigma^2)]}. \tag{25.5}$$

The velocity for a wave (u_y) with oscillations perpendicular to the direction of propagation (but still in the plane of the plate) is equal to the velocity c_t of transverse waves in an infinite medium.

Thus we see that longitudinal waves in rods and plates are of the same nature as in an infinite medium, only the velocity being different; as before, it is independent of the frequency. Entirely different results are obtained for *bending waves* in rods and plates, for which the oscillations are in a direction perpendicular to the axis of the rod or the plane of the plate, i.e. involve bending.

The equations for free oscillations of a plate can be written down at once from the equilibrium equation (12.5). To do so, we must replace $-P$ by the acceleration $\ddot{\zeta}$ multiplied by the mass ρh per unit area of the plate. This gives

$$\rho\frac{\partial^2\zeta}{\partial t^2}+(D/h)\triangle^2\zeta=0, \tag{25.6}$$

where $\triangle$ is the two-dimensional Laplacian.

Let us consider a monochromatic elastic wave, and accordingly seek a solution of equation (25.6) in the form

$$\zeta = \text{constant}\times e^{i(\mathbf{k}\cdot\mathbf{r}-\omega t)}, \tag{25.7}$$

where the wave vector $\mathbf{k}$ has, of course, only two components, k_x and k_y. Substituting in (25.6), we obtain the equation

$$-\rho\omega^2+Dk^4/h=0.$$

Hence we have the following relation between the frequency and the wave number:

$$\omega = k^2\sqrt{(D/\rho h)} = k^2\sqrt{\{Eh^2/12\rho(1-\sigma^2)\}}. \tag{25.8}$$

Thus the frequency is proportional to the square of the wave number, whereas in waves in an infinite medium it is proportional to the wave number itself.

Knowing the relation between the frequency and the wave number, we can determine the velocity of propagation of the wave from the formula

$$\mathbf{U} = \partial\omega/\partial\mathbf{k}.$$

The derivatives of k^2 with respect to the components k_x, k_y are respectively $2k_x$, $2k_y$. The velocity of propagation of the wave is therefore

$$\mathbf{U} = \mathbf{k}\sqrt{\{Eh^2/3\rho(1-\sigma^2)\}}. \tag{25.9}$$

It is proportional to the wave vector, and not a constant as it is for waves in a medium infinite in three dimensions.†

Similar results are obtained for bending waves in thin rods. The bending deflections of the rod are supposed small. The equations of motion are obtained by replacing $-K_x$ and $-K_y$ in the equations of equilibrium for a slightly bent rod (20.4) by the product of the acceleration $\ddot{X}$ or $\ddot{Y}$ and the mass ρS per unit length of the rod (S being its cross-sectional area). Thus

$$\rho S\ddot{X} = EI_y\partial^4 X/\partial z^4, \qquad \rho S\ddot{Y} = EI_x\partial^4 Y/\partial z^4. \tag{25.10}$$

We again seek solutions of these equations in the form

$$X = \text{constant} \times e^{i(kz-\omega t)}, \qquad Y = \text{constant} \times e^{i(kz-\omega t)}.$$

Substituting in (25.10), we obtain the following relations between the frequency and the wave number:

$$\omega = k^2\sqrt{(EI_y/\rho S)}, \qquad \omega = k^2\sqrt{(EI_x/\rho S)}, \tag{25.11}$$

for vibrations in the x and y directions respectively. The corresponding velocities of propagation are

$$U^{(x)} = 2k\sqrt{(EI_y/\rho S)}, \qquad U^{(y)} = 2k\sqrt{(EI_x/\rho S)}. \tag{25.12}$$

Finally, there is a particular case of vibration of rods called *torsional vibration*. The corresponding equations of motion are derived by equating $C\partial\tau/\partial z$ (see §18) to the time derivative of the angular momentum of the rod per unit length. This angular momentum is $\rho I\partial\phi/\partial t$, where $\partial\phi/\partial t$ is the angular velocity (ϕ being the angle of rotation of the cross-section considered) and $I = \int(x^2+y^2)\,\mathrm{d}f$ is the moment of inertia of the cross-section about its centre of mass; for pure torsional vibration each cross-section of the rod performs rotary vibrations about its centre of mass, which remains at rest. Putting $\tau = \partial\phi/\partial z$, we obtain the equation of motion in the form

$$C\partial^2\phi/\partial z^2 = \rho I\partial^2\phi/\partial t^2. \tag{25.13}$$

Hence we see that the velocity of propagation of torsional oscillations along the rod is

$$\sqrt{(C/\rho I)}. \tag{25.14}$$

† The wave number $k = 2\pi/\lambda$, where λ is the wavelength. Hence the velocity of propagation should increase without limit as λ tends to zero. This physically impossible result is obtained because formula (25.9) is not valid for short waves.

PROBLEMS

PROBLEM 1. Determine the characteristic frequencies of longitudinal vibrations of a rod with length l, with one end fixed and the other free.

SOLUTION. At the fixed end ($z = 0$) we must have $u_z = 0$, and at the free end ($z = l$) $\sigma_{zz} = Eu_{zz} = 0$, i.e. $\partial u_z/\partial z = 0$. We seek a solution of equation (25.1) in the form

$$u_z = A \cos(\omega t + \alpha) \sin kz,$$

where $k = \omega\sqrt{(\rho/E)}$. From the condition at $z = l$ we have $\cos kl = 0$, whence the characteristic frequencies are

$$\omega = \sqrt{(E/\rho)}(2n+1)\pi/2l,$$

n being any integer.

PROBLEM 2. The same as Problem 1, but for a rod with both ends free or both fixed.

SOLUTION. In either case $\omega = \sqrt{(E/\rho)}\, n\pi/l$.

PROBLEM 3. Determine the characteristic frequencies of vibration of a string with length l.

SOLUTION. The equation of motion of the string is

$$\frac{\partial^2 X}{\partial z^2} - \frac{\rho S}{T}\frac{\partial^2 X}{\partial t^2} = 0;$$

cf. the equilibrium equation (20.17). The boundary conditions are that $X = 0$ for $z = 0$ and l. The characteristic frequencies are $\omega = \sqrt{(\rho S/T)}\, n\pi/l$.

PROBLEM 4. Determine the characteristic transverse vibrations of a rod (with length l) with clamped ends.

SOLUTION. Equation (25.10), on substituting $X = X_0(z)\cos(\omega t + \alpha)$, becomes

$$\mathrm{d}^4 X_0/\mathrm{d}z^4 = \kappa^4 X_0,$$

where $\kappa^4 = \omega^2 \rho S/EI_y$. The general integral of this equation is

$$X_0 = A\cos\kappa z + B\sin\kappa z + C\cosh\kappa z + D\sinh\kappa z.$$

The constants A, B, C and D are determined from the boundary conditions that $X = \mathrm{d}X/\mathrm{d}z = 0$ for $z = 0$ and l. The result is

$$X_0 = A\{(\sin\kappa l - \sinh\kappa l)(\cos\kappa z - \cosh kz) - $$
$$ - (\cos\kappa l - \cosh\kappa l)(\sin\kappa z - \sinh\kappa z)\},$$

and the equation $\cos\kappa l \cosh\kappa l = 1$, the roots of which give the characteristic frequencies. The smallest characteristic frequency is

$$\omega_{\min} = \frac{22{\cdot}4}{l^2}\sqrt{\frac{EI_y}{\rho S}}.$$

PROBLEM 5. The same as Problem 4, but for a rod with supported ends.

SOLUTION. In the same way as in Problem 4, we obtain $X_0 = A\sin\kappa z$, and the frequencies are given by $\sin\kappa l = 0$, i.e. $\kappa = n\pi/l$ ($n = 1, 2, \ldots$). The smallest frequency is

$$\omega_{\min} = \frac{9{\cdot}87}{l^2}\sqrt{\frac{EI_y}{\rho S}}.$$

PROBLEM 6. The same as Problem 4, but for a rod with one end clamped and the other free.

SOLUTION. We have for the displacement

$$X_0 = A\{(\cos\kappa l + \cosh\kappa l)(\cos\kappa z - \cosh\kappa z) + $$
$$ + (\sin\kappa l - \sinh\kappa l)(\sin\kappa z - \sinh\kappa z)\}$$

(the clamped end being at $z = 0$ and the free end at $z = l$), and for the characteristic frequencies the equation $\cos\kappa l \cosh\kappa l + 1 = 0$. The smallest frequency is

$$\omega_{\min} = \frac{3{\cdot}52}{l^2}\sqrt{\frac{EI_y}{\rho S}}.$$

Problem 7. Determine the characteristic vibrations of a rectangular plate with sides a and b, with its edges supported.

Solution. Equation (25.6), on substituting $\zeta = \zeta_0(x,y)\cos(\omega t + \alpha)$, becomes

$$\Delta^2\zeta_0 - \kappa^4\zeta_0 = 0,$$

where $\kappa^4 = 12\rho(1-\sigma^2)\omega^2/Eh^2$. We take the coordinate axes along the sides of the plate. The boundary conditions (12.11) become $\zeta = \partial^2\zeta/\partial x^2 = 0$ for $x = 0$ and a,

$$\zeta = \partial^2\zeta/\partial y^2 = 0$$

for $y = 0$ and b. The solution which satisfies these conditions is

$$\zeta_0 = A\,\sin(m\pi x/a)\,\sin(n\pi y/b),$$

where m and n are integers. The frequencies are given by

$$\omega = h\sqrt{\frac{E}{12\rho(1-\sigma^2)}}\pi^2\left[\frac{m^2}{a^2}+\frac{n^2}{b^2}\right].$$

Problem 8. Determine the characteristic frequencies for the vibration of a rectangular membrane with sides a and b.

Solution. The equation for the vibration of a membrane is $T\Delta\zeta = \rho h\ddot{\zeta}$; cf. the equilibrium equation (14.9). The edges of the membrane must be fixed, so that $\zeta = 0$. The corresponding solution for a rectangular membrane is

$$\zeta = A\,\sin(m\pi x/a)\,\sin(n\pi y/b)\,\cos\,\omega t,$$

where the characteristic frequencies are given by

$$\omega^2 = \frac{T\pi^2}{\rho h}\left(\frac{m^2}{a^2}+\frac{n^2}{b^2}\right),$$

m and n being integers.

Problem 9. Determine the velocity of propagation of torsional vibrations in a rod whose cross-section is a circle, an ellipse, or an equilateral triangle, and in a rod in the form of a long thin rectangular plate.

Solution. For a circular cross-section with radius R, the moment of inertia is $I = \frac{1}{2}\pi R^4$; C is given in §16, Problem 1, and we find the velocity to be $\sqrt{(\mu/\rho)}$, which is the same as the velocity c_t.

Similarly (using the results of §16, Problems 2 to 4), we find for a rod with an elliptical cross-section the velocity $[2ab/(a^2+b^2)]c_t$, for one with an equilateral triangular cross-section $\sqrt{(3/5)}c_t$, and for one which is a long rectangular plate $(2h/d)c_t$. All these are less than c_t.

Problem 10. The surface of an incompressible fluid of infinite depth is covered by a thin elastic plate. Determine the relation between the wave number and the frequency for waves which are simultaneously propagated in the plate and near the surface of the fluid.

Solution. We take the plane of the plate as $z = 0$, and the x-axis in the direction of propagation of the wave; let the fluid be in $z < 0$. The equation of motion of the plate alone would be

$$\rho_0 h\frac{\partial^2\zeta}{\partial t^2} = -D\frac{\partial^4\zeta}{\partial x^4},$$

where ρ_0 is the volume density of the plate. When the fluid is present, the right-hand side of this equation must also include the force exerted by the fluid on unit area of the plate, i.e. the pressure p of the fluid. The pressure in the wave, however, can be expressed in terms of the velocity potential by $p = -\rho\partial\phi/\partial t$ (we neglect gravity). Hence we obtain

$$\rho_0 h\frac{\partial^2\zeta}{\partial t^2} = -D\frac{\partial^4\zeta}{\partial x^4} - \left[\rho\frac{\partial\phi}{\partial t}\right]_{z=0}. \tag{1}$$

Next, the normal component of the fluid velocity at the surface must be equal to that of the plate, whence

$$\partial\zeta/\partial t = [\partial\phi/\partial z]_{z=0}. \tag{2}$$

The potential ϕ must satisfy everywhere in the fluid the equation

$$\frac{\partial^2\phi}{\partial x^2}+\frac{\partial^2\phi}{\partial z^2} = 0. \tag{3}$$

We seek ζ in the form of a travelling wave $\zeta = \zeta_0 e^{ikx - i\omega t}$; accordingly, we take as the solution of equation (3) the surface wave $\phi = \phi_0 e^{i(kx-\omega t)}e^{kz}$, which is damped in the interior of the fluid. Substituting these expressions in (1) and (2), we obtain two equations for ϕ_0 and ζ_0, and the compatibility condition is

$$\omega^2 = \frac{Dk^5}{\rho + h\rho_0 k}.$$

§26. Anharmonic vibrations

The whole of the theory of elastic vibrations given above is approximate to the extent that any theory of elasticity is so which is based on Hooke's law. It should be recalled that the theory begins from an expansion of the elastic energy as a power series with respect to the strain tensor, which includes terms up to and including the second order. The components of the stress tensor are then linear functions of those of the strain tensor, and the equations of motion are linear.

The most characteristic property of elastic waves in this approximation is that any wave can be obtained by simple superposition (i.e. as a linear combination) of separate monochromatic waves. Each of these is propagated independently, and could exist by itself without involving any other motion. We may say that the various monochromatic waves which are simultaneously propagated in a single medium do not interact with one another.

These properties, however, no longer hold in subsequent approximations. The effects which appear in these approximations, though small, may be of importance as regards certain phenomena. They are usually called *anharmonic effects*, since the corresponding equations of motion are non-linear and do not admit simple periodic (harmonic) solutions.

We shall consider here anharmonic effects of the third order, arising from terms in the elastic energy which are cubic in the strains. It would be too cumbersome to write out the corresponding equations of motion in their general form. However, the nature of the resulting effects can be ascertained as follows. The cubic terms in the elastic energy give quadratic terms in the stress tensor, and therefore in the equations of motion. Let us suppose that all the linear terms in these equations are on the left-hand side, and all the quadratic terms on the right-hand side. Solving these equations by the method of successive approximations, we omit the quadratic terms in the first approximation. This leaves the ordinary linear equations, whose solution $\mathbf{u}_0$ can be put in the form of a superposition of monochromatic travelling waves: constant $\times\, e^{i(\mathbf{k}\cdot\mathbf{r}-\omega t)}$, with definite relations between ω and $\mathbf{k}$. On going to the second approximation, we must put $\mathbf{u} = \mathbf{u}_0 + \mathbf{u}_1$ and retain only the terms in $\mathbf{u}_0$ on the right-hand sides of the equations (the quadratic terms). Since $\mathbf{u}_0$, by definition, satisfies the homogeneous linear equations obtained by putting the right-hand sides equal to zero, the terms in $\mathbf{u}_0$ on the left-hand sides will cancel. The result is a set of inhomogeneous linear equations for the components of the vector $\mathbf{u}_1$, where the right-hand sides contain only known functions of the coordinates and time. These functions, which are obtained by substituting $\mathbf{u}_0$ for $\mathbf{u}$ in the right-hand sides of the original equations, are sums of terms each of which is proportional to

$$e^{i[(\mathbf{k}_1 - \mathbf{k}_2)\cdot\mathbf{r} - (\omega_1 - \omega_2)t]}$$

or

$$e^{i[(\mathbf{k}_1 + \mathbf{k}_2)\cdot\mathbf{r} - (\omega_1 + \omega_2)t]},$$

where ω_1, ω_2, $\mathbf{k}_1$, $\mathbf{k}_2$ are the frequencies and wave vectors of any two monochromatic waves in the first approximation.

A particular integral of linear equations of this type is a sum of terms containing similar exponential factors to those in the free terms (the right-hand sides) of the equations, with suitably chosen coefficients. Each such term corresponds to a travelling wave with frequency $\omega_1 \pm \omega_2$ and wave vector $\mathbf{k}_1 \pm \mathbf{k}_2$. Frequencies equal to the sum or difference of the frequencies of the original waves are called *combination frequencies*.

Thus the anharmonic effects in the third order have the result that the set of fundamental monochromatic waves (with frequencies $\omega_1, \omega_2, \ldots$ and wave vectors $\mathbf{k}_1, \mathbf{k}_2, \ldots$) has superposed on it other "waves" of small intensity, whose frequencies are the combination frequencies such as $\omega_1 \pm \omega_2$, and whose wave vectors are such as $\mathbf{k}_1 \pm \mathbf{k}_2$. We call these "waves" in quotation marks because they are a correction effect and cannot exist alone except in certain special cases (see below). The values $\omega_1 \pm \omega_2$ and $\mathbf{k}_1 \pm \mathbf{k}_2$ do not in general satisfy the relations which hold between the frequencies and wave vectors for ordinary monochromatic waves.

It is clear, however, that there may happen to be particular values of ω_1, $\mathbf{k}_1$ and ω_2, $\mathbf{k}_2$ such that one of the relations for monochromatic waves in the medium considered also holds for $\omega_1 + \omega_2$ and $\mathbf{k}_1 + \mathbf{k}_2$ (for definiteness, we shall discuss sums and not differences). Putting $\omega_3 = \omega_1 + \omega_2$, $\mathbf{k}_3 = \mathbf{k}_1 + \mathbf{k}_2$, we can say that, mathematically, ω_3 and $\mathbf{k}_3$ then correspond to waves which satisfy the homogeneous linear equations of motion (with zero on the right-hand side) in the first approximation. If the right-hand sides in the second approximation contain terms proportional to $e^{i(\mathbf{k}_3 \cdot \mathbf{r} - \omega_3 t)}$, then a particular integral will be a wave with the same frequency and an amplitude which increases indefinitely with time.

Thus the superposition of two monochromatic waves with values of ω_1, $\mathbf{k}_1$ and ω_2, $\mathbf{k}_2$ whose sum ω_3, $\mathbf{k}_3$ satisfies the above condition leads, by the anharmonic effects, to resonance: a new monochromatic wave (with parameters ω_3, $\mathbf{k}_3$) is formed, whose amplitude increases with time and eventually is no longer small. It is evident that, if a wave with ω_3, $\mathbf{k}_3$ is formed on superposition of those with ω_1, $\mathbf{k}_1$ and ω_2, $\mathbf{k}_2$, then the superposition of waves with ω_1, $\mathbf{k}_1$ and ω_3, $\mathbf{k}_3$ will also give a resonance with $\omega_2 = \omega_3 - \omega_1$, $\mathbf{k}_2 = \mathbf{k}_3 - \mathbf{k}_1$, and similarly ω_2, $\mathbf{k}_2$ and ω_3, $\mathbf{k}_3$ lead to ω_1, $\mathbf{k}_1$.

In particular, for an isotropic body ω and $\mathbf{k}$ are related by $\omega = c_t k$ or $\omega = c_l k$, with $c_l > c_t$. It is easy to see in which cases either of these relations can hold for each of the three combinations

$$\omega_1, \mathbf{k}_1; \quad \omega_2, \mathbf{k}_2; \quad \omega_3 = \omega_1 + \omega_2, \quad \mathbf{k}_3 = \mathbf{k}_1 + \mathbf{k}_2.$$

If $\mathbf{k}_1$ and $\mathbf{k}_2$ are not in the same direction, $k_3 < k_1 + k_2$, and so it is clear that resonance can then occur only in the following two cases: (1) the waves with ω_1, $\mathbf{k}_1$ and ω_2, $\mathbf{k}_2$ are transverse and that with ω_3, $\mathbf{k}_3$ longitudinal; (2) one of the waves with ω_1, $\mathbf{k}_1$ and ω_2, $\mathbf{k}_2$ is transverse and the other longitudinal, and that with ω_3, $\mathbf{k}_3$ is longitudinal. If the vectors $\mathbf{k}_1$ and $\mathbf{k}_2$ are in the same direction, however, resonance is possible when all three waves are longitudinal or all three are transverse.

The anharmonic effect involving resonance occurs not only when several monochromatic waves are superposed, but also when there is only one wave, with parameters ω_1, $\mathbf{k}_1$. In this case the right-hand sides of the equations of motion contain terms proportional to $e^{2i(\mathbf{k}_1 \cdot \mathbf{r} - \omega_1 t)}$. If ω_1 and $\mathbf{k}_1$ satisfy the usual condition, however, then $2\omega_1$ and $2\mathbf{k}_1$ do so too, since this condition is homogeneous and of degree one. Thus the anharmonic effect results in the appearance, besides the monochromatic waves with ω_1, $\mathbf{k}_1$ previously obtained, of waves with $2\omega_1$, $2\mathbf{k}_1$, i.e. with twice the frequency and twice the wave vector, and amplitude increasing with time.

Finally, we may briefly discuss how we can set up the equations of motion, allowing for the anharmonic terms. The strain tensor must now be given by the complete expression (1.3):

$$u_{ik} = \frac{1}{2}\left(\frac{\partial u_i}{\partial x_k} + \frac{\partial u_k}{\partial x_i} + \frac{\partial u_l}{\partial x_i}\frac{\partial u_l}{\partial x_k}\right), \tag{26.1}$$

in which the terms quadratic in u_i can not be neglected. Next, the general expression for the energy density† $\mathscr{E}$, in bodies having a given symmetry, must be written as a scalar formed from the components of the tensor u_{ik} and some constant tensors characteristic of the substance involved; this scalar will contain terms up to a given power of u_{ik}. Substituting the expression (26.1) for u_{ik} and omitting terms in u_i of higher orders than that power, we find the energy $\mathscr{E}$ as a function of the derivatives $\partial u_i/\partial x_k$ to the required accuracy.

In order to obtain the equations of motion, we notice the following result. The variation $\delta\mathscr{E}$ may be written

$$\delta\mathscr{E} = \frac{\partial\mathscr{E}}{\partial(\partial u_i/\partial x_k)}\delta\frac{\partial u_i}{\partial x_k},$$

or, putting

$$\sigma_{ik} = \frac{\partial\mathscr{E}}{\partial(\partial u_i/\partial x_k)}, \tag{26.2}$$

$$\delta\mathscr{E} = \sigma_{ik}\frac{\partial\delta u_i}{\partial x_k} = \frac{\partial}{\partial x_k}(\sigma_{ik}\delta u_i) - \delta u_i\frac{\partial\sigma_{ik}}{\partial x_k}.$$

The coefficients of $-\delta u_i$ are the components of the force per unit volume of the body. They formally appear the same as before, and so the equations of motion can again be written

$$\rho_0\ddot{u}_i = \partial\sigma_{ik}/\partial x_k, \tag{26.3}$$

where ρ_0 is the density of the undeformed body, and the components of the tensor σ_{ik} are now given by (26.2), with $\mathscr{E}$ correct to the required accuracy. The tensor σ_{ik} is no longer symmetrical.

It should be emphasized that σ_{ik} is no longer the momentum flux density (the stress tensor). In the ordinary theory this interpretation was derived by integrating the body force density $\partial\sigma_{ik}/\partial x_k$ over the volume of the body. This derivation depended on the fact that, in performing the integration, we made no distinction between the coordinates of points in the body before and after the deformation. In subsequent approximations, however, this distinction must be made, and the surface bounding the region of integration is not the same as the actual surface of the region considered after the deformation.

It has been shown in §2 that the symmetry of the tensor σ_{ik} is due to the conservation of angular momentum. This result no longer holds, since the angular momentum density is not $x_i\dot{u}_k - x_k\dot{u}_i$ but $(x_i+u_i)\dot{u}_k - (x_k+u_k)\dot{u}_i$.

PROBLEM

Write down the general expression for the elastic energy of an isotropic body in the third approximation.

SOLUTION. From the components of a symmetrical tensor of rank two we can form two quadratic scalars (u_{ik}^2 and u_{ll}^2) and three cubic scalars (u_{ll}^3, $u_{ll}u_{ik}^2$ and $u_{ik}u_{il}u_{kl}$). Hence the most general scalar containing terms

† We here use the internal energy $\mathscr{E}$, and not the free energy F, since adiabatic vibrations are involved.

quadratic and cubic in u_{ik}, with scalar coefficients (since the body is isotropic), is

$$\mathscr{E} = \mu u_{ik}^2 + (\tfrac{1}{2}K - \tfrac{1}{3}\mu)u_{ll}^2 + \tfrac{1}{3}A u_{ik}u_{il}u_{kl} + B u_{ik}^2 u_{ll} + \tfrac{1}{3}C u_{ll}^3;$$

the coefficients of u_{ik}^2 and u_{ll}^2 have been expressed in terms of the moduli of compression and rigidity, and A, B, C are three new constants. Substituting the expression (26.1) for u_{ik} and retaining terms up to and including the third order, we find the elastic energy to be

$$\mathscr{E} = \tfrac{1}{4}\mu\left(\frac{\partial u_i}{\partial x_k} + \frac{\partial u_k}{\partial x_i}\right)^2 + (\tfrac{1}{2}K - \tfrac{1}{3}\mu)\left(\frac{\partial u_l}{\partial x_l}\right)^2 +$$
$$+ (\mu + \tfrac{1}{4}A)\frac{\partial u_i}{\partial x_k}\frac{\partial u_l}{\partial x_i}\frac{\partial u_l}{\partial x_k} + (\tfrac{1}{2}B + \tfrac{1}{2}K - \tfrac{1}{3}\mu)\frac{\partial u_l}{\partial x_l}\left(\frac{\partial u_i}{\partial x_k}\right)^2 +$$
$$+ \tfrac{1}{12}A\frac{\partial u_i}{\partial x_k}\frac{\partial u_k}{\partial x_l}\frac{\partial u_l}{\partial x_i} + \tfrac{1}{2}B\frac{\partial u_i}{\partial x_k}\frac{\partial u_k}{\partial x_i}\frac{\partial u_l}{\partial x_l} + \tfrac{1}{3}C\left(\frac{\partial u_l}{\partial x_l}\right)^3.$$

CHAPTER IV

DISLOCATIONS†

§27. Elastic deformations in the presence of a dislocation

ELASTIC deformations in a crystal may arise not only by the action of external forces on it but also because of internal structural defects present in the crystal. The principal type of defect that influences the mechanical properties of crystals is called a *dislocation*. The study of the properties of dislocations on the atomic or microscopic scale is not, of course, within the scope of this book; we shall here consider only purely macroscopic aspects of the phenomenon as it affects elasticity theory. For a better understanding of the physical significance of the relations obtained, however, we shall first give two simple examples to show what is the nature of dislocation defects as regards the structure of the crystal lattice.

Let us imagine that an "extra" half-plane is put into a crystal lattice of which a cross-section is shown in Fig. 22; in this diagram, the added half-plane is the upper half of the yz-plane. The edge of this half-plane (the z-axis, at right angles to the plane of the diagram) is then called an *edge dislocation*. In the immediate neighbourhood of the dislocation the crystal lattice is greatly distorted, but even at a distance of a few lattice periods the crystal planes fit together in an almost regular manner. The deformation nevertheless exists even far from the dislocation. It is clearly seen on going round a closed circuit of lattice points in the xy-plane, with the origin within the circuit: if the displacement of each point from its position in the ideal lattice is denoted by the vector $\mathbf{u}$, the total increment of this vector around the circuit will not be zero, but equals one lattice period in the x-direction.

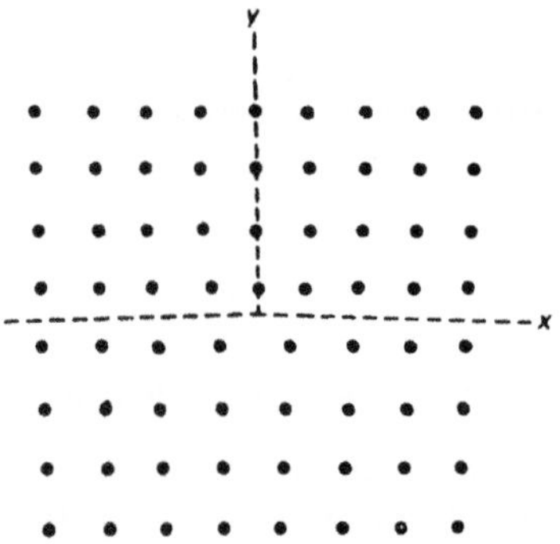

FIG. 22

Another type of dislocation may be visualized as the result of "cutting" the lattice along a half-plane and then shifting the parts of the lattice on either side of the cut in opposite directions to a distance of one lattice period parallel to the edge of the cut (then called a

† This chapter was written jointly with A. M. Kosevich.

screw dislocation). Such a dislocation converts the lattice planes into a helicoidal surface, like a spiral staircase without the steps. In a complete circuit round the dislocation line (the axis of the helicoidal surface) the lattice point displacement vector increment is one lattice period along that axis. Figure 23 shows a diagram of such a cut.

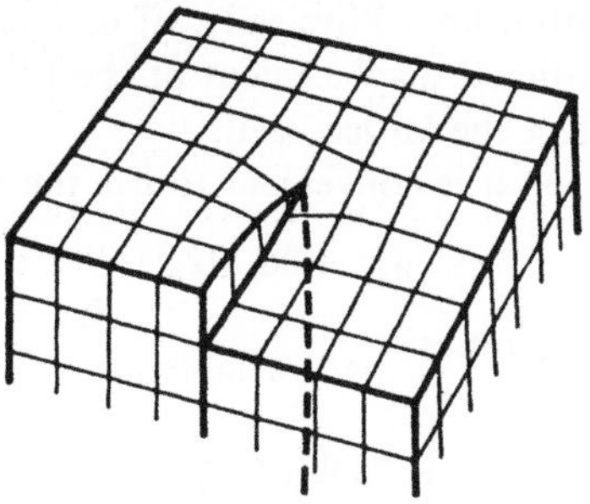

FIG. 23

Macroscopically, a dislocation deformation of a crystal regarded as a continuous medium has the following general property; after a passage round any closed contour L which encloses the dislocation line D, the elastic displacement vector $\mathbf{u}$ receives a certain finite increment $\mathbf{b}$ which is equal to one of the lattice vectors in magnitude and direction; the constant vector $\mathbf{b}$ is called the *Burgers vector* of the dislocation concerned. This property may be expressed as

$$\oint_L du_i = \oint \frac{\partial u_i}{\partial x_k} dx_k = -b_i, \tag{27.1}$$

where the direction in which the contour is traversed and the chosen direction of the tangent vector τ to the dislocation line are assumed to be related by the corkscrew rule (Fig. 24). The dislocation line itself is a line of singularities of the deformation field.

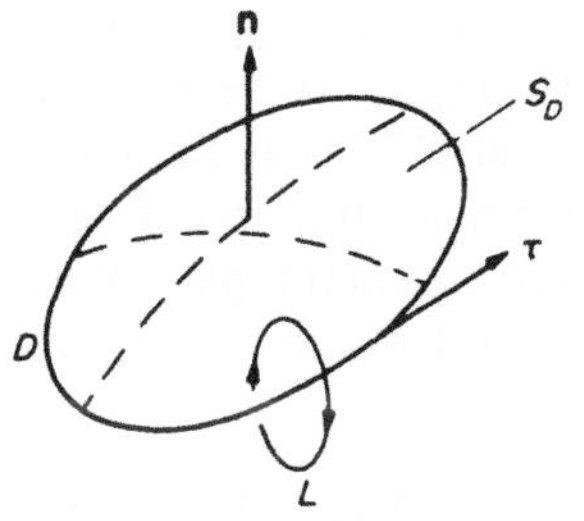

FIG. 24

The simple cases of edge and screw dislocations mentioned above correspond to straight lines D with $\tau \perp \mathbf{b}$ and $\tau \parallel \mathbf{b}$. We may also note that in the representation given by Fig. 22 edge dislocations with opposite directions of $\mathbf{b}$ differ in that the "extra" crystal half-plane lies above or below the xy-plane; such dislocations are said to have opposite *signs*.

In the general case, the dislocation is a curve, along which the angle between τ and $\mathbf{b}$ varies. The Burgers vector $\mathbf{b}$ itself is always constant along the dislocation line. It is also evident that this line cannot simply terminate within the crystal (see the next-but-one footnote): it must either reach the surface of the crystal at both ends or (as usually happens in actual cases) form a closed loop.

The condition (27.1) thus signifies that in the presence of a dislocation the displacement vector is not a single-valued function of the coordinates, but receives a certain increment in a passage round the dislocation line. Physically, of course, there is no ambiguity: the increment **b** denotes an additional displacement of the lattice points equal to a lattice vector, and this does not affect the lattice itself.

In the subsequent discussion it is convenient to use the notation

$$w_{ik} = \partial u_k/\partial x_i, \tag{27.2}$$

so that the condition (27.1) becomes

$$\oint_L w_{ik}\,dx_i = -b_k. \tag{27.3}$$

The (unsymmetrical) tensor w_{ik} is called the *distortion tensor*. Its symmetrical part gives the ordinary strain tensor:

$$u_{ik} = \tfrac{1}{2}(w_{ik} + w_{ki}). \tag{27.4}$$

According to the foregoing discussion the tensors w_{ik} and u_{ik}, and therefore the stress tensor σ_{ik}, are single-valued functions of the coordinates, unlike the function $\mathbf{u}(\mathbf{r})$.

The condition (27.3) may also be written in a differential form. To do so, we transform the integral round the contour L into one over a surface S_L spanning this contour:†

$$\oint_L w_{mk}\,dx_m = \int_{S_L} e_{ilm}\frac{\partial w_{mk}}{\partial x_l}\,df_i. \tag{27.5}$$

Since the tensor e_{ilm} is antisymmetrical in the suffixes l and m, and the tensor $\partial w_{mk}/\partial x_l = \partial^2 u_k/\partial x_l \partial x_m$ is symmetrical in them, the integrand is identically zero everywhere except where the line D meets the surface S_L; on the dislocation line itself, which is a line of singularities, the representation of the w_{mk} as the derivatives (27.2) is no longer meaningful.‡ At these points, the w_{ik} are to be determined by means of the appropriate delta function so that the integral (27.5) has the required value $-b_k$. Let $\boldsymbol{\xi}$ be a two-dimensional position vector from a given point on the dislocation axis, in a plane perpendicular to $\boldsymbol{\tau}$. The element of area in this plane is expressed in terms of the element $d\mathbf{f}$ of the surface S_L as $\boldsymbol{\tau}\cdot d\mathbf{f}$. From this definition of the two-dimensional delta function $\delta(\boldsymbol{\xi})$,

$$\int \delta(\boldsymbol{\xi})\boldsymbol{\tau}\cdot d\mathbf{f} = \tau_i \int_{S_L} \delta(\boldsymbol{\xi})\,df_i = 1.$$

---

† The transformation is made, according to Stokes' theorem, by replacing dx_m by the operator $df_i e_{ilm}\partial/\partial x_l$, where e_{ilm} is the antisymmetric unit tensor. It should be recalled also that any expression have the form $e_{ilm}a_i b_l$ is the m component of the vector product $\mathbf{a}\times\mathbf{b}$.

‡ If the dislocation line ended at some point within the body, the surface S_L could be chosen so as to enclose that point and thus nowhere intersect the line D. The integral (27.5) would then be zero, contrary to the condition stated.

It is therefore clear that to achieve the necessary result we must put

$$e_{ilm}\,\partial w_{mk}/\partial x_l = -\tau_i b_k \delta(\xi). \tag{27.6}$$

This is the required differential form.

The displacement field $\mathbf{u}(\mathbf{r})$ around the dislocation can be expressed in a general form if we know the Green's tensor $G_{ik}(\mathbf{r})$ of the equations of equilibrium of the anisotropic medium considered, i.e. the function which determines the displacement component u_i produced in an infinite medium by a unit force applied at the origin along the x_k-axis (see §8). This can easily be done by using the following formal device.

Instead of seeking many-valued solutions of the equations of equilibrium, we shall regard $\mathbf{u}(\mathbf{r})$ as a single-valued function, which undergoes a fixed discontinuity $\mathbf{b}$ on some arbitrarily chosen surface S_D spanning the dislocation loop D. If $\mathbf{u}_+$ and $\mathbf{u}_-$ are the values of the function on the upper and lower edges of the discontinuity S_D, then

$$\mathbf{u}_+ - \mathbf{u}_- = \mathbf{b}. \tag{27.7}$$

The upper and lower edges are defined as shown in Fig. 24: the normal $\mathbf{n}$ to the surface S_D in the direction indicated relative to τ is from the lower edge to the upper. The integration along L from the upper to the lower edge then gives (27.3) with the correct sign. The tensors w_{ik} and u_{ik}, which are formally defined by (27.3) and (27.4), have a delta-function singularity on the discontinuity surface:

$$w_{ik}^{(S)} = n_i b_k \delta(\zeta), \quad u_{ik}^{(S)} = \tfrac{1}{2}(n_i b_k + n_k b_i)\delta(\zeta), \tag{27.8}$$

where ζ is the coordinate measured from the surface S_D along the normal $\mathbf{n}$; $\mathrm{d}\zeta = \mathbf{n}\cdot\mathrm{d}\mathbf{l}$, where $\mathrm{d}\mathbf{l}$ is an element of length of L.

Since there is no actual physical singularity in the space around the dislocation, the stress tensor σ_{ik} must, as already mentioned, be a single-valued and everywhere continuous function. The strain tensor (27.8), however, is formally related to a stress tensor $\sigma_{ik}^{(S)} = \lambda_{iklm} u_{lm}^{(S)}$, which also has a singularity on the surface S_D. In order to eliminate this we must define fictitious body forces distributed over the surface S_D with a certain density $f^{(S)}$. The equations of equilibrium in the presence of body forces are $\partial\sigma_{ik}/\partial x_k + f_i^{(S)} = 0$ (cf. (2.8)). Hence it is clear that we must put

$$f_i^{(S)} = -\frac{\partial \sigma_{ik}^{(S)}}{\partial x_k} = -\lambda_{iklm}\frac{\partial u_{lm}^{(S)}}{\partial x_k}. \tag{27.9}$$

Thus the problem of finding the many-valued function $\mathbf{u}(\mathbf{r})$ is equivalent to that of finding a single-valued but discontinuous function in the presence of body forces given by formulae (27.7) and (27.9). We can now use the formula

$$u_i(\mathbf{r}) = \int G_{ij}(\mathbf{r}-\mathbf{r}') f_j^{(S)}(\mathbf{r}')\,\mathrm{d}V'.$$

We substitute (27.9) and integrate by parts; the integration with the delta function is then trivial, giving

$$u_i(\mathbf{r}) = -\lambda_{jklm} b_m \int_{S_D} n_l \frac{\partial}{\partial x_k} G_{ij}(\mathbf{r}-\mathbf{r}')\,\mathrm{d}f'. \tag{27.10}$$

This solves the problem.†

---

† The tensor G_{ij} for an anisotropic medium has been derived in the paper by I. M. Lifshitz and L. N. Rozentsveĭg quoted in §8, Problem. This tensor is in general very complicated. For a straight dislocation, which corresponds to a two-dimensional problem of elasticity theory, it may be simpler to solve the equations of equilibrium directly.

The deformation (27.9) has its simplest form far from the closed dislocation loop. If we imagine the loop to be situated near the origin, then at distances r large compared with the linear dimensions of the loop we have

$$u_i(\mathbf{r}) = -\lambda_{jklm} d_{lm} \partial G_{ij}(\mathbf{r})/\partial x_k, \tag{27.11}$$

where

$$d_{ik} = S_i b_k, \; S_i = \int_{S_D} n_i \mathrm{d}f = \tfrac{1}{2} e_{ikl} \oint_D x_k \mathrm{d}x_l, \tag{27.12}$$

and e_{ikl} is the antisymmetric unit tensor. The axial vector $\mathbf{S}$ has components equal to the areas bounded by the projections of the loop D on planes perpendicular to the corresponding coordinate axes; the tensor d_{ik} may be called the *dislocation moment tensor*. The components of the tensor G_{ij} are first-order homogeneous functions of the coordinates x, y, z (see §8, Problem). We therefore see from (27.11) that $u_i \propto 1/r^2$, and the corresponding stress field $\sigma_{ik} \propto 1/r^3$.

It is also easy to ascertain the way in which the elastic stresses vary with distance near a straight dislocation. In cylindrical polar coordinates z, r, ϕ (with the z-axis along the dislocation line) the deformation will depend only on r and ϕ. The integral (27.3) must, in particular, be unchanged by an arbitrary change in the size of any contour in the xy-plane which leaves the shape of the contour the same. It is clear that this can be true only if all the $w_{ik} \propto 1/r$. The tensor u_{ik}, and therefore the stresses σ_{ik}, will be proportional to the same power, $1/r$.†

Although we have hitherto spoken only of dislocations, the formulae derived are applicable also to deformations caused by other kinds of defect in the crystal structure. Dislocations are linear defects; there exist also defects in which the regular structure is interrupted through a region near a given surface.‡ Such a defect can be macroscopically described as a surface of discontinuity on which the displacement vector $\mathbf{u}$ is discontinuous but the stresses σ_{ik} are continuous, by virtue of the equilibrium conditions. If the discontinuity $\mathbf{b}$ is the same everywhere on the surface, the resulting strain is just the same as that due to a dislocation along the edge of the surface. The only difference is that the vector $\mathbf{b}$ is not equal to a lattice vector. However, the position of the surface S_D discussed above is no longer arbitrary; it must coincide with the actual physical discontinuity. Such a surface of discontinuity involves a certain additional energy which may be described by means of an appropriate surface-tension coefficient.

PROBLEMS

PROBLEM 1. Derive the differential equations of equilibrium for a dislocation deformation in an isotropic medium, expressed in terms of the displacement vector.§

† Attention is drawn to a certain analogy between the elastic deformation field round a dislocation line and the magnetic field of constant line currents. The current is replaced by the Burgers vector, which must be constant along the dislocation line, like the current. Similar analogies will also be readily seen in the relations given below. However, quite apart from the entirely different nature of the two physical effects, these analogies are not far-reaching, because the tensor character of the corresponding quantities is different.

‡ A well-known example of a defect of this type is a narrow twinned layer in a crystal.

§ The physical meaning of this and other problems relating to an isotropic medium is purely conventional, since actual dislocations by their nature occur only in crystals, i.e. in anisotropic media. Such problems have illustrative value, however.

SOLUTION. In terms of the stress tensor or strain tensor the equations of equilibrium have the usual form $\partial\sigma_{ik}/\partial x_k = 0$ or, substituting σ_{ik} from (5.11),

$$\frac{\partial u_{ik}}{\partial x_k} + \frac{\sigma}{1-2\sigma}\frac{\partial u_{ll}}{\partial x_i} = 0. \tag{1}$$

To convert to the vector $\mathbf{u}$ we must use the differential condition (27.6). Multiplying (27.6) by e_{ikn} and summing over i and k, we obtain†

$$\frac{\partial w_{nk}}{\partial x_k} - \frac{\partial w_{kk}}{\partial x_n} = -(\boldsymbol{\tau}\times\mathbf{b})_n\delta(\boldsymbol{\xi}). \tag{2}$$

Writing (1) in the form

$$\tfrac{1}{2}\frac{\partial w_{ik}}{\partial x_k} + \tfrac{1}{2}\frac{\partial w_{ki}}{\partial x_k} + \frac{\sigma}{1-2\sigma}\frac{\partial w_{ll}}{\partial x_i} = 0$$

and substituting (2), we find

$$\frac{\partial w_{ki}}{\partial x_k} + \frac{1}{1-2\sigma}\frac{\partial w_{ll}}{\partial x_i} = (\boldsymbol{\tau}\times\mathbf{b})_i\delta(\boldsymbol{\xi}).$$

Now changing to $\mathbf{u}$ in accordance with (27.2), we find the required equation for the multi-valued function $\mathbf{u}(\mathbf{r})$:

$$\triangle\mathbf{u} + \frac{1}{1-2\sigma}\,\mathbf{grad}\operatorname{div}\mathbf{u} = \boldsymbol{\tau}\times\mathbf{b}\,\delta(\boldsymbol{\xi}). \tag{3}$$

The solution of this equation must satisfy the condition (27.1).

PROBLEM 2. Determine the deformation near a straight screw dislocation in an isotropic medium.

SOLUTION. We take cylindrical polar coordinates z, r, ϕ, with the z-axis along the dislocation line; the Burgers vector is $b_x = b_y = 0$, $b_z = b$. It is evident from symmetry that the displacement $\mathbf{u}$ is parallel to the z-axis and is independent of the coordinate z. The equation of equilibrium (3), Problem 1, reduces to $\triangle u_z = 0$. The solution which satisfies the condition (27.1) is‡ $u_z = b\phi/2\pi$. The only non-zero components of the tensors u_{ik} and σ_{ik} are $u_{z\phi} = b/4\pi r$, $\sigma_{z\phi} = \mu b/2\pi r$, and the deformation is therefore a pure shear.

The free energy of the dislocation (per unit length) is given by the integral

$$F = \tfrac{1}{2}\int 2u_{z\phi}\sigma_{z\phi}\,dV = \frac{\mu b^2}{4\pi}\int\frac{dr}{r},$$

which diverges logarithmically at both limits. As the lower limit we must take the order of magnitude of the interatomic distances ($\sim b$), at which the deformation is large and the macroscopic theory is inapplicable. The upper limit is determined by a dimension of the order of the length L of the dislocation. Then $F = (\mu b^2/4\pi)\log(L/b)$. The energy of the deformation in the "core" of the dislocation near its axis (in a region of cross-sectional area $\sim b^2$) can be estimated as $\sim \mu b^2$. When $\log(L/b) \gg 1$ this energy is small in comparison with that of the elastic deformation field.§

PROBLEM 3. Determine the internal stresses in an anisotropic medium near a screw dislocation which is perpendicular to a plane of symmetry of the crystal.

SOLUTION. We take coordinates x, y, z so that the z-axis is along the dislocation line, and again write $b_z = b$. The vector $\mathbf{u}$ again has only the component $u_z = u(x, y)$. Since the xy-plane is a plane of symmetry, all the

---

† Using also the formula $e_{ilm}e_{ikn} = \delta_{lk}\delta_{mn} - \delta_{ln}\delta_{mk}$.

‡ In all the problems on straight dislocations we take the vector $\boldsymbol{\tau}$ in the negative z-direction.

§ These estimates are general ones and are valid in order of magnitude for any dislocation (and not only for a screw dislocation).

It should be noted that in practice the values of $\log(L/b)$ are usually not very large, and the energy of the core is therefore a considerable fraction of the total energy of the dislocation.

components of the tensor λ_{iklm} are zero which contain the suffix z an odd number of times. Thus only two components of the tensor σ_{ik} are non-zero:

$$\sigma_{xz} = \lambda_{xzxz}\frac{\partial u}{\partial x} + \lambda_{xzyz}\frac{\partial u}{\partial y},$$

$$\sigma_{yz} = \lambda_{yzxz}\frac{\partial u}{\partial x} + \lambda_{yzyz}\frac{\partial u}{\partial y}.$$

We define a two-dimensional vector $\boldsymbol{\sigma}$ and a two-dimensional tensor $\lambda_{\alpha\beta}$: $\sigma_\alpha = \sigma_{\alpha z}$, $\lambda_{\alpha\beta} = \lambda_{\alpha z\beta z}$ $(\alpha = 1, 2)$. Then $\sigma_\alpha = \lambda_{\alpha\beta}\partial u/\partial x_\beta$, and the equation of equilibrium becomes div $\boldsymbol{\sigma} = 0$. The required solution of this equation must satisfy the condition (27.1): $\oint \mathbf{grad}\, u \cdot d\mathbf{l} = b$.

In this form, the problem is the same as that of finding the magnetic induction and magnetic field (represented by $\boldsymbol{\sigma}$ and **grad** u) is an anisotropic medium with magnetic permeability $\lambda_{\alpha\beta}$ near a straight current of strength $I = cb/4\pi$. Using the solution derived in electrodynamics, we obtain (see *ECM*, §30, Problem 5)

$$\sigma_{\alpha z} = -\frac{b}{2\pi}\frac{\lambda_{\alpha\beta}e_{\beta\gamma z}x_\gamma}{\sqrt{|\lambda|\cdot\lambda^{-1}_{\alpha'\beta'}x_{\alpha'}x_{\beta'}}},$$

where $|\lambda|$ is the determinant of the tensor $\lambda_{\alpha\beta}$.

PROBLEM 4. Determine the deformation near a straight edge dislocation in an isotropic medium.

SOLUTION. Let the z-axis be along the dislocation line, and the Burgers vector be $b_x = b$, $b_y = b_z = 0$. It is evident from the symmetry of the problem that the displacement vector lies in the xy-plane and is independent of z, so that the problem is a two-dimensional one. In the rest of this solution all vectors and vector operations are two-dimensional in the xy-plane.

We shall seek a solution of the equation

$$\triangle\mathbf{u} + \frac{1}{1-2\sigma}\,\mathbf{grad}\,\mathrm{div}\,\mathbf{u} = -b\mathbf{j}\delta(\mathbf{r})$$

(see Problem 1; $\mathbf{j}$ is a unit vector along the y-axis) in the form $\mathbf{u} = \mathbf{u}^{(0)} + \mathbf{w}$, where $\mathbf{u}^{(0)}$ is a vector with components $u^{(0)}{}_x = b\phi/2\pi$, $u^{(0)}{}_y = (b/2\pi)\log r$; these are the imaginary and real parts of $(b/2\pi)\log(x+iy)$, r and ϕ being polar coordinates in the xy-plane. This vector satisfies the condition (27.1). The problem therefore reduces to finding the single-valued function $\mathbf{w}$. Since, as is easily verified, div $\mathbf{u}^{(0)} = 0$, $\triangle\mathbf{u}^{(0)} = b\mathbf{j}\delta(\mathbf{r})$, it follows that $\mathbf{w}$ satisfies the equation

$$\triangle\mathbf{w} + \frac{1}{1-2\sigma}\,\mathbf{grad}\,\mathrm{div}\,\mathbf{w} = -2b\mathbf{j}\delta(\mathbf{r}).$$

This is the equation of equilibrium under forces concentrated along the z-axis with volume density $Eb\mathbf{j}\delta(\mathbf{r})/2(1+\sigma)$; cf. §8, Problem, equation (1). By means of the Green's tensor found in that problem for an infinite medium, the calculation of $\mathbf{w}$ is reduced to that of the integral

$$\mathbf{w} = \frac{b}{8\pi(1-\sigma)}\cdot 2\int_0^\infty\left[\frac{(3-4\sigma)\mathbf{j}}{R} + \frac{\mathbf{r}y}{R^3}\right]dz',$$

$$R = \sqrt{(r^2+z'^2)}.$$

The result is

$$u_x = \frac{b}{2\pi}\left\{\tan^{-1}\frac{y}{x} + \frac{1}{2(1-\sigma)}\frac{xy}{x^2+y^2}\right\},$$

$$u_y = -\frac{b}{2\pi}\left\{\frac{1-2\sigma}{2(1-\sigma)}\log\sqrt{(x^2+y^2)} + \frac{1}{2(1-\sigma)}\frac{x^2}{x^2+y^2}\right\}.$$

The stress tensor calculated from this has Cartesian components

$$\sigma_{xx} = -bB\frac{y(3x^2+y^2)}{(x^2+y^2)^2},$$

$$\sigma_{yy} = bB\frac{y(x^2-y^2)}{(x^2+y^2)^2},$$

$$\sigma_{xy} = bB\frac{x(x^2-y^2)}{(x^2+y^2)^2},$$

and polar components

$$\sigma_{rr} = \sigma_{\phi\phi} = -(bB/r)\sin\phi,$$

$$\sigma_{r\phi} = (bB/r)\cos\phi,$$

where $B \equiv \mu/2\pi(1-\sigma)$.

PROBLEM 5. An infinity of identical parallel straight edge dislocations in an isotropic medium lie in one plane perpendicular to their Burgers vectors and at equal distances h apart. Find the shear stresses due to such a "dislocation wall" at distances large compared with h.

SOLUTION. Let the dislocations be in the yz-plane and parallel to the z-axis. According to the results of Problem 4, the total stress due to all the dislocations at the point (x, y) is given by the sum

$$\sigma_{xy}(x, y) = bBx \sum_{n=-\infty}^{\infty} \frac{x^2-(y-nh)^2}{[x^2+(y-nh)^2]^2}.$$

This may be written in the form

$$\sigma_{xy} = -bB\frac{\alpha}{h}\left[J(\alpha,\beta)+\alpha\frac{\partial J(\alpha,\beta)}{\partial\alpha}\right],$$

where

$$J(\alpha,\beta) = \sum_{n=-\infty}^{\infty} \frac{1}{\alpha^2+(\beta-n)^2}, \quad \alpha = x/h,\ \beta = y/h.$$

According to Poisson's summation formula

$$\sum_{n=-\infty}^{\infty} f(n) = \sum_{k=-\infty}^{\infty} \int_{-\infty}^{\infty} f(x)e^{2\pi ikx}\,dx,$$

we find

$$J(\alpha,\beta) = \int_{-\infty}^{\infty} \frac{d\xi}{\alpha^2+\xi^2} + 2\,\mathrm{re} \sum_{k=1}^{\infty} e^{2\pi ik\beta} \int_{-\infty}^{\infty} \frac{e^{2\pi ik\xi}\,d\xi}{\alpha^2+\xi^2}$$

$$= \frac{\pi}{\alpha} + \frac{2\pi}{\alpha}\sum_{k=1}^{\infty} e^{-2\pi k\alpha}\cos 2\pi k\beta.$$

When $\alpha = x/h \gg 1$ only the first term need be retained in the sum over k, and the result is

$$\sigma_{xy} = 4\pi^2 B\frac{bx}{h^2}e^{-2\pi x/h}\cos(2\pi y/h).$$

Thus the stresses decrease exponentially away from the wall.

PROBLEM 6. Determine the deformation of an isotropic medium around a dislocation loop (J. M. Burgers 1939).

SOLUTION. We start from (27.10). The tensor λ_{iklm} for an isotropic medium may be written as

$$\lambda_{iklm} = \mu\left\{\frac{2\sigma}{1-2\sigma}\delta_{ik}\delta_{lm} + \delta_{il}\delta_{km} + \delta_{im}\delta_{kl}\right\}.$$

The Green's tensor for an isotropic medium has been derived in §8, Problem, and may be written as

$$G_{ik}(\mathbf{R}) = \frac{1}{16\pi\mu(1-\sigma)R}\left\{(3-4\sigma)\delta_{ik} + \nu_i\nu_k\right\}.$$

Here $\mathbf{R} = \mathbf{r}-\mathbf{r}'$ is the radius vector from the element $d\mathbf{f}'$ at $\mathbf{r}'$ to the point $\mathbf{r}$ at which the deformation is considered; $\boldsymbol{\nu} = \mathbf{R}/R$ is a unit vector in that direction. Substituting these expressions in (27.11) and carrying out the differentiations in the integral, we obtain finally

$$\mathbf{u}(\mathbf{r}) = \frac{1-2\sigma}{8\pi(1-\sigma)}\int_{S_D}\frac{1}{R^2}\{\mathbf{b}(\boldsymbol{\nu}\cdot d\mathbf{f}') + (\mathbf{b}\cdot\boldsymbol{\nu})d\mathbf{f}' - \boldsymbol{\nu}(\mathbf{b}\cdot d\mathbf{f}')\} + \frac{3}{8\pi(1-\sigma)}\int_{S_D}\frac{1}{R^2}\boldsymbol{\nu}(\mathbf{b}\cdot\boldsymbol{\nu})(\boldsymbol{\nu}\cdot d\mathbf{f}'). \quad (1)$$

The integrals here can be expressed in terms of integrals along D, i.e. along the dislocation loop. To do so, we use the formulae

$$\oint_D \frac{1}{R}\mathbf{b}\times d\mathbf{l}' = \int_{S_D} \frac{1}{R^2}\left\{(\mathbf{b}\cdot\boldsymbol{\nu})d\mathbf{f}' - \boldsymbol{\nu}(\mathbf{b}\cdot d\mathbf{f}')\right\},$$

$$\oint_D (\mathbf{b}\times\boldsymbol{\nu})\cdot d\mathbf{l}' = -\int_{S_D} \frac{1}{R}\left\{\mathbf{b}\cdot d\mathbf{f}' + (\mathbf{b}\cdot\boldsymbol{\nu})(\boldsymbol{\nu}\cdot d\mathbf{f}')\right\}.$$

The integrals on the right are derived from the contour integrals on the left by means of Stokes' theorem, according to which the transformation is made by the change $d\mathbf{l}' \to d\mathbf{f}\times\nabla'$ (where $\nabla' = \partial/\partial\mathbf{r}'$); since the integrand depends only on $\mathbf{r}-\mathbf{r}'$, this transformation is equivalent to $d\mathbf{l}' \to d\mathbf{f}'\times\nabla$ (where $\nabla = \partial/\partial\mathbf{r}$). We also define the solid angle Ω subtended by the loop D at the point considered:

$$\Omega = \int \frac{1}{R^2}\boldsymbol{\nu}\cdot d\mathbf{f}'.$$

The displacement field is then

$$\mathbf{u}(\mathbf{r}) = \mathbf{b}\frac{\Omega}{4\pi} + \frac{1}{4\pi}\oint_D \frac{1}{R}\mathbf{b}\times d\mathbf{l}' + \frac{1}{8\pi(1-\sigma)}\nabla\oint_D (\mathbf{b}\times\boldsymbol{\nu})\cdot d\mathbf{l}'.$$

The non-uniqueness of this function lies in the first term: the angle Ω changes by 4π on passing round D.

Far from the loop, the expression (1) becomes

$$\mathbf{u}(\mathbf{r}) = \frac{1-2\sigma}{8\pi(1-\sigma)R^2}\left\{\mathbf{S}(\mathbf{b}\cdot\boldsymbol{\nu}) + \mathbf{b}(\mathbf{S}\cdot\boldsymbol{\nu}) - \boldsymbol{\nu}(\mathbf{S}\cdot\mathbf{b})\right\} + \frac{3}{8\pi(1-\sigma)R^2}(\mathbf{S}\cdot\boldsymbol{\nu})(\mathbf{b}\cdot\boldsymbol{\nu})\boldsymbol{\nu}.$$

This could also be obtained directly from (27.11) and (27.12).

§28. The action of a stress field on a dislocation

Let us consider a dislocation loop D in a field of elastic stresses $\sigma_{ik}^{(e)}$ created by given external loads, and calculate the force on the loop in such a field. According to the general rules, this must be done by finding the work δR_D done on the dislocation when it undergoes an infinitesimal displacement.

Let us return to the concept of the dislocation loop D (§27) as the line spanned by the displacement vector discontinuity surface S_D; the amount of the discontinuity is given by (27.7). The displacement of D changes the surface S_D. Let $\delta\mathbf{x}$ be the displacement vector of points on D. Under this displacement, the line element $d\mathbf{l}$ sweeps out an area $\delta\mathbf{f} = \delta\mathbf{x}\times\delta\mathbf{l} = \delta\mathbf{x}\times\boldsymbol{\tau}dl$, and this gives the increase in the area of the surface S_D. Since we are here considering an actual physical displacement of the dislocation, we have to take into account the fact that the operation mentioned is accompanied by a change in the physical volume of the medium. Since the displacements $\mathbf{u}$ of the points in the medium on either side of the surface differ by $\mathbf{b}$, the change is given by the product

$$\delta V = \mathbf{b}\cdot d\mathbf{f} = (\delta\mathbf{x}\times\boldsymbol{\tau})\cdot\mathbf{b}\,df = \delta\mathbf{x}\cdot(\boldsymbol{\tau}\times\mathbf{b})\,df. \tag{28.1}$$

Two physically different situations are therefore possible. In one, $\delta V \equiv 0$, and the displacement of the dislocation line involves no change in volume. This will happen if the displacement occurs in the plane defined by the vectors $\boldsymbol{\tau}$ and $\mathbf{b}$, called the *glide plane* or *slip plane* of the dislocation element concerned. The envelope of the family of glide planes of all the elements of length in the loop D is called the *glide surface* of the dislocation; it is a cylinder with its generators parallel to the Burgers vector $\mathbf{b}$.† The physically distinctive

† The possible systems of glide planes in an anisotropic medium are actually governed by its crystal lattice structure.

feature of the glide plane is that it is the only one in which a comparatively easy mechanical movement of the dislocation is possible (usually referred to in this case as a *glide*).†

As the area of the surface S_D changes during the movement of the dislocation, so does the deformation singularity (27.8) concentrated on the line D. This change may be expressed as

$$\delta u_{ik}^{(\mathrm{pl})} = \tfrac{1}{2}\{b_i(\delta \mathbf{x}\times\boldsymbol{\tau})_k + b_k(\delta\mathbf{x}\times\boldsymbol{\tau})_i\}\,\delta(\boldsymbol{\xi}), \tag{28.2}$$

where $\delta(\boldsymbol{\xi})$ is the two-dimensional delta function defined in §27. We must emphasize that this value is uniquely determined by the shape of D and the displacement $\delta\mathbf{x}$, in contrast to (27.8), which depends on the arbitrary choice of the surface S_D.

The expression (28.2) describes a local inelastic residual or *plastic* strain not associated with elastic stresses. The corresponding work done ultimately by external sources is given by the integral

$$\int \sigma_{ik}^{(e)}\,\delta u_{ik}\,dV$$

(cf. (3.2)), where δu_{ik} is the total geometrical change in the deformation. This consists of elastic and plastic parts; we are concerned here only with the work related to the plastic part.‡ After substituting $\delta u_{ik}^{(\mathrm{pl})}$ from (28.2), there remains (because of the delta function) only an integration along the dislocation loop D:

$$\delta R_D = \oint_D \sigma_{ik}^{(e)}\,e_{ilm}\,\delta x_l\,\tau_m\,dl. \tag{28.3}$$

The coefficient of δx_l in the integrand is the force f_l acting on unit length of the dislocation line. Thus

$$f_i = e_{ikl}\,\tau_k\sigma_{lm}^{(e)}\,b_m \tag{28.4}$$

(M. O. Peach and J. S. Koehler 1950). This force $\mathbf{f}$ is perpendicular to $\boldsymbol{\tau}$, i.e. to the dislocation line.

Formula (28.3) has an intuitive interpretation. From the above discussion, the displacement of the dislocation line element amounts to the cutting of an area $d\mathbf{f}$ and a shift of the upper edge through $\mathbf{b}$ relative to the lower edge. The internal stress force applied to $d\mathbf{f}$ is $\sigma_{ik}^{(e)}df_k$, and the work done by this force in the shift is $b_i\sigma_{ik}^{(e)}df_k$.

Since (28.4) in that form relates only to movement in the glide plane, the component of the force $\mathbf{f}$ in that plane may be introduced immediately. Let $\boldsymbol{\kappa}$ be a unit vector along the normal to the dislocation line in the glide plane. Then

$$f_\perp = \mathbf{f}\cdot\boldsymbol{\kappa} = e_{ikl}\kappa_i\tau_k b_m\sigma_{lm}^{(e)}$$

or

$$f_\perp = \nu_l\sigma_{lm}^{(e)}\,b_m, \tag{28.5}$$

where $\boldsymbol{\nu} = \boldsymbol{\kappa}\times\boldsymbol{\tau}$ is a vector normal to the glide plane. Since $\mathbf{b}$ and $\boldsymbol{\nu}$ are at right angles, we

† For example, for the movement of the edge dislocation shown in Fig. 22 in its glide plane (the xz-plane), comparatively slight atomic displacements are sufficient, which make crystal half-planes farther and farther from the yz-plane into "extra" half-planes.

‡ In deriving the equations of motion, virtual plastic and elastic strains are to be regarded as independent variables. Since the equation of motion of the dislocation is under consideration, only the plastic strain need be taken into account.

can take two of the coordinate axes along these vectors and find that $f_\perp$ is determined by only one component of the tensor $\sigma_{lm}^{(e)}$.

If, however, the displacement of the dislocation is not in the glide plane, $\delta V \neq 0$. This means that the shift of the edges of the cut would give rise to an excess of material (when one edge encroaches on the other) or a deficiency of material (when a gap is formed between the edges as they move apart). This is not acceptable if we suppose that the medium remains continuous with constant density (apart from the elastic strains) as the dislocation moves. In an actual crystal, the excess material is removed, or the deficiency made good, by diffusion, the dislocation axis becoming a source or sink for diffusional fluxes of matter.† Movement of dislocations accompanied by the healing of defects in the continuous medium by diffusion is called *climb*.‡

It is clear from the above that, allowing dislocation climb as a possible virtual displacement, one must suppose that climb, like glide, occurs without any local change in the volume of the medium. This means that from the strain (28.2) we have to subtract the part $\frac{1}{3}\delta_{ik}u_{ll}^{(pl)}$ which accounts for the change in volume, representing the plastic strain by the tensor

$$\delta u_{ik}^{(pl)} = \{\tfrac{1}{2}b_i(\delta\mathbf{x}\times\boldsymbol{\tau})_k + \tfrac{1}{2}b_k(\delta\mathbf{x}\times\boldsymbol{\tau})_i - \tfrac{1}{3}\delta_{ik}\mathbf{b}\cdot(\delta\mathbf{x}\times\boldsymbol{\tau})\}\,\delta(\boldsymbol{\xi}). \tag{28.6}$$

Accordingly (28.4) is replaced by the following expression for the force acting on the dislocation:§

$$f_i = e_{ikl}\tau_k b_m(\sigma_{lm}^{(e)} - \tfrac{1}{3}\delta_{lm}\sigma_{nn}^{(e)}) \tag{28.7}$$

(J. Weertman 1965).

The total force on the entire dislocation loop is

$$F_i = e_{ikl}b_m \oint_D (\sigma_{lm}^{(e)} - \tfrac{1}{3}\delta_{lm}\sigma_{nn}^{(e)})\,dx_k. \tag{28.8}$$

This is zero except in an inhomogeneous stress field: when $\sigma_{lm}^{(e)}$ = constant, the integral reduces to $\oint dx_k \equiv 0$. If the stress field varies only slightly along the loop, we can write

$$F_i = e_{ikl}b_m \frac{\partial}{\partial x_p}(\sigma_{ml}^{(e)} - \tfrac{1}{3}\delta_{lm}\sigma_{nn}^{(e)}) \oint_D x_p\,dx_k;$$

the loop is assumed to be near the origin. The integrals here form an antisymmetric tensor:

$$\oint x_p\,dx_k = -\oint x_k\,dx_p.$$

We can then easily express the force in terms of the dislocation moment d_{kl} in (27.12)¶:

$$F_i = d_{lm}\frac{\partial\sigma_{lm}^{(e)}}{\partial x_i} + \tfrac{1}{3}\left(d_{il}\frac{\partial\sigma_{nn}^{(e)}}{\partial x_l} - d_{ll}\frac{\partial\sigma_{nn}^{(e)}}{\partial x_i}\right). \tag{28.9}$$

† For example, the dislocation shown in Fig. 22 can move in the yz-plane only through a loss of material from the "extra" half-plane by diffusion.

‡ Since such a process is limited by diffusion, it can be important in practice only at sufficiently high temperatures.

§ Evidently, a uniform compression of the crystal cannot produce a force **f**, and the expression (28.7) has this property.

¶ The derivation uses also the formula $e_{ikl}e_{imn} = \delta_{km}\delta_{ln} - \delta_{kn}\delta_{lm}$, and the equilibrium equation $\partial\sigma_{lm}^{(e)}/\partial x_m = 0$.

In a homogeneous stress field this force is zero, as already mentioned. In that case, however, the loop is acted on by a torque

$$K_i = e_{ilm}\oint x_l f_m \, \mathrm{d}l,$$

which can also be expressed in terms of the dislocation moment:

$$K_i = e_{ikl} d_{km}(\sigma_{lm}^{(e)} - \tfrac{1}{3}\delta_{lm}\sigma_{nn}^{(e)}). \tag{28.10}$$

PROBLEMS

PROBLEM 1. Find the force of interaction between two parallel screw dislocations in an isotropic medium.

SOLUTION. The force per unit length acting on one dislocation in the stress field due to the other dislocation is determined from formula (28.5), using the results of §27, Problem 2. It is a radial force of magnitude $f = \mu b_1 b_2/2\pi r$. Dislocations of like sign ($b_1 b_2 > 0$) repel, while those of unlike sign ($b_1 b_2 < 0$) attract.

PROBLEM 2. A straight screw dislocation lies parallel to the plane free surface of an isotropic medium. Find the force acting on the dislocation.

SOLUTION. Let the yz-plane be the surface of the body, and let the dislocation be parallel to the z-axis with coordinates $x = x_0$, $y = 0$.

The stress field which leaves the surface of the medium a free surface is described by the sum of the fields of the dislocation and its image in the yz-plane, considered to lie in an infinite medium:

$$\sigma_{xz} = \frac{\mu b}{2\pi}\left[\frac{y}{(x-x_0)^2+y^2} - \frac{y}{(x+x_0)^2+y^2}\right],$$

$$\sigma_{yz} = -\frac{\mu b}{2\pi}\left[\frac{x-x_0}{(x-x_0)^2+y^2} - \frac{x+x_0}{(x+x_0)^2+y^2}\right].$$

Such a field exerts a force on the dislocation considered which is equal to the attraction exerted by its image, i.e. the dislocation is attracted to the surface of the medium by a force $f = \mu b^2/4\pi x_0$.

PROBLEM 3. Find the force of interaction between two parallel edge dislocations in an isotropic medium which are in parallel glide planes.

SOLUTION. Let the glide planes be parallel to the xz-plane and let the z-axis be parallel to the dislocation lines; as in §27, Problem 4, we put $\tau_z = -1$, $b_x = b$. Then the force on unit length of the dislocation in the field of the elastic stresses σ_{ik} has components $f_x = b\sigma_{xy}$, $f_y = -b\sigma_{xx}$. In the case considered, σ_{ik} is determined by the expressions derived in §27, Problem 4. If one dislocation is along the z-axis, it exerts on the other dislocation (passing through the point $(x, y, 0)$) a force whose polar components are $f_r = b_1 b_2 D/r$, $f_\phi = (b_1 b_2 D/r)\sin 2\phi$, $D = \mu/2\pi(1-\sigma)$. The component of this force in the glide plane is $f_x = (b_1 b_2 D/r)\cos\phi\cos 2\phi$, which is zero when $\phi = \frac{1}{2}\pi$ or $\frac{1}{4}\pi$. The former position corresponds to stable equilibrium when $b_1 b_2 > 0$, the latter when $b_1 b_2 < 0$.

§29. A continuous distribution of dislocations

If a crystal contains several dislocations at the same time which are at relatively short distances apart (although far apart compared with the lattice constant, of course), it is useful to treat them by means of an averaging process: we consider "physically infinitesimal" volume elements in the crystal with a large number of dislocation lines through each.

An equation which expresses a fundamental property of dislocation deformations can be formulated by a natural generalization of equation (27.6). We define a tensor ρ_{ik} (the *dislocation density tensor*) such that its integral over a surface spanning any contour L is equal to the sum **b** of the Burgers vectors of all the dislocation lines embraced by the contour:

$$\int_{S_L} \rho_{ik}\, df_i = b_k. \tag{29.1}$$

The continuous functions ρ_{ik} describe the distribution of dislocations in the crystal. This tensor now replaces the expression on the right of equation (27.6):

$$e_{ilm}\partial w_{mk}/\partial x_l = -\rho_{ik}. \tag{29.2}$$

This equation shows that the tensor ρ_{ik} must satisfy the condition

$$\partial \rho_{ik}/\partial x_i = 0; \tag{29.3}$$

for a single dislocation, this condition simply states that the Burgers vector is constant along the dislocation line.

When the dislocations are treated in this way, the tensor w_{ik} becomes a primary quantity describing the deformation and determining the strain tensor through (27.4). A displacement vector **u** related to w_{ik} by the definition (27.2) cannot exist; this is clear from the fact that with such a definition the left-hand side of equation (29.2) would be identically zero throughout the crystal.

So far we have assumed the dislocations to be at rest. Let us now see how a set of equations may be formulated so as to allow in principle elastic deformations and stresses in a medium where dislocations are moving in a given manner† (E. Kröner and G. Rieder 1956).

Equation (29.2) is independent of whether the dislocations are at rest or in motion. The tensor w_{ik} still determines the elastic deformation; its symmetrical part is the elastic strain tensor, which is related to the stress tensor in the usual way, by Hooke's law.

This equation, however, is now insufficient for a complete formulation of the problem. The full set of equations must also determine the velocity **v** of the points in the medium.

It must be borne in mind that the movement of dislocations causes not only a change in the elastic deformation but also a change in the shape of the crystal which does not involve stresses, i.e. a *plastic deformation*. The motion of dislocations is, as already mentioned, a mechanism of plastic deformation. This is clearly illustrated by Fig. 25, where the passage of the edge dislocation from left to right causes the part of the crystal above the glide plane to be shifted to the right by one lattice period; since the lattice is then regular, the crystal remains unstressed. Unlike an elastic deformation, which is uniquely defined by the thermodynamic state of the body, a plastic deformation depends on the process which occurs. In considering dislocations at rest we have no need to distinguish elastic and plastic deformations, since we are concerned only with stresses which are independent of the previous history of the crystal.

Let **u** be the geometrical displacement vector of points in the medium, measured, say, from their position before the deformation process begins; its time derivative $\dot{\mathbf{u}} = \mathbf{v}$. If the "total distortion" tensor $W_{ik} = \partial u_k/\partial x_i$ is formed from the vector **u**, its "plastic part" $w_{ik}^{(pl)}$

† We shall not discuss here the problem of determining this motion itself from the forces applied to the body. The solution of such a problem requires a detailed study of the microscopic mechanism of the motion of dislocations and their retardation by various defects, which must take account of the conditions occurring in actual crystals.

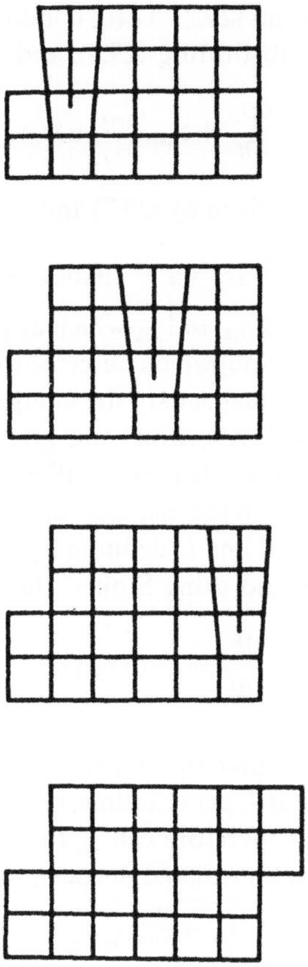

FIG. 25

is obtained by subtracting from W_{ik} the "elastic distortion" tensor, which is the same as the tensor w_{ik} in (29.2). We use the notation

$$-j_{ik} = \partial w_{ik}^{(\mathrm{pl})}/\partial t; \tag{29.4}$$

the symmetrical part of j_{ik} gives the rate of variation of the plastic strain tensor: the change in $u_{ik}^{(\mathrm{pl})}$ in an infinitesimal time interval δt is

$$\delta u_{ik}^{(\mathrm{pl})} = -\tfrac{1}{2}(j_{ik}+j_{ki})\delta t. \tag{29.5}$$

We may note, in particular, that, if a plastic deformation occurs without destroying the continuity of the body, the trace of the tensor j_{ik} is zero: a plastic deformation causes no extension or compression of the body (which would always involve the appearance of internal stresses), i.e. $u_{kk}^{(\mathrm{pl})} = 0$, and therefore $j_{kk} = -\partial u_{kk}^{(\mathrm{pl})}/\partial t = 0$.

Substituting in the definition (29.4) $w_{ik}^{(\mathrm{pl})} = W_{ik} - w_{ik}$, we can write it as

$$\frac{\partial w_{ik}}{\partial t} = \frac{\partial v_k}{\partial x_i} + j_{ik}, \tag{29.6}$$

an equation which relates the rates of change of the elastic and plastic deformations. Here the j_{ik} must be regarded as given quantities which must satisfy conditions ensuring the compatibility of equations (29.6) and (29.2). These conditions are found by differentiating (29.2) with respect to time and substituting (29.6), and are

$$\frac{\partial \rho_{ik}}{\partial t} + e_{ilm}\frac{\partial j_{mk}}{\partial x_l} = 0. \tag{29.7}$$

The complete set of equations is given by (29.2) and (29.6), together with the dynamical equations

$$\rho \dot{v}_i = \partial \sigma_{ik}/\partial x_k, \quad \sigma_{ik} = \lambda_{iklm} u_{lm} = \lambda_{iklm} w_{lm} \tag{29.8}$$

(A. M. Kosevich 1962). The tensors ρ_{ik} and j_{ik} which appear in these equations are given functions of the coordinates (and time) which describe the distribution and movement of the dislocations. These functions must satisfy the compatibility conditions of equations (29.2) with one another and with (29.6), which are given by (29.3) and (29.7).

The condition (29.7) may be regarded as a differential expression of the "law of conservation of the Burgers vector" in the medium: integrating both sides of this equation over a surface spanning some closed line L, defining by (29.1) the total Burgers vector $\mathbf{b}$ of the dislocations embraced by L, and using Stokes' theorem, we obtain

$$\frac{\mathrm{d}b_k}{\mathrm{d}t} = -\oint_L j_{ik}\mathrm{d}x_i. \tag{29.9}$$

The form of this equation shows that the integral on the right gives the "flux" of the Burgers vector through the contour L per unit time, i.e. the Burgers vector carried across L by moving dislocations. We may therefore call j_{ik} the *dislocation flux density tensor.*

In particular, it is clear that for an isolated dislocation loop the tensor j_{ik} has the form

$$\begin{aligned} j_{ik} &= e_{ilm}\rho_{lk}V_m \\ &= e_{ilm}\tau_l V_m b_k \delta(\xi), \end{aligned} \tag{29.10}$$

in accordance with the expression (28.2) for the plastic strain when the dislocation moves; $\mathbf{V}$ is the velocity of the dislocation line at a particular point on it. The flux vector through the element $\mathrm{d}\mathbf{l}$ of the contour L is $j_{ik}dl_i$ and is proportional to $\mathrm{d}\mathbf{l}\cdot\boldsymbol{\tau}\times\mathbf{V} = \mathbf{V}\cdot\mathrm{d}\mathbf{l}\times\boldsymbol{\tau}$, i.e. the component of $\mathbf{V}$ in a direction perpendicular to both $\mathrm{d}\mathbf{l}$ and $\boldsymbol{\tau}$; from geometrical considerations it is evident that this is correct, since only that velocity component causes the dislocation to intersect the element $\mathrm{d}\mathbf{l}$.

We may note that the trace of the tensor (29.10) is proportional to the component of the velocity of the dislocation along the normal to its glide plane. It has been mentioned above that the absence of any inelastic change in density of the medium is ensured by the condition $j_{ii} = 0$. We see that for an individual dislocation this condition signifies motion in the glide plane, in accordance with the previous discussion of the physical nature of the movement of dislocations; see the second footnote to §28.

Finally, let us consider the case where dislocation loops are distributed in the crystal in such a way that their total Burgers vector (denoted by $\mathbf{B}$) is zero.† This condition signifies that integration over any cross-section of the body gives

$$\int \rho_{ik}\mathrm{d}f_i = 0. \tag{29.11}$$

† The presence of a dislocation involves a certain bending of the crystal, as shown schematically in Fig. 26 (greatly exaggerated). The condition $\mathbf{B} = 0$ means that there is no macroscopic bending of the crystal as a whole.

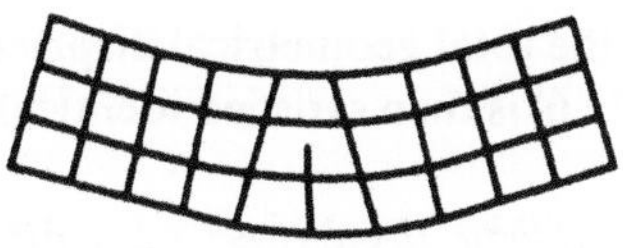

FIG. 26

From this it follows that the dislocation density in this case can be written as

$$\rho_{ik} = e_{ilm}\partial P_{mk}/\partial x_i \qquad (29.12)$$

(F. Kroupa 1962); then the integral (29.11) becomes an integral along a contour outside the body, and is zero. It may also be noted that the expression (29.12) necessarily satisfies the condition (29.3).

It is easy to see that the tensor P_{ik} thus defined represents the dislocation moment density in the deformed crystal, and may therefore be called the "dislocation polarization": the total dislocation moment D_{ik} of the crystal is, by definition,

$$D_{ik} = \sum S_i b_k = \tfrac{1}{2} e_{ilm} \sum b_k \oint_D x_l \mathrm{d}x_m$$

$$= \tfrac{1}{2} \int e_{ilm}\, x_l \rho_{mk}\, \mathrm{d}V,$$

where the summation is over all dislocation loops and the integration is over the whole volume of the crystal. Substituting (29.12), we obtain

$$D_{ik} = \tfrac{1}{2} \int e_{ilm} e_{mpq} x_l \frac{\partial P_{qk}}{\partial x_p} \mathrm{d}V$$

$$= \tfrac{1}{2} \int x_m \left(\frac{\partial P_{mk}}{\partial x_i} - \frac{\partial P_{ik}}{\partial x_m} \right) \mathrm{d}V$$

and, after integrating by parts in each term,

$$D_{ik} = \int P_{ik}\, \mathrm{d}V. \qquad (29.13)$$

The dislocation flux density is given in terms of the same tensor P_{ik} by

$$j_{ik} = -\partial P_{ik}/\partial t. \qquad (29.14)$$

This is easily seen, for example, by calculating the integral $\int j_{ik}\, \mathrm{d}V$ over an arbitrary part of the volume of the body, using the expression (29.10), to give a sum over all dislocation loops within that volume. We may note that the expression (29.14) together with (29.12) automatically satisfies the condition (29.7).

A comparison of (29.14) and (29.4) shows that $\delta w_{ik}^{(\mathrm{pl})} = \delta P_{ik}$. If we agree to regard the plastic deformation as absent in the state with $P_{ik} = 0$, then $w_{ik}^{(\mathrm{pl})} = P_{ik}$,† and

$$w_{ik} = W_{ik} - w_{ik}^{(\mathrm{pl})} = \partial u_k/\partial x_i - P_{ik}, \qquad (29.15)$$

† It is assumed that the entire deformation process occurs with $\mathbf{B} = 0$. This point must be emphasized, since there is a fundamental difference between the tensors P_{ik} and $w_{ik}^{(\mathrm{pl})}$: whereas P_{ik} is a function of the state of the body, the tensor $w_{ik}^{(\mathrm{pl})}$ is not, but depends on the process which has brought the body into that state.

where u_k is again the vector of the total geometrical displacement from the position in the undeformed state. Equation (29.6) is then satisfied identicaly, and the dynamical equation (29.8) becomes

$$\rho \ddot{u}_i - \lambda_{iklm}\partial^2 u_m/\partial x_k \partial x_l = -\lambda_{iklm}\partial P_{lm}/\partial x_k. \tag{29.16}$$

Thus the determination of the elastic deformation due to moving dislocations with $\mathbf{B} = 0$ reduces to a problem of ordinary elasticity theory with body forces distributed in the crystal with density $-\lambda_{iklm}\partial P_{lm}/\partial x_k$.

§30. Distribution of interacting dislocations

Let us consider a large number of similar straight dislocations lying parallel in the same glide plane, and derive an equation to determine their equilibrium distribution. Let the z-axis be parallel to the dislocations, and the xz-plane be the glide plane.

We shall suppose for definiteness that the Burgers vectors of the dislocations are in the x-direction. Then the force in the glide plane on unit length of a dislocation is $b\sigma_{xy}$, where σ_{xy} is the stress at the position of the dislocation.

The stresses created by one straight dislocation (and acting on another dislocation) decrease inversely as the distance from it. The stress at a point x due to a dislocation at a point x' is therefore $bD/(x-x')$, where D is a constant of the order of the elastic moduli of the crystal. It may be shown that this constant D is positive, i.e. two like dislocations in the same glide plane repel each other.†

Let $\rho(x)$ be the line density of dislocations on a segment (a_1, a_2) of the x-axis; $\rho(x)\,dx$ is the sum of the Burgers vectors of dislocations passing through points in the interval dx. Then the total stress at a point x on the x-axis due to all the dislocations is given by the integral

$$\sigma_{xy}(x) = -D\int_{a_1}^{a_2} \frac{\rho(\xi)d\xi}{\xi - x}. \tag{30.1}$$

For points in the segment (a_1, a_2) this integral must be taken as a principal value in order to exclude the physically meaningless action of a dislocation on itself.

If the crystal is also subjected to a two-dimensional stress field $\sigma_{xy}^{(e)}(x, y)$ in the xy-plane, caused by given external loads, each dislocation will be subjected to a force $b(\sigma_{xy} + p(x))$, where for brevity $p(x)$ denotes $\sigma_{xy}^{(e)}(x, 0)$. The condition of equilibrium is that this force be zero: $\sigma_{xy} + p = 0$, i.e.

$$P\int_{a_1}^{a_2} \frac{\rho(\xi)d\xi}{\xi - x} = \frac{p(x)}{D} \equiv \omega(x), \tag{30.2}$$

where P denotes, as usual, the principal value. This is an integral equation to determine the equilibrium distribution $\rho(x)$. It is a singular equation with a Cauchy kernel.

The solution of such an equation is equivalent to a problem in the theory of functions of a complex variable which may be formulated as follows.

† For an isotropic medium this has been proved in §28, Problem 3.

Let $\Omega(z)$ denote a function defined throughout the complex z-plane (cut from a_1 to a_2) as the integral

$$\Omega(z) = \int_{a_1}^{a_2} \frac{\rho(\xi)\mathrm{d}\xi}{\xi - z}. \tag{30.3}$$

Let $\Omega^+(x)$ and $\Omega^-(x)$ denote the limiting values $\Omega(z)$ on the upper and lower edges of the cut. They are equal to similar integrals along the segment (a_1, a_2) with an indentation in the form of an infinitesimal semicircle below or above the point $z = x$ respectively, i.e.

$$\Omega^{\pm}(x) = P\int_{a_1}^{a_2} \frac{\rho(\xi)\mathrm{d}\xi}{\xi - x} \pm i\pi\rho(x). \tag{30.4}$$

If $\rho(\xi)$ satisfies equation (30.2), the principal value of the integral is $\omega(x)$, and we therefore have

$$\Omega^+(x) + \Omega^-(x) = 2\omega(x), \tag{30.5}$$

$$\Omega^+(x) - \Omega^-(x) = 2i\pi\rho(x). \tag{30.6}$$

Thus the problem of solving equation (30.2) is equivalent to that of finding an analytic function $\Omega(z)$ with the property (30.5); $\rho(x)$ is then given by (30.6). The physical conditions of the problem in question also require that $\Omega(\infty) = 0$; this follows because far from the dislocations ($x \to \pm\infty$) the stresses σ_{xy} must be zero (by the definition (30.3), $\sigma_{xy}(x) = -D\Omega(x)$ outside the segment (a_1, a_2)).

Let us first consider the case where there are no external stresses ($p(x) \equiv 0$), and the dislocations are constrained by some obstacles (lattice defects) at the ends of the segment (a_1, a_2). When $\omega(x) = 0$ we have from (30.5) $\Omega^+(x) = -\Omega^-(x)$, i.e. the function $\Omega(z)$ must change sign in a passage round each of the points a_1, a_2. This condition is satisfied by any function of the form

$$\Omega(z) = \frac{P(z)}{\sqrt{[(a_2 - z)(z - a_1)]}}, \tag{30.7}$$

where $P(z)$ is a polynomial. The condition $\Omega(\infty) = 0$ means that we must take $P(z) = 1$ (apart from a constant coefficient), so that

$$\Omega(z) = \frac{1}{\sqrt{[(a_2 - z)(z - a_1)]}}. \tag{30.8}$$

The required function $\rho(x)$ will, according to (30.6), have the same form. The coefficient is determined from the condition

$$\int_{a_1}^{a_2} \rho(\xi)\mathrm{d}\xi = B, \tag{30.9}$$

where B is the sum of the Burgers vectors of all the dislocations, and so we have

$$\rho(x) = \frac{B}{\pi\sqrt{[(a_2 - x)(x - a_1)]}}. \tag{30.10}$$

We see that the dislocations pile up towards the obstacles at the ends of the segment, with density inversely proportional to the square root of the distance from the obstacle. The stress outside the segment (a_1, a_2) increases in the same manner as the ends of the segment are approached, e.g. for $x > a_2$

$$\sigma_{xy} \cong \frac{BD}{\sqrt{[(x-a_2)(a_2-a_1)]}}.$$

In other words, the concentration of dislocations at the boundary leads to a stress concentration beyond the boundary.

Let us now suppose that under the same conditions (obstacles at the fixed ends of the segment) there is also an external stress field $p(x)$. Let $\Omega_0(z)$ denote a function of the form (30.7), and let us rewrite equation (30.5) divided by $\Omega_0^+ = -\Omega_0^-$ as

$$\frac{\Omega^+(x)}{\Omega_0^+(x)} - \frac{\Omega^-(x)}{\Omega_0^-(x)} = \frac{2\omega(x)}{\Omega_0^+(x)}.$$

A comparison of this with (30.6) shows that

$$\frac{\Omega(z)}{\Omega_0(z)} = \frac{1}{i\pi}\int_{a_1}^{a_2} \frac{\omega(\xi)}{\Omega_0^+(\xi)}\frac{d\xi}{\xi - z} + i\pi P(z), \tag{30.11}$$

where $P(z)$ is a polynomial. A solution which satisfies the condition $\Omega(\infty) = 0$ is obtained by taking as $\Omega_0(z)$ the function (30.8) and putting $P(z) = C$, a constant. The required function $\rho(x)$ is hence found by means of (30.6), and the result is

$$\rho(x) = -\frac{1}{\pi^2\sqrt{[(a_2-x)(x-a_1)]}} P\int_{a_1}^{a_2} \omega(\xi)\sqrt{[(a_2-\xi)(\xi-a_1)]}\frac{d\xi}{\xi - x} + \frac{C}{\sqrt{[(a_2-x)(x-a_1)]}}. \tag{30.12}$$

The constant C is determined by the condition (30.9). Here also $\rho(x)$ increases as $(a_2 - x)^{-1/2}$ when $x \to a_2$ (and similarly when $x \to a_1$), and a similar concentration of stresses occurs on the other side of the boundary.

If there is an obstacle only on one side (at a_2, say) the required solution must satisfy the condition of finite stress for all $x < a_2$, including the point $x = a_1$; the position of the latter point is not known beforehand and must be determined by solving the problem. With respect to $\Omega(z)$ this means that $\Omega(a_1)$ must be finite. Such a function (satisfying also the condition $\Omega(\infty) = 0$) is obtained from the same formula (30.11) by taking for $\Omega_0(z)$ the function $\sqrt{[(z-a_1)/(a_2-z)]}$, which is also of the form (30.7), and putting $P(z) = 0$ in (30.11). The result is

$$\rho(x) = -\frac{1}{\pi^2}\sqrt{\frac{x-a_1}{a_2-x}}\, P\int_{a_1}^{a_2} \sqrt{\frac{a_2-\xi}{\xi-a_1}}\frac{\omega(\xi)d\xi}{\xi-x}. \tag{30.13}$$

When $x \to a_1$, $\rho(x)$ tends to zero as $\sqrt{(x-a_1)}$. The total stress $\sigma_{xy}(x) + p(x)$ tends to zero according to a similar law on the other side of the point a_1.

Finally, let there be no obstacle at either end of the segment, and let the dislocations be constrained only by external stresses $p(x)$. The corresponding $\Omega(z)$ is obtained by putting in (30.11) $\Omega_0(z) = \sqrt{[(a_2-z)(z-a_1)]}$, $P(z) = 0$. The condition $\Omega(\infty) = 0$, however, here requires the fulfilment of a further condition: taking the limit as $z \to \infty$ in (30.11), we find

$$\int_{a_1}^{a_2} \frac{\omega(\xi)d\xi}{\sqrt{[(a_2-\xi)(\xi-a_1)]}} = 0. \tag{30.14}$$

The function $\rho(x)$ is given by

$$\rho(x) = -\frac{1}{\pi^2}\sqrt{[(a_2-x)(x-a_1)]}P\int_{a_1}^{a_2} \frac{\omega(\xi)}{\sqrt{[(a_2-\xi)(\xi-a_1)]}}\frac{d\xi}{\xi-x}, \tag{30.15}$$

the coordinates a_1 and a_2 of the ends of the segment being determined by the conditions (30.9) and (30.14).

PROBLEM

Find the distribution of dislocations in a uniform stress field $p(x) = p_0$ over a segment with obstacles at one or both ends.

SOLUTION. When there is an obstacle at one end (a_2) the calculation of the integral (30.13) gives

$$\rho(x) = \frac{p_0}{\pi D}\sqrt{\frac{x-a_1}{a_2-x}}.$$

The condition (30.9) determines the length of the segment occupied by dislocations: $a_2 - a_1 = 2BD/P_0$. Beyond the obstacle there is a concentration of stresses near it according to

$$\sigma_{xy} \cong p_0\sqrt{\frac{a_2-a_1}{x-a_2}}.$$

For a segment of length $2L$ bounded by two obstacles we take the origin of x at the midpoint and obtain from (30.12)

$$\rho(x) = \frac{1}{\pi\sqrt{(L^2-x^2)}}\left(\frac{p_0}{D}x + B\right).$$

§31. Equilibrium of a crack in an elastic medium

The problem of the equilibrium of a crack is somewhat distinctive among the problems of elasticity theory. From the point of view of that theory, a crack is a cavity in an elastic medium, which exists when internal stresses are present in the medium and closes up when the load is removed. The shape and size of the crack depend considerably on the stresses acting on it. The mathematical feature of the problem is therefore that the boundary conditions are given on a surface which is initially unknown and must itself be determined in solving the problem.†

Let us consider a crack in an isotropic medium, with infinite length and uniform in the z-direction and in a plane stress field $\sigma_{ik}^{(e)}(x, y)$; this is a two-dimensional problem of elasticity theory. We shall suppose that the stresses are symmetrical about the centre of the cross-section of the crack. Then the outline of the cross-section will also be symmetrical (Fig. 27). Let its length be $2L$ and its variable width $h(x)$; since the crack is symmetrical, $h(-x) = h(x)$.

† The quantitative theory of cracks discussed here is due to G. I. Barenblatt (1959).

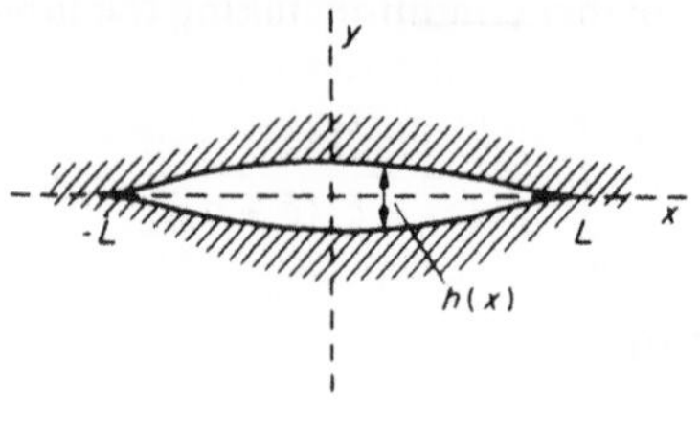

FIG. 27

We shall assume the crack to be thin ($h \ll L$). Then the boundary conditions on its surface can be applied to the corresponding segment of the x-axis. Thus the crack is regarded as a line of discontinuity (in the xy-plane) on which the normal component of the displacement $u_y = \pm\frac{1}{2}h$ is discontinuous.

Instead of $h(x)$ we define a new unknown function $\rho(x)$ by the formulae

$$h(x) = \int_x^L \rho(x)dx, \qquad \rho(-x) = -\rho(x). \tag{31.1}$$

The function $\rho(x)$ may be conveniently, though purely formally, interpreted as a density of straight dislocations lying in the z-direction and continuously distributed along the x-axis, with their Burgers vectors in the y-direction.† It has been shown in §27 that a dislocation line may be regarded as the edge of a surface of discontinuity on which the displacement $\mathbf{u}$ has a discontinuity $\mathbf{b}$. In the form (31.1) the discontinuity h of the normal displacement at the point x is regarded as the sum of the Burgers vectors of all the dislocations lying to the right of that point; the equation $\rho(-x) = -\rho(x)$ signifies that the dislocations to the right and to the left of the point $x = 0$ have opposite signs.

By means of this representation we can write down immediately an expression for the normal stresses (σ_{yy}) on the x-axis. These consist of the stresses $\sigma_{yy}{}^{(e)}(x, 0)$ resulting from the external loads (which for brevity we denote by $p(x)$) and the stresses $\sigma_{yy}{}^{(cr)}(x)$ due to the deformation caused by the crack. Regarding the latter stresses as being due to dislocations distributed over the segment $(-L, L)$, we obtain (similarly to (30.1))

$$\sigma_{yy}{}^{(cr)}(x) = -D\int_{-L}^{L} \frac{\rho(\xi)d\xi}{\xi - x}; \tag{31.2}$$

for points in the segment $(-L, L)$ itself, the integral must be taken as a principal value. For an isotropic medium,

$$D = \frac{\mu}{2\pi(1-\sigma)} = \frac{E}{4\pi(1-\sigma^2)}; \tag{31.3}$$

see §28, Problem 3. The stresses σ_{xy} due to such dislocations in an isotropic medium are zero on the x-axis.

The boundary condition on the free surface of the crack, applied (as already mentioned) to the corresponding segment of the x-axis, requires that the normal stresses $\sigma_{yy} =$

† It is for this reason that the theory of cracks is described here in the chapter on dislocations, although physically the phenomena are quite different.

$\sigma_{yy}^{(cr)} + p(x)$ be zero. This condition, however, needs to be made more precise, for the following reason.

Let us make the assumption (which will be confirmed by the result) that the edges of the crack join smoothly near its ends, so that the surfaces approach very closely. Then it is necessary to take into account the forces of molecular attraction between the surfaces; the action of these forces extends to a distance r_0 large compared with interatomic distances. These forces will be of importance in a narrow region near the end of the crack where $h \lesssim r_0$; the length of this region will be denoted by d in order of magnitude, and will be estimated later.

Let G be the force of molecular cohesion per unit area of the crack; it depends on the distance h between the surfaces.† When these forces are taken into account, the boundary condition becomes

$$\sigma_{yy}^{(cr)} + p(x) - G = 0. \tag{31.4}$$

It is reasonable to suppose that the shape of the crack near its end is determined by the nature of the cohesion forces and does not depend on the external loads applied to the body. Then, in finding the shape of the main part of the crack from the external forces $p(x)$, the quantity G becomes a given function $G(x)$ independent of $p(x)$ (over the region d, outside which it is unimportant).

Substituting $\sigma_{yy}^{(cr)}$ from (31.2) in (31.4), we thus obtain the following integral equation for $\rho(x)$:

$$P \int_{-L}^{L} \frac{\rho(\xi)\mathrm{d}\xi}{\xi - x} = \frac{1}{D} p(x) - \frac{1}{D} g(x) \equiv \omega(x). \tag{31.5}$$

Since the ends of the crack are assumed not fixed, the stresses must remain finite there. This means that, in solving the integral equation (31.5), we now have the last of the cases discussed in §30, for which the solution is given by (30.15). With the origin at the midpoint of the segment $(-L, L)$ this formula becomes

$$\rho(x) = -\frac{1}{\pi^2}\sqrt{(L^2 - x^2)}P \int_{-L}^{L} \frac{\omega(\xi)}{\sqrt{(L^2 - \xi^2)}} \frac{\mathrm{d}\xi}{\xi - x}. \tag{31.6}$$

The condition (30.14) must be satisfied, which in this case gives

$$\int_0^L \frac{p(x)\,\mathrm{d}x}{\sqrt{(L^2 - x^2)}} - \int_0^L \frac{G(x)\,\mathrm{d}x}{\sqrt{(L^2 - x^2)}} = 0 \tag{31.7}$$

(where the integrals from $-L$ to L have been replaced by integrals from 0 to L, using the symmetry of the problem). Since $G(x)$ is zero except in the range $L - x \sim d$, in the second integral we can put $L^2 - x^2 \cong 2L(L - x)$; the condition (31.7) then becomes

$$\int_0^L \frac{p(x)\,\mathrm{d}x}{\sqrt{(L^2 - x^2)}} = \frac{M}{\sqrt{(2L)}}, \tag{31.8}$$

† In the macroscopic theory, the function $G(x)$ is to be regarded as increasing smoothly, as $L - x$ decreases, up to a maximum value at the end of the crack.

where M denotes the constant

$$M = \int_0^d \frac{G(\xi)\,d\xi}{\sqrt{\xi}}, \tag{31.9}$$

which depends on the medium concerned. This constant can be expressed in terms of the ordinary macroscopic properties of the body, its elastic moduli and surface tension α; as will be shown later, the relation is

$$M = \sqrt{[\pi\alpha E/(1-\sigma^2)]}. \tag{31.10}$$

The equation (31.8) determines the length $2L$ of the crack from the given stress distribution $p(x)$. For example, for a crack widened by concentrated forces f applied to the midpoints of the sides ($p(x) = f\delta(x)$) we find

$$\begin{aligned} 2L &= f^2/M^2 \\ &= f^2(1-\sigma^2)/\pi\alpha E. \end{aligned} \tag{31.11}$$

It must be remembered, however, that stable equilibrium of a crack is not possible for every distribution $p(x)$. For instance, with uniform widening stresses ($p(x) =$ constant $\equiv p_0$) (31.8) gives

$$\begin{aligned} 2L &= 4M^2/\pi^2 {p_0}^2 \\ &= 4\alpha E/\pi(1-\sigma^2){p_0}^2. \end{aligned} \tag{31.12}$$

This inverse relation (L decreasing when p_0 increases) shows that the state is unstable. The value of L determined by (31.12) corresponds to unstable equilibrium and gives the "critical" crack length: longer cracks grow spontaneously, but shorter ones close up, a result first derived by A. A. Griffith (1920).

Let us now return to the consideration of the shape of the crack. When $L - x \lesssim d$, the region $L - \xi \sim d$ is the most important in the integral in (31.6). The integral can then be replaced by its limiting value as $x \to L$; the result is $\rho =$ constant $\times \sqrt{(L-x)}$, whence†

$$h(x) = \text{constant} \times (L-x)^{3/2} \qquad (L - x \sim d). \tag{31.13}$$

We see that over the terminal region d the two sides of the crack in fact join smoothly. The value of the coefficient in (31.13) depends on the properties of the cohesion forces and can not be expressed in terms of the ordinary macroscopic parameters.‡

For the part farther from the end, where $d \ll L - x \ll L$, the region $L - \xi \sim d$ is again the most important in the integral in (31.6), and $\omega(\xi) \cong -G(\xi)/D$. In addition to putting $L^2 - x^2 \cong 2L(L-x)$, $L^2 - \xi^2 \cong 2L(L-\xi)$, we can here replace $\xi - x$ by $L - x$, obtaining $\rho = M/\pi^2 D\sqrt{(L-x)}$, where M is the same constant as in (31.9), (31.10). Hence

$$h(x) = 2M\sqrt{(L-x)}/\pi^2 D \qquad (d \ll L - x \ll L). \tag{31.14}$$

Thus the end of the crack has a shape independent of the applied forces (and therefore of

† In order to proceed to the limit we must first divide the integral in (31.6) into two integrals with numerators $\omega(\xi) - \omega(L)$ and $\omega(L)$; the second integral makes no contribution to the limiting value.

‡ An estimate of the coefficient in (31.13) gives a value of the order of $\sqrt{a/d}$, where a is the dimension of an atom (using $\alpha \sim aE$, $M \sim E\sqrt{a}$). An estimate of the length d is obtained from the condition $h(d) \sim r_0$, whence $d \sim {r_0}^2/a \gg r_0$. It should be mentioned, however, that in practice the required inequalities are satisfied only by a small margin, so that the resulting shape of the terminal projection of the crack is not to be taken as exact.

the length of the crack) throughout the range $L-x \ll L$: when $L-x \gg d$ the shape is given by (31.14), and when $L-x \sim d$ it has an infinitely sharp projection (31.13) (Fig. 28). The shape of the remainder of the crack does depend on the applied forces.

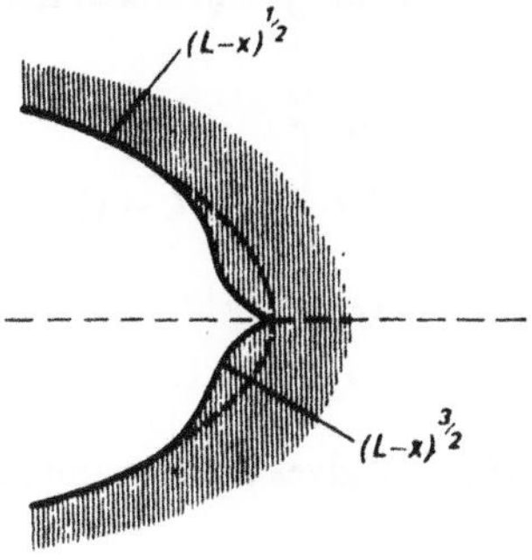

FIG. 28

If we ignore details, of the order of the radius of action of the cohesion forces, the crack therefore has a smooth outline with ends rounded according to the parabolas (31.14), and this shape is entirely determined by the applied forces and the ordinary macroscopic parameters. The small ($\sim d$) terminal projections which actually occur are of fundamental significance, however, since they ensure that the stresses remain finite at the ends of the crack.

The stresses caused by the crack on the continuation of the x-axis are given by formula (31.2). At distances $x-L$ such that $d \ll x-L \ll L$, we have†

$$\sigma_{yy} \cong \sigma_{yy}^{(cr)} \cong M/\pi\sqrt{(x-L)}. \tag{31.15}$$

The increase in the stresses as the edge of the crack is approached continues according to this law up to distances $x-L \sim d$, and σ_{yy} then drops to zero at the point $x = L$.

It remains to derive the formula (31.10) already given above, which relates the constant M to the ordinary macroscopic quantities. To do this, we write down the condition for the total free energy to be a minimum by equating to zero its variation under a change in the length L.

Firstly, when the length of the crack increases by δL the surface energy at its two free surfaces increases by $\delta F_{\text{surf}} = 2\alpha\delta L$. Secondly, the "opening" of the crack end reduces the elastic energy F_{el} by $\frac{1}{2}\int\sigma_{yy}(x)\eta(x)\,dx$, where $\eta(x)$ is the difference in width between the displaced and undisplaced crack shapes. Since the shape of the crack end is independent of its length, $\eta(x) = h(x-\delta L) - h(x)$. The stress $\sigma_{yy} = 0$ for $x < L$, and $h(x) = 0$ for $x > L$. Hence

$$\delta F_{\text{el}} = -\tfrac{1}{2}\int_{L}^{L+\delta L} \sigma_{yy}(x)h(x-\delta L)\,dx.$$

† The integral is easily calculated directly, but it is not necessary to do this if we use the relation between the functions $\rho(x)$ for $x < L$ and $\sigma_{yy}^{(cr)}$ for $x > L$, which is evident from the results of §30.

Substituting (31.14) and (31.15), we find

$$\delta F_{el} = -\frac{M^2}{\pi^3 D}\int_{L}^{L+\delta L}\sqrt{\frac{L+\delta L-x}{x-L}}\,dx$$

$$= -\frac{M^2}{\pi^3 D}\int_{0}^{\delta L}\frac{\sqrt{y}\,dy}{\sqrt{(\delta L-y)}}$$

$$= -\frac{M^2}{2\pi^2 D}\delta L.$$

Finally, the condition $\delta F_{surf} + \delta F_{el} = 0$ gives the relation $M^2 = 4\pi^2\alpha D$, and hence we have (31.10).†

---

† It may be noted that the theory described above, including the relation (31.10), is in fact applicable as it stands only to ideally brittle bodies, i.e. those which remain linearly elastic up to fracture, such as glass and fused quartz. In bodies which exhibit plasticity the formation of the crack may be accompanied by plastic deformation at its ends.

CHAPTER V

THERMAL CONDUCTION AND VISCOSITY IN SOLIDS

§32. The equation of thermal conduction in solids

NON-UNIFORM heating of a solid does not cause convection as it generally does in fluids. Hence the transfer of heat is effected in solids by thermal conduction alone. The processes of thermal conduction in solids are therefore described by somewhat simpler equations than those for fluids, where they are complicated by convection.

The equation of thermal conduction in a solid can be derived immediately from the law of conservation of energy in the form of an "equation of continuity for heat". The amount of heat absorbed per unit time in unit volume of the body is $T\partial S/\partial t$, where S is the entropy per unit volume. This must be put equal to $-\operatorname{div}\mathbf{q}$, where $\mathbf{q}$ is the heat flux density. This flux can almost always be written as $-\kappa\,\mathbf{grad}\,T$, i.e. it is proportional to the temperature gradient (κ being the thermal conductivity). Thus

$$T\partial S/\partial t = \operatorname{div}(\kappa\,\mathbf{grad}\,T). \tag{32.1}$$

According to formula (6.4), the entropy can be written as

$$S = S_0(T) + K\alpha u_{ii},$$

where α is the thermal expansion coefficient and S_0 the entropy in the undeformed state. We shall suppose that, as usually happens, the temperature differences in the body are so small that quantities such as κ, α, etc. may be regarded as constants. Then equation (32.1), after substitution of the above expression for S, becomes

$$T\frac{\partial S_0}{\partial t} + \alpha K T\frac{\partial u_{ii}}{\partial t} = \kappa \triangle T.$$

According to a well-known formula of thermodynamics, we have

$$C_p - C_v = K\alpha^2 T,$$

whence

$$\alpha K T = (C_p - C_v)/\alpha.$$

The time derivative of S_0 can be written as $(\partial S_0/\partial T)\cdot(\partial T/\partial t)$, where the derivative $\partial S_0/\partial T$ is taken for $u_{ii} \equiv \operatorname{div}\mathbf{u} = 0$, i.e. at constant volume, and therefore is equal to C_v/T.

The resulting equation of thermal conduction is

$$C_v\frac{\partial T}{\partial t} + \frac{C_p - C_v}{\alpha}\frac{\partial}{\partial t}\operatorname{div}\mathbf{u} = \kappa \triangle T. \tag{32.2}$$

In order to obtain a complete system of equations, it is necessary to add an equation describing the deformation of a non-uniformly heated body. This is the equilibrium equation (7.8):

$$2(1-\sigma)\,\mathbf{grad}\operatorname{div}\mathbf{u} - (1-2\sigma)\,\mathbf{curl}\,\mathbf{curl}\,\mathbf{u} = \tfrac{2}{3}\alpha(1+\sigma)\,\mathbf{grad}\,T. \tag{32.3}$$

From equation (32.3) we can in principle determine the deformation of the body for any given temperature distribution. Substituting the expression for div **u** thus obtained in equation (32.2), we derive an equation giving the temperature distribution, in which the only unknown function is $T(x, y, z, t)$.

For example, let us consider thermal conduction in a infinite solid in which the temperature distribution satisfies only one condition: at infinity, the temperature tends to a constant value T_0, and there is no deformation. In such a case equation (32.3) leads to the following relation between div **u** and T (see §7, Problem 8):

$$\operatorname{div} \mathbf{u} = \frac{1+\sigma}{3(1-\sigma)}\alpha(T-T_0).$$

Substituting this expression in (32.2), we obtain

$$\frac{(1+\sigma)C_p + 2(1-2\sigma)C_v}{3(1-\sigma)}\frac{\partial T}{\partial t} = \kappa \triangle T, \tag{32.4}$$

which is the ordinary equation of thermal conduction.

An equation of this type also describes the temperature distribution along a thin straight rod, if one (or both) of its ends is free. The temperature may be assumed constant over any transverse cross-section, so that T is a function only of the coordinate x along the rod and of the time. The thermal expansion of such a rod causes a change in its length, but no departure from straightness and no internal stresses. Hence it is clear that the derivative $\partial S/\partial t$ in the general equation (32.1) must be taken at constant pressure and, since $(\partial S/\partial t)_p = C_p/T$, the temperature distribution will satisfy the one-dimensional thermal conduction equation $C_p \partial T/\partial t = \kappa \partial^2 T/\partial x^2$.

It should be mentioned, however, that the temperature distribution in a solid can in practice always be determined, with sufficient accuracy, by a simple thermal conduction equation. The reason is that the second term on the left-hand side of equation (32.2) is a correction of order $(C_p - C_v)/C_v$ relative to the first term. In solids, however, the difference between the two specific heats is usually very small, and if it is neglected the equation of thermal conduction in solids can always be written

$$\partial T/\partial t = \chi \Delta T, \tag{32.5}$$

where χ is the thermometric conductivity, defined as the ratio of the thermal conductivity κ to some mean specific heat per unit volume C.

§33. Thermal conduction in crystals

In an anisotropic body, the direction of the heat flux **q** is not in general that of the temperature gradient. Hence, instead of the formula

$$\mathbf{q} = -\kappa \, \mathbf{grad} \, T$$

relating **q** to the temperature gradient, we have in a crystal the more general relation

$$q_i = -\kappa_{ik} \partial T/\partial x_k. \tag{33.1}$$

The tensor κ_{ik}, of rank two, is called the *thermal conductivity tensor* of the crystal. In accordance with (33.1), the equation of thermal conduction (32.5) has also a more general form,

$$C\frac{\partial T}{\partial t} = \kappa_{ik}\frac{\partial^2 T}{\partial x_i \partial x_k}. \tag{33.2}$$

A general theorem can be stated: the thermal conductivity tensor is symmetrical, i.e.

$$\kappa_{ik} = \kappa_{ki}. \tag{33.3}$$

This relation, which we shall now prove, is a consequence of the symmetry of the kinetic coefficients (see *SP* 1, §120).

The rate of increase of the total entropy of the body by irreversible processes of thermal conduction is

$$\dot{S}_{\text{tot}} = -\int \frac{\operatorname{div}\mathbf{q}}{T}\,\mathrm{d}V = -\int \operatorname{div}\frac{\mathbf{q}}{T}\,\mathrm{d}V + \int \mathbf{q}\cdot\mathbf{grad}\frac{1}{T}\,\mathrm{d}V.$$

The first integral, on being transformed into a surface integral, is seen to be zero. Thus

$$\dot{S}_{\text{tot}} = \int \mathbf{q}\cdot\mathbf{grad}\frac{1}{T}\,\mathrm{d}V = -\int \frac{\mathbf{q}\cdot\mathbf{grad}\,T}{T^2}\,\mathrm{d}V,$$

or

$$\dot{S}_{\text{tot}} = -\int \frac{1}{T^2} q_i \frac{\partial T}{\partial x_i}\,\mathrm{d}V. \tag{33.4}$$

In accordance with the general definition of the kinetic coefficients,† we can deduce from (33.4) that in the case considered the coefficients $T^2\kappa_{ik}$ in

$$q_i = -T^2\kappa_{ik}\left(\frac{1}{T^2}\frac{\partial T}{\partial x_k}\right)$$

are kinetic coefficients. Hence the result (33.3) follows immediately from the symmetry of the kinetic coefficients.

The quadratic form

$$-q_i\frac{\partial T}{\partial x_i} = \kappa_{ik}\frac{\partial T}{\partial x_i}\frac{\partial T}{\partial x_k}$$

must be positive, since the time derivative (33.4) of the entropy must be positive. The condition for a quadratic form to be positive is that the eigenvalues of the matrix of its coefficients be positive. Hence all the principal values of the thermal conductivity tensor κ_{ik} are always positive; this is evident also from simple considerations regarding the direction of the heat flux.

The number of independent components of the tensor κ_{ik} depends on the symmetry of the crystal. Since the tensor κ_{ik} is symmetrical, this number is evidently the same as the number for the thermal expansion tensor (§10), which is also a symmetrical tensor of rank two.

§34. Viscosity of solids

In discussing motion in elastic bodies, we have so far assumed that the deformation is reversible. In reality, this process is thermodynamically reversible only if it occurs with infinitesimal speed, so that thermodynamic equilibrium is established in the body at every instant. An actual motion, however, has finite velocities; the body is not in equilibrium at every instant, and therefore processes will take place in it which tend to return it to

---

† We here use the definition given in *FM*, §59.

equilibrium. The existence of these processes has the result that the motion is irreversible, and, in particular, mechanical energy† is dissipated, ultimately into heat.

The dissipation of energy occurs by two means. Firstly, when the temperature at different points in the body is different, irreversible processes of thermal conduction take place in it. Secondly, if any internal motion occurs in the body, there are irreversible processes arising from the finite velocity of that motion. This means of energy dissipation may be referred to, as in fluids, as internal friction or *viscosity*.

In most cases the velocity of macroscopic motions in the body is so small that the energy dissipation is not considerable. Such "almost irreversible" processes can be described by means of what is called the *dissipative function*.‡ If we have a mechanical system whose motion involves the dissipation of energy, this motion can be described by the ordinary equations of motion, with the forces acting on the system augmented by the *dissipative forces* or *frictional forces*, which are linear functions of the velocities. These forces can be written as the velocity derivatives of a certain quadratic function R of the velocities, called the dissipative function. The frictional force f_a corresponding to a generalized coordinate q_a of the system is then given by $f_a = -\partial R/\partial \dot{q}_a$. The dissipative function R is a positive quadratic form in the velocities $\dot{q}_a$. The above relation is equivalent to

$$\delta R = -\sum_a f_a \delta \dot{q}_a, \tag{34.1}$$

where δR is the change in the dissipative function caused by an infinitesimal change in the velocities. It can also be shown that the dissipative function is half the decrease in the mechanical energy of the system per unit time.

It is easy to generalize equation (34.1) to the case of motion with friction in a continuous medium. The state of the system is then defined by a continuum of generalized coordinates. These are the displacement vector $\mathbf{u}$ at each point in the body. Accordingly, the relation (34.1) can be written in the integral form

$$\delta \int R \, \mathrm{d}V = -\int f_i \delta v_i \, \mathrm{d}V \tag{34.2}$$

where $\mathbf{v} = \dot{\mathbf{u}}$ and the f_i are the components of the dissipative force vector $\mathbf{f}$ per unit volume of the body; we write the total dissipative function for the body as $\int R \, \mathrm{d}V$, where R is the dissipative function per unit volume.

Let us now determine the general form of the dissipative function R for deformed bodies. The function R, which describes the internal friction, must be zero if there is no internal friction, and in particular if the body executes only a general translatory or rotary motion. In other words, the dissipative function must be zero if $\dot{\mathbf{u}} = \text{constant}$ or $\dot{\mathbf{u}} = \mathbf{\Omega} \times \mathbf{r}$. This means that it must depend not on the velocity itself but on its gradient, and can contain only such combinations of the derivatives as vanish when $\dot{\mathbf{u}} = \mathbf{\Omega} \times \mathbf{r}$. These are the sums

$$v_{ik} = \frac{1}{2}\left(\frac{\partial v_i}{\partial x_k} + \frac{\partial v_k}{\partial x_i}\right),$$

i.e. the time derivatives of the components of the strain tensor.† Thus the dissipative function must be a quadratic function of v_{ik}. The most general form of such a function is

† By *mechanical energy* we here mean the sum of the kinetic energy of the macroscopic motion in the elastic body and its (elastic) potential energy arising from the deformation.

‡ See *SP* 1, §121.

$$R = \tfrac{1}{2}\eta_{iklm}v_{ik}v_{lm}. \tag{34.3}$$

The tensor η_{iklm}, of rank four, may be called the *viscosity tensor*. It has the following evident symmetry properties:‡

$$\eta_{iklm} = \eta_{lmik} = \eta_{kilm} = \eta_{ikml}. \tag{34.4}$$

The expression (34.3) is exactly analogous to the expression (10.1) for the free energy of a crystal: the elastic modulus tensor is replaced by the tensor η_{iklm}, and u_{ik} by v_{ik}. Hence the results obtained in §10 for the tensor λ_{iklm} in crystals of various symmetries are wholly valid for the tensor η_{iklm} also.

In particular, the tensor η_{iklm} in an isotropic body has only two independent components, and R can be written in a form analogous to the expression (4.3) for the elastic energy of an isotropic body:

$$R = \eta(v_{ik} - \tfrac{1}{3}\delta_{ik}v_{ll})^2 + \tfrac{1}{2}\zeta v_{ll}^{\,2}, \tag{34.5}$$

where η and ζ are the two coefficients of viscosity. Since R is a positive function, the coefficients η and ζ must be positive.

The relation (34.2) is entirely analogous to that for the elastic free energy, $\delta\int F\,\mathrm{d}V = -\int F_i\delta u_i\,\mathrm{d}V$, where $F_i = \partial\sigma_{ik}/\partial x_k$ is the force per unit volume. Hence the expression for the dissipative force f_i in terms of the tensor v_{ik} can be written down at once by analogy with the expression for F_i in terms of u_{ik}. We have

$$f_i = \partial\sigma'_{ik}/\partial x_k, \tag{34.6}$$

where the dissipative stress tensor σ'_{ik} is defined by

$$\sigma'_{ik} = \partial R/\partial v_{ik} = \eta_{iklm}v_{lm}. \tag{34.7}$$

The viscosity can therefore be taken into account in the equations of motion by simply replacing the stress tensor σ_{ik} in those equations by the sum $\sigma_{ik} + \sigma'_{ik}$.

In an isotropic body,

$$\sigma'_{ik} = 2\eta(v_{ik} - \tfrac{1}{3}\delta_{ik}v_{ll}) + \zeta v_{ll}\delta_{ik}. \tag{34.8}$$

This expression is, as we should expect, formally identical with that for the viscosity stress tensor in a fluid.

§35. The absorption of sound in solids

The absorption of sound in solids can be calculated in a manner entirely analogous to that used for fluids (see *FM*, §79). Here we shall give the calculations for an isotropic body. The total energy dissipated in the body is the sum

$$\dot{E}_{\text{mech}} = -(\kappa/T)\int(\nabla T)^2\,\mathrm{d}V - 2\int R\,\mathrm{d}V,$$

† Cf. the entirely analogous arguments on the viscosity of fluids in *FM*, §15.

‡ The existence of the dissipative function is a consequence of Onsager's principle of the symmetry of the kinetic coefficients. This principle leads to the first equation (34.4) for the coefficients in the linear relations (34.7), whereby the dissipative function can be defined. This will be shown directly in a similar context in §42.

the first term being due to thermal conduction and the second to viscosity. Using the expression (34.5), we therefore have

$$\dot{E}_{\text{mech}} = -\frac{\kappa}{T}\int (\mathbf{grad}\, T)^2\, dV - 2\eta \int (v_{ik} - \tfrac{1}{3}\delta_{ik}v_{ll})^2\, dV - \zeta \int v_{ll}^2\, dV. \qquad (35.1)$$

To calculate the temperature gradient, we use the fact that sound oscillations are adiabatic in the first approximation. Using the expression (6.4) for the entropy, we can write the adiabatic condition as $S_0(T) + K\alpha u_{ii} = S_0(T_0)$, where T_0 is the temperature in the undeformed state. Expanding the difference $S_0(T) - S_0(T_0)$ in powers of $T - T_0$, we have as far as the first-order terms $S_0(T) - S_0(T_0) = (T - T_0)\,\partial S_0/\partial T_0 = C_v(T - T_0)/T_0$. The derivative of the entropy is taken for $u_{ii} = 0$, i.e. at constant volume. Thus

$$T - T_0 = -T\alpha K u_{ii}/C_v.$$

Using also the relations $K \equiv K_{\text{iso}} = C_v K_{\text{ad}}/C_p$ and $K_{\text{ad}}/\rho = c_l^2 - 4c_t^2/3$, we can rewrite this result as

$$T - T_0 = -\frac{T\alpha\rho(c_l^2 - 4c_t^2/3)}{C_p} u_{ii}. \qquad (35.2)$$

Let us first consider the absorption of transverse sound waves. The thermal conduction cannot result in the absorption of these waves (in the approximation considered). For, in a transverse wave, we have $u_{ii} = 0$, and therefore the temperature in it is constant, by (35.2). Let the wave be propagated along the x-axis; then

$$u_x = 0, \quad u_y = u_{0y}\cos(kx - \omega t), \quad u_z = u_{0z}\cos(kx - \omega t),$$

and the only non-zero components of the strain tensor are $u_{xy} = -\frac{1}{2}ku_{0y}\sin(kx - \omega t)$, $u_{xz} = -\frac{1}{2}ku_{0z}\sin(kx - \omega t)$.

We shall consider the energy dissipation per unit volume of the body; the (time) average value of this quantity is, from (35.1),

$$\overline{\dot{E}}_{\text{mech}} = -\tfrac{1}{2}\eta\omega^4(u_{0y}^2 + u_{0z}^2)/c_t^2,$$

where we have put $k = \omega/c_t$. The total mean energy of the wave is twice the mean kinetic energy, i.e.

$$\bar{E} = \rho\int \overline{\dot{u}^2}\, dV;$$

for unit volume we have

$$\bar{E} = \tfrac{1}{2}\rho\omega^2(u_{0y}^2 + u_{0z}^2).$$

The sound absorption coefficient is defined as the ratio of the mean energy dissipation to twice the mean energy flux in the wave; this quantity gives the manner of variation of the wave amplitude with distance. The amplitude decreases as $e^{-\gamma x}$. Thus we find the following expression for the absorption coefficient for transverse waves:

$$\gamma_t = \tfrac{1}{2}|\overline{\dot{E}}_{\text{mech}}|/c_t\bar{E} = \eta\omega^2/2\rho c_t^3. \qquad (35.3)$$

In a longitudinal sound wave $u_x = u_0\cos(kx - \omega t)$, $u_y = u_z = 0$. A similar calculation, using formulae (35.1) and (35.2), gives

$$\gamma_l = \frac{\omega^2}{2\rho c_l^3}\left[\left(\frac{4}{3}\eta + \zeta\right) + \frac{\kappa T\alpha^2\rho^2 c_l^2}{C_p^2}\left(1 - \frac{4c_t^2}{3c_l^2}\right)^2\right]. \qquad (35.4)$$

These formulae relate, strictly speaking, only to a completely isotropic and amorphous body. They give, however, the correct order of magnitude for the absorption of sound in anisotropic single crystals also.

The absorption of sound in polycrystalline bodies exhibits peculiar properties. If the wavelength λ of the sound is small in comparison with the dimensions a of the individual crystallites (grains), then the sound is absorbed in each crystallite in the same way as in a large crystal, and the absorption coefficient is proportional to ω^2.

If $\lambda \gg a$, however, the nature of the absorption is different. In such a wave we can assume that each crystallite is subject to a uniformly distributed pressure. However, since the crystallites are anisotropic, and so are the boundary conditions at their surfaces of contact, the resulting deformation is not uniform. It varies considerably (by an amount of the same order as itself) over the dimension of a crystallite, and not over one wavelength as in a homogeneous body. When sound is absorbed, the rates of change of the deformation (v_{ik}) and the temperature gradients are of importance. Of these, the former are still of the usual order of magnitude. The temperature gradients within each crystallite are anomalously large, however. Hence the absorption due to thermal conduction will be large compared with that due to viscosity, and only the former need be calculated.

Let us consider two limiting cases. The time during which the temperature is equalized by thermal conduction over distances $\sim a$ (the relaxation time for thermal conduction) is of the order of a^2/χ. Let us first assume that $\omega \ll \chi/a^2$. This means that the relaxation time is small compared with the period of the oscillations in the wave, and so thermal equilibrium is nearly established in each crystallite; in this case we have almost isothermal oscillations.

Let T' be the temperature difference in a crystallite, and T_0' the corresponding difference in an adiabatic process. The heat transferred by thermal conduction per unit volume is $-\operatorname{div}\mathbf{q} = \kappa\triangle T' \sim \kappa T'/a^2$. The amount of heat evolved in the deformation is of the order of $\dot{T}_0'C \sim \omega T_0'C$, where C is the specific heat. Equating the two, we obtain $T' \sim T_0'\omega a^2/\chi$. The temperature varies by an amount of the order of T' over the dimension of the crystallite, and so its gradient is of magnitude $\sim T'/a$. Finally, T_0' is found from (35.2), with $u_{ii} \sim ku \sim \omega u/c$ (u being the amplitude of the displacement vector):

$$T_0' \sim T\alpha\rho c\omega u/C; \tag{35.5}$$

in obtaining orders of magnitude, we naturally neglect the difference between the various velocities of sound. Using these results, we can calculate the energy dissipated per unit volume:

$$\bar{\dot{E}}_{\text{max}} \sim \frac{\kappa}{T}(\mathbf{grad}\, T)^2 \sim \frac{\kappa}{T}\left(\frac{T'}{a}\right)^2.$$

Dividing this by the energy flux $c\bar{E} \sim c\rho\omega^2u^2$, we find the damping coefficient to be

$$\gamma \sim T\alpha^2\rho c a^2\omega^2/\chi C \quad \text{for} \quad \omega \ll \chi/a^2 \tag{35.6}$$

(C. Zener 1938). Comparing this expression with the general expressions (35.3) and (35.4), we can say that, in the case considered, the absorption of sound by a polycrystalline body is the same as if it had a viscosity

$$\eta \sim T\alpha^2\rho^2c^4a^2/\chi C,$$

which is much larger than the actual viscosity of the component crystallites.

Next, let us consider the opposite limiting case, where $\omega \gg \chi/a^2$. In other words, the relaxation time is large compared with the period of oscillations in the wave, and no noticeable equalization of the temperature differences due to the deformation can occur in one period. It would be incorrect, however, to suppose that the temperature gradients which determine the absorption of sound are of the order of T_0'/a. This assumption would take into account only thermal conduction in each crystallite, whereas heat exchange between neighbouring crystallites must be of importance in the case in question (M. A. Isakovich 1948). If the crystallites were thermally insulated the temperature differences occurring at their boundaries would be of the same order T_0' as those within each individual crystallite. In reality, however, the boundary conditions require the continuity of the temperature across the surface separating two crystallites. We therefore have "temperature waves" propagated away from the boundary into the crystallite; these are damped at a distance† $\delta \sim \sqrt{(\chi/\omega)}$. In the case under consideration $\delta \ll a$, i.e. the main temperature gradient is of the order of T_0'/δ and occurs over distances small compared with the total dimension of a crystallite. The corresponding fraction of the volume of the crystallite is $\sim a^2\delta$; taking the ratio of this to the total volume $\sim a^3$, we find the mean energy dissipation

$$\bar{\dot{E}}_{\text{mech}} \sim \frac{\kappa}{T}\left(\frac{T_0'}{\delta}\right)^2 \frac{a^2\delta}{a^3} \approx \frac{\kappa T_0'^2}{Ta\delta}.$$

Substituting for T_0' the expression (35.5) and dividing by $c\bar{E} \sim c\rho\omega^2 u^2$, we obtain the required absorption coefficient:

$$\gamma \sim T\alpha^2 \rho c\sqrt{(\chi\omega)}/aC \quad \text{for} \quad \omega \gg \chi/a^2 \tag{35.7}$$

It is proportional to the square root of the frequency.‡

Thus the sound absorption coefficient in a polycrystalline body varies as ω^2 at very low frequencies ($\omega \ll \chi/a^2$); for $\chi/a^2 \ll \omega \ll c/a$ it varies as $\sqrt{\omega}$, and for $\omega \gg c/a$ it again varies as ω^2.

Similar considerations hold for the damping of transverse waves in thin rods and plates (C. Zener 1938). If h is the thickness of the rod or plate, then for $\lambda \gg h$ the transverse temperature gradient is important, and the damping is mainly due to thermal conduction (see the Problems). If also $\omega \ll \chi/h^2$, the oscillations may be regarded as isothermal, and therefore, in determining (for example) the characteristic frequencies of vibrations of the rod or plate, the isothermal values of the moduli of elasticity must be used.

PROBLEMS

PROBLEM 1. Determine the damping coefficient for longitudinal vibrations of a rod.

SOLUTION. The damping coefficient for the vibrations is defined as $\beta = |\bar{\dot{E}}_{\text{mech}}|/2\bar{E}$; the amplitude of the vibrations diminishes with time as $e^{-\beta t}$.

In a longitudinal wave, any short section of the rod is subject to simple extension or compression; the components of the strain tensor are $u_{zz} = \partial u_z/\partial z$, $u_{xx} = u_{yy} = -\sigma_{\text{ad}}\partial u_z/\partial z$. We put $u_z = u_0 \cos kz \cos \omega t$, where

† It may be recalled that, if a thermally conducting medium is bounded by the plane $x = 0$, at which the excess temperature varies periodically according to $T' = T_0' e^{-i\omega t}$, then the temperature distribution in the medium is given by the "temperature wave" $T' = T_0' e^{-i\omega t} \exp[-(1+i)x\sqrt{(\omega/2\chi)}]$; see *FM*, §52.

‡ The same frequency dependence is found for the absorption of sound propagated in a fluid near a solid wall (in a pipe, for instance); see *FM*, §79, Problems.

$k = \omega/\sqrt{(E_{ad}/\rho)}$. Calculations similar to those given above lead to the following expression for the damping coefficient:

$$\beta = \frac{\omega^2}{2\rho}\left\{\frac{\eta}{3}\frac{3c_l^2 - 4c_t^2}{(c_l^2 - c_t^2)c_t^2} + \frac{\zeta c_t^2}{(c_l^2 - c_t^2)(3c_l^2 - 4c_t^2)} + \frac{\kappa T\alpha^2\rho^2}{9C_p^2}\right\}.$$

Here we have written E_{ad} and σ_{ad} in terms of the velocities c_l, c_t by means of formulae (22.4).

PROBLEM 2. The same as Problem 1, but for longitudinal oscillations of a plate.

SOLUTION. For waves whose direction of oscillation is parallel to that of their propagation (the x-axis, say) the non-zero components of the strain tensor are

$$u_{xx} = \partial u_x/\partial x, \qquad u_{zz} = -[\sigma_{ad}/(1-\sigma_{ad})]\,\partial u_x/\partial x;$$

see (13.1). The velocity of propagation of these waves is $\sqrt{[E_{ad}/\rho(1-\sigma_{ad}^2)]}$. A calculation gives

$$\beta = \frac{\omega^2}{2\rho}\left\{\frac{\eta}{3}\frac{3c_l^4 + 4c_t^4 - 6c_l^2c_t^2}{c_t^2c_l^2(c_l^2 - c_t^2)} + \frac{\zeta c_t^2}{c_l^2(c_l^2 - c_t^2)} + \frac{\kappa T\alpha^2\rho^2(1+\sigma_{ad})^2}{9C_p^2}\right\}.$$

For waves whose direction of oscillation is perpendicular to the direction of propagation, $u_{ll} = 0$, and the damping is caused only by the viscosity η. In this case the damping coefficient is $\beta = \eta\omega^2/2\rho c_t^2$. This applies also to the damping of torsional vibrations of rods.

PROBLEM 3. Determine the damping coefficient for transverse vibrations of a rod (with frequencies such that $\omega \gg \chi/h^2$, where h is the thickness of the rod).

SOLUTION. The damping is due mainly to thermal conduction. According to §17, we have for each volume element in the rod $u_{zz} = x/R$, $u_{xx} = u_{yy} = -\sigma_{ad}x/R$ (for bending in the xz-plane); for $\omega \gg \chi/h^2$, the vibrations are adiabatic. For small deflections the radius of curvature $R = 1/X''$, so that $u_{ii} = (1-2\sigma_{ad})xX''$, the prime denoting differentiation with respect to z. The temperature varies most rapidly across the rod, and so $(\mathbf{grad}\,T)^2 \cong (\partial T/\partial x)^2$. Using (35.1) and (35.2), we obtain for the total mean energy dissipation in the rod $-(\kappa T\alpha^2 E_{ad}^2 S/9C_p^2)\int \overline{X''^2}\,dz$, where S is the cross-sectional area of the rod. The mean total energy is twice the potential energy $E_{ad}I_y\int X''^2\,dz$. The damping coefficient is

$$\beta = \kappa T\alpha^2 S E_{ad}/18 I_y C_p^2.$$

PROBLEM 4. The same as Problem 3, but for transverse vibrations of a plate.

SOLUTION. According to (11.4), we have for any volume element in the plate

$$u_{ii} = -\frac{1-2\sigma_{ad}}{1-\sigma_{ad}}z\frac{\partial^2\zeta}{\partial x^2}$$

for bending in the xz-plane. The energy dissipation is found from formulae (35.1) and (35.2) and the mean total energy is twice the expression (11.6). The damping coefficient is

$$\beta = \frac{2\kappa T\alpha^2 E_{ad}}{3C_p^2h^2}\cdot\frac{1+\sigma_{ad}}{1-\sigma_{ad}} = \frac{2\kappa T\alpha^2\rho}{3C_p^2h^2}\cdot\frac{(3c_l^2 - 4c_t^2)^2c_t^2}{(c_l^2 - c_t^2)c_l^2}.$$

PROBLEM 5. Determine the change in the characteristic frequencies of transverse vibrations of a rod due to the fact that the vibrations are not adiabatic. The rod is in the form of a long plate of thickness h. The surface of the rod is supposed thermally insulated.

SOLUTION. Let $T_{ad}(x, t)$ be the temperature distribution in the rod for adiabatic vibrations, and $T(x, t)$ the actual temperature distribution; x is a coordinate across the thickness of the rod, and the temperature variation in the yz-plane is neglected. Since, for $T = T_{ad}$, there is no heat exchange between various parts of the body, it is clear that the thermal conduction equation must be

$$\frac{\partial}{\partial t}(T - T_{ad}) = \chi\frac{\partial^2 T}{\partial x^2}.$$

For periodic vibrations of frequency ω, the differences $\tau_{ad} = T_{ad} - T_0$, $\tau = T - T_0$ from the equilibrium temperature T_0 are proportional to $e^{-i\omega t}$, and we have $\tau'' + i\omega\tau/\chi = i\omega\tau_{ad}/\chi$, the prime denoting differentiation with respect to x. Since, by (35.2), τ_{ad} is proportional to u_{ll}, and the components u_{ik} are proportional to x (see §17), it follows that $\tau_{ad} = Ax$, where A is a constant which need not be calculated, since it does not appear in the final result. The solution of the equation $\tau'' + i\omega\tau/\chi = i\omega Ax/\chi$, with the boundary condition $\tau' = 0$ for $x = \pm\frac{1}{2}h$ (the surface of the rod being insulated), is

$$\tau = A\left(x - \frac{\sin kx}{k\cos\frac{1}{2}kh}\right), \qquad k = (1+i)\sqrt{(\omega/2\chi)}.$$

The moment M_y of the internal stress forces in a rod bent in the xz-plane is composed of the isothermal part $M_{y,\,\mathrm{iso}}$ (i.e. the value for isothermal bending) and the part due to the non-uniform heating of the rod. If $M_{y,\,\mathrm{ad}}$ is the moment in adiabatic bending, the second part of the moment is reduced from $M_{y,\,\mathrm{ad}} - M_{y,\,\mathrm{iso}}$ in the ratio

$$1 + f(\omega) = \int_{-\frac{1}{2}h}^{\frac{1}{2}h} z\tau\,\mathrm{d}z \bigg/ \int_{-\frac{1}{2}h}^{\frac{1}{2}h} z\tau_{\mathrm{ad}}\,\mathrm{d}z.$$

Defining the Young's modulus E_ω for any frequency ω as the coefficient of proportionality between M_y and I_y/R (see (17.8)), and noticing that $E_{\mathrm{ad}} - E = E^2T\alpha^2/9C_p$ (see (6.8); E is the isothermal Young's modulus), we can put

$$E_\omega = E + [1 + f(\omega)]\,E^2T\alpha^2/9C_p.$$

A calculation shows that $f(\omega) = (24/k^3h^3)(\frac{1}{2}kh - \tan\frac{1}{2}kh)$. For $\omega \to \infty$ we obtain $f = 1$, which is correct, since $E_\infty = E_{\mathrm{ad}}$, and for $\omega \to 0$, $f = 0$ and $E_0 = E$.

The frequencies of the characteristic vibrations are proportional to the square root of the Young's modulus (see §25, Problems 4–6). Hence

$$\omega = \omega_0\left[1 + f(\omega_0)\frac{ET\alpha^2}{18C_p}\right],$$

where ω_0 are the characteristic frequencies for adiabatic vibrations. This value of ω is complex. Separating the real and imaginary parts ($\omega = \omega' + i\beta$), we find the characteristic frequencies

$$\omega' = \omega_0\left[1 - \frac{ET\alpha^2}{3C_p}\cdot\frac{1}{\xi^3}\cdot\frac{\sinh\xi - \sin\xi}{\cosh\xi + \cos\xi}\right]$$

and the damping coefficient

$$\beta = \frac{2ET\alpha^2\chi}{3C_ph^2}\left[1 - \frac{1}{\xi}\cdot\frac{\sinh\xi + \sin\xi}{\cosh\xi + \cos\xi}\right],$$

where $\xi = h\sqrt{(\omega_0/2\chi)}$.

For large ξ the frequency ω tends to ω_0, as it should, and the damping coefficient to $2ET\alpha^2\chi/3C_ph^2$, in accordance with the result of Problem 3.

Small values of ξ correspond to almost isothermal conditions; in this case

$$\omega \cong \omega_0\left(1 - \frac{ET\alpha^2}{18C_p}\right) \cong \omega_0\sqrt{(E/E_{\mathrm{ad}})},$$

and the damping coefficient $\beta = ET\alpha^2h^2\omega_0^{\,2}/180C_p\chi$.

§36. Highly viscous fluids

For typical fluids, the Navier–Stokes equations are valid if the periods of the motion are large compared with times characterizing the molecules. This, however, is not true for very viscous fluids. In such fluids, the usual equations of fluid mechanics become invalid for much larger periods of the motion. There are viscous fluids which, during short intervals of time (though these are long compared with molecular times), behave as solids (for instance, glycerine and resin). Amorphous solids (for instance, glass) may be regarded as a limiting case of such fluids having a very large viscosity.

The properties of these fluids can be described by the following method, due to Maxwell. They are elastically deformed during short intervals of time. When the deformation ceases, shear stresses remain in them, although these are damped in the course of time, so that after a sufficiently long time almost no internal stress remains in the fluid. Let τ be of the order of the time during which the stresses are damped (sometimes called the *Maxwellian relaxation time*). Let us suppose that the fluid is subjected to some variable external forces, which vary periodically in time with frequency ω. If the period $1/\omega$ is large compared with the relaxation time τ, i.e. $\omega\tau \ll 1$, the fluid under consideration will behave as an ordinary viscous fluid. If, however, the frequency ω is sufficiently large (so that $\omega\tau \gg 1$), the fluid will behave as an amorphous solid.

In accordance with these "intermediate" properties, the fluids in question can be characterized by both a viscosity coefficient η and a modulus of rigidity μ. It is easy to obtain a relation between the orders of magnitude of η, μ and the relaxation time τ. When periodic forces of sufficiently small frequency act, and so the fluid behaves like an ordinary fluid, the stress tensor is given by the usual expression for viscosity stresses in a fluid, i.e.

$$\sigma_{ik} = 2\eta \dot{u}_{ik} = -2i\omega\eta u_{ik}.$$

In the opposite limit of large frequencies, the fluid behaves like a solid, and the internal stresses must be given by the formulae of the theory of elasticity, i.e. $\sigma_{ik} = 2\mu u_{ik}$; we are speaking of pure shear deformations, i.e. we assume that $u_{ii} = \sigma_{ii} = 0$. For frequencies $\omega \sim 1/\tau$, the stresses given by these two expressions must be of the same order of magnitude. Thus $\eta u/\lambda\tau \sim \mu u/\lambda$, whence

$$\eta \sim \tau\mu. \tag{36.1}$$

This is the required relation.

Finally, let us derive the equation of motion which qualitatively describes the behaviour of these fluids. To do so, we make a very simple assumption concerning the damping of the internal stresses (when motion ceases): namely, that they are damped exponentially, i.e. $d\sigma_{ik}/dt = -\sigma_{ik}/\tau$. In a solid, however, we have $\sigma_{ik} = 2\mu u_{ik}$, and so $d\sigma_{ik}/dt = 2\mu du_{ik}/dt$. It is easy to see that the equation

$$\frac{d\sigma_{ik}}{dt} + \frac{\sigma_{ik}}{\tau} = 2\mu\frac{du_{ik}}{dt} \tag{36.2}$$

gives the correct result in both limiting cases of slow and rapid motions, and may therefore serve as an interpolatory equation for intermediate cases.

For example, in periodic motion, where u_{ik} and σ_{ik} depend on the time through a factor $e^{-i\omega t}$, we have from (36.2) $-i\omega\sigma_{ik} + \sigma_{ik}/\tau = -2i\omega\mu u_{ik}$, whence

$$\sigma_{ik} = \frac{2\mu u_{ik}}{1 + i/\omega\tau}. \tag{36.3}$$

For $\omega\tau \gg 1$, this formula gives $\sigma_{ik} = 2\mu u_{ik}$, i.e. the usual expression for solid bodies, while for $\omega\tau \ll 1$ we have $\sigma_{ik} = -2i\omega\mu\tau u_{ik} = 2\mu\tau\dot{u}_{ik}$, the usual expression for a fluid of viscosity $\mu\tau$.

CHAPTER VI

MECHANICS OF LIQUID CRYSTALS†

§37. Static deformations of nematics

LIQUID crystals are, macroscopically, anisotropic fluids. The mechanics of these substances has features characteristic of ordinary liquids and of elastic media, and is in this respect intermediate between fluid mechanics and elasticity theory.

There are various types of liquid crystals. The *nematic liquid crystals* or *nematics* are substances which in the undeformed state are both macroscopically and microscopically homogeneous; the anisotropy of the medium is due only to the anisotropic spatial orientation of the molecules (see *SP* 1, §§ 139, 140). The great majority of known nematics belong to the simplest type, in which the anisotropy is fully defined by specifying at each point in the medium a unit vector **n** along one particular direction; **n** is called the *director*. The quantities **n** and −**n** are physically equivalent, and so only a particular axis is distinguished, the two opposite directions along it being equivalent. The properties of these nematics in each volume element are invariant under inversion (a change in sign of all three coordinates).‡ Only this type of nematic liquid crystals will be discussed here.

The state of a nematic substance is thus described by specifying at each point, together with the usual quantities for a liquid (density ρ, pressure p and velocity **v**), the director **n**. All these appear as unknown functions of coordinates and time, in the equation of motion of a nematic.

In equilibrium, a nematic at rest under no external forces (which includes forces exerted by the boundary walls) is homogeneous, with **n** constant throughout its volume. In a deformed nematic, the direction of **n** varies slowly in space, the word "slowly" being taken in the sense usual in macroscopic theory: the characteristic dimensions of the deformation are much greater than molecular dimensions, so that the derivatives $\partial n_i/\partial x_k$ are to be regarded as small quantities.

In this chapter, it will be more convenient to relate all thermodynamic quantities to unit volume of the deformed body, not of the undeformed body as in previous chapters. Then the free energy density F of a nematic substance is made up of the free energy $F_0(\rho, T)$ of the undeformed nematic and the deformation energy F_d. The latter is given by an expression quadratic in the derivatives of **n**, its general form being

$$\begin{aligned} F_d &= F - F_0 \\ &= \tfrac{1}{2}K_1(\operatorname{div}\mathbf{n})^2 + \tfrac{1}{2}K_2(\mathbf{n}\cdot\mathbf{curl}\,\mathbf{n})^2 + \tfrac{1}{2}K_3(\mathbf{n}\times\mathbf{curl}\,\mathbf{n})^2; \end{aligned} \tag{37.1}$$

see *SP* 1, § 140. For the unit vector **n**(**r**), since $\nabla\mathbf{n}^2 \equiv 0$,

---

† This chapter was written jointly with L. P. Pitaevskiĭ.

‡ Nematics not invariant under inversion are unstable with respect to a deformation which converts them into cholesterics; see *SP* 1, §140, and §44 below.

$$\mathbf{n}\times\mathbf{curl}\,\mathbf{n} = -(\mathbf{n}\cdot\nabla)\mathbf{n}, \tag{37.2}$$

and so the last term in (37.1) can be written in the equivalent form $\frac{1}{2}K_3[(\mathbf{n}\cdot\nabla)\mathbf{n}]^2$.

The energy (37.1) in nematic mechanics has a role similar to the elastic energy of a deformed solid, and its presence gives this mechanics some of the features of elasticity theory.†

The three quadratic combinations of derivatives in (37.1) are independent; each can be different from zero when the other two are not. The condition for the undeformed state to be stable is therefore that all three coefficients K_1, K_2, K_3 (functions of density and temperature) be positive. We call these the *elastic moduli* or *Frank's moduli* of the nematic.

Deformations in which only one of the quantities div $\mathbf{n}$, $\mathbf{n}\cdot\mathbf{curl}\,\mathbf{n}$ and $\mathbf{n}\times\mathbf{curl}\,\mathbf{n}$ is not zero are called respectively *splays*, *twists* and *bends*. In general, of course, the deformation of a nematic includes all three kinds simultaneously. To illustrate their nature, some simple examples will be given. Let a nematic medium occupy the space between two coaxial cylindrical surfaces; r, ϕ, z be cylindrical polar coordinates with the z-axis along the axis of the cylinders. If the director $\mathbf{n}$ is radial at every point in the medium ($n_r = 1$, $n_\phi = n_z = 0$), the deformation is a splay (div $\mathbf{n} = 1/r$). If it is everywhere along a circle whose centre is on the z-axis ($n_\phi = 1$, $n_r = n_z = 0$), we have a pure bend ($\mathrm{curl}_z\mathbf{n} = 1/r$). If the director changes across a plane parallel slab of nematic at right angles to the z-axis according to $n_x = \cos\phi(z)$, $n_y = \sin\phi(z)$, $n_z = 0$, we have a pure twist ($\mathbf{n}\cdot\mathbf{curl}\,\mathbf{n} = -\phi'(z)$).

The walls which form the boundary of the volume occupied by a liquid crystal, and the free surface, tend to orient the medium, as will be discussed more fully below. Hence the mere presence of a boundary causes in general a deformation of a liquid crystal at rest. The question arises of finding the equations which describe this deformation, that is, which determine the equilibrium distribution $\mathbf{n}(\mathbf{r})$ for given boundary conditions (J. L. Ericksen 1966).

To do so, we start from the general thermodynamic condition of equilibrium: a minimum of the total free energy $\int F\,dV$, which is a functional of $\mathbf{n}(\mathbf{r})$. Since $\mathbf{n}$ is a unit vector, this functional is to be minimized with the auxiliary condition $\mathbf{n}^2 = 1$. With the familiar method of undetermined Lagrange multipliers, we must equate to zero the variation

$$\delta\int\{F - \tfrac{1}{2}\lambda(\mathbf{r})\mathbf{n}^2\}\,dV, \tag{37.3}$$

where $\lambda(\mathbf{r})$ is an arbitrary function.

The integrand depends on the functions $n_i(\mathbf{r})$ and on their derivatives. We have‡

$$\delta\int F\,dV = \int\left\{\frac{\partial F}{\partial n_i}\delta n_i + \frac{\partial F}{\partial(\partial_k n_i)}\partial_k\,\delta n_i\right\}dV$$

$$= \int\left\{\frac{\partial F}{\partial n_i} - \partial_k\frac{\partial F}{\partial(\partial_k n_i)}\right\}\delta n_i\,dV + \oint\frac{\partial F}{\partial(\partial_k n_i)}\delta n_i\,df_k. \tag{37.4}$$

† Deformation of a liquid crystal causes in general dielectric polarization of it and accordingly the presence of an electric field (the *piezoelectric* or *flexoelectric effect*; see *ECM*, §17). This effect is usually weak, and its influence on the mechanical properties of the substance will here be neglected. We shall also ignore the influence of an external magnetic field on the properties of liquid crystals; because of the anisotropy of the magnetic (in practice diamagnetic) susceptibility of the nematic, a magnetic field has an orienting effect on it, which may be considerable.

‡ In this chapter, to simplify the notation, we shall use the abbreviation $\partial_i = \partial/\partial x_i$ for the operator of differentiation with respect to a coordinate, which is common in recent literature.

The second term, an integral over the surface of the body, is important only in determining the boundary conditions. Putting at present $\delta\mathbf{n} = 0$ at the boundaries, we thus find as the variation of the total free energy

$$\delta \int F \, dV = -\int \mathbf{H}\cdot\delta\mathbf{n} \, dV, \tag{37.5}$$

where $\mathbf{H}$ is a vector whose components are

$$H_i = \partial_k \Pi_{ki} - \partial F/\partial n_i, \quad \Pi_{ki} = \partial F/\partial(\partial_k n_i). \tag{37.6}$$

This $\mathbf{H}$ acts as a field which tends to "straighten out" the direction of $\mathbf{n}$ throughout the liquid crystal, and is called the *molecular field*.

Equation (37.3) becomes

$$\int (\mathbf{H} + \lambda\mathbf{n})\cdot\delta\mathbf{n} \, dV = 0,$$

from which, since the variation $\delta\mathbf{n}$ is arbitrary, we find the equilibrium equation $\mathbf{H} = -\lambda\mathbf{n}$. Hence $\lambda = -\mathbf{H}\cdot\mathbf{n}$, so that the longitudinal component of this equation is satisfied by choosing λ. The equilibrium equation thus reduces in practice to the condition that $\mathbf{H}$ and $\mathbf{n}$ be collinear at every point in the medium; the longitudinal component of $\mathbf{H}$ has no physical significance. The equilibrium condition may therefore be written as

$$\mathbf{h} \equiv \mathbf{H} - \mathbf{n}(\mathbf{n}\cdot\mathbf{H}) = 0, \tag{37.7}$$

with a vector $\mathbf{h}$ which is such that $\mathbf{n}\cdot\mathbf{h} = 0$.

An explicit expression can be found for the molecular field corresponding to the free energy (37.1). To differentiate with respect to $\partial_k n_i$, we note that

$$\operatorname{div}\mathbf{n} = \partial_l n_l, \quad \operatorname{curl}_l \mathbf{n} = e_{lki}\,\partial_k n_i$$

(where e_{ikl} is the antisymmetric unit tensor), and so

$$\frac{\partial \operatorname{div}\mathbf{n}}{\partial(\partial_k n_i)} = \delta_{ik}, \quad \frac{\partial}{\partial(\partial_k n_i)}\operatorname{curl}_l \mathbf{n} = e_{lki}.$$

The resulting expression for the tensor Π_{ki} is

$$\Pi_{ki} = K_1\,\delta_{ik}\operatorname{div}\mathbf{n} + K_2(\mathbf{n}\cdot\mathbf{curl}\,\mathbf{n})n_l e_{lki} + K_3[(\mathbf{n}\times\mathbf{curl}\,\mathbf{n})\times\mathbf{n}]_l e_{lki}. \tag{37.8}$$

The further differentiation in accordance with the definition (37.6) yields the following fairly complicated formula for the molecular field:

$$\mathbf{H} = \nabla(K_1 \operatorname{div}\mathbf{n}) - \{K_2(\mathbf{n}\cdot\mathbf{curl}\,\mathbf{n})\,\mathbf{curl}\,\mathbf{n} + \mathbf{curl}[K_2(\mathbf{n}\cdot\mathbf{curl}\,\mathbf{n})\mathbf{n}]\} +$$
$$+ K_3[(\mathbf{n}\times\mathbf{curl}\,\mathbf{n})\times\mathbf{curl}\,\mathbf{n}] + \mathbf{curl}[K_3\,\mathbf{n}\times(\mathbf{n}\times\mathbf{curl}\,\mathbf{n})]. \tag{37.9}$$

The boundary conditions on the equations of equilibrium cannot be derived in a general form: they depend not only on the elastic energy (37.1) but also on the specific form of interaction between the liquid and the boundary wall. This surface energy would have to be included in the free energy which is minimized in order to obtain the equilibrium conditions. In practice, the surface forces are usually so great as to determine the direction of $\mathbf{n}$ at the boundary regardless of the type of deformation within the sample. If the solid boundary surface is anisotropic, this direction is entirely definite or is one of the few such directions. If it is isotropic, however, including the case of a free surface, only the angle between **n** and the normal to the surface is specified. If this angle is zero, the direction of **n** is definite, along the normal to the surface. If the angle is not zero, the possible directions form a conical surface with a definite vertex angle.

In the latter case, an additional boundary condition has to be imposed. This is determined by the requirement that the surface integral in (37.4) be zero for variations $\delta\mathbf{n}$ that are rotations of **n** about the normal at each point on the surface without change in the angle to the normal (that is, variations which do not affect the surface energy). Such variations have the form $\delta\mathbf{n} = \boldsymbol{\nu}\times\mathbf{n}\,\delta\phi$, where $\boldsymbol{\nu}$ is a unit vector along the normal, and $\delta\phi$ is the angle of rotation, which is arbitrary (at each point on the surface). Writing the surface element as $d\mathbf{f} = \boldsymbol{\nu}\,df$, we find

$$\oint \Pi_{ki}\, e_{imn} n_n \nu_m \nu_k\, \delta\phi\, df = 0,$$

from which, since $\delta\phi$ is arbitrary, we get the boundary condition

$$\Pi_{ki}\, e_{imn} n_n \nu_m \nu_k = 0, \tag{37.10}$$

or, if the z-axis is along $\boldsymbol{\nu}$,

$$\Pi_{zx} n_y - \Pi_{zy} n_x = 0. \tag{37.11}$$

Lastly, the following comment may be made regarding the elastic moduli occurring in (37.1). Since they are defined as coefficients in the free energy, they determine the isothermal deformations of the body. It is easy to see, however, that the same coefficients in nematics determine the adiabatic deformations also. We have seen in §6 that for a solid the difference between the isothermal and adiabatic moduli results from a term in the free energy that is linear in the strain tensor. A term linear in the derivatives $\partial_k n_i$ might play a similar role for nematics. Such a term would have to be a scalar and also invariant under a change in the sign of **n**. It is evident that no such term can be constructed: the product $\mathbf{n}\cdot\mathbf{curl}\,\mathbf{n}$ is a pseudoscalar, and the only true scalar is div **n**, which changes sign with **n**. For this reason, the isothermal and adiabatic moduli of a nematic are the same, just as for the shear modulus of an isotropic solid (§6). These arguments can also be expressed in a slightly different manner. In the absence of a linear term, the quadratic elastic energy (37.1) is a first "small correction" to the thermodynamic quantities for an undeformed body. The theorem of small increments (*SP* 1, §15) shows that, when expressed in terms of the corresponding thermodynamic variables (temperature or entropy), this correction is the same for the free energy and the internal energy.

§38. Straight disclinations in nematics

The equilibrium state of a nematic substance with given boundary conditions does not necessarily have at all points a continuous distribution **n**(**r**) with a definite direction of the vector **n** everywhere. In the mechanics of nematics, it is necessary to consider also deformations where **n**(**r**) may have singular points or lines at which the direction of **n** is not definite. The linear singularities are called *disclinations*.

The necessary occurrence of disclinations may be illustrated by means of simple examples. Let us consider a nematic in a long cylindrical vessel, the boundary conditions requiring **n** to be perpendicular to the surface of the vessel. It is reasonable to expect that in equilibrium the vector **n** at each point lies radially in the cross-section of the cylinder (Fig. 29a); the direction of **n** is then obviously indeterminate on the cylinder axis, which is

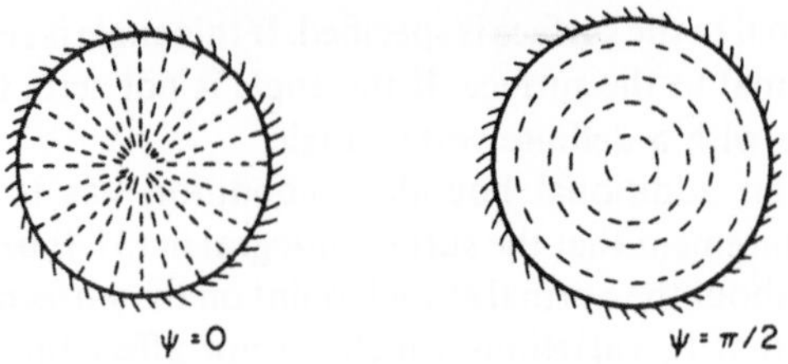

FIG. 29

therefore a disclination. If, on the other hand, the boundary conditions require **n** to be parallel to the vessel wall and in the cross-section plane, we get a distribution in which the vectors **n** are everywhere along concentric circles in the cross-section planes, with the centre on the cylinder axis (Fig. 29b); here again, the direction of **n** on the axis is indeterminate.

These are two simple cases of straight disclinations. Let us now take the general problem of the possible distributions **n**(**r**) in straight disclinations in an infinite nematic medium. The distribution **n**(**r**) in such a disclination is evidently independent of the coordinate along its length, and we need therefore consider the distribution only in planes perpendicular to the disclination axis. We shall suppose that **n** itself is everywhere in such a plane. Thus we have a two-dimensional problem in the mechanics of nematics. Some general properties of the solution can be derived from general arguments without looking at any specific equilibrium equations.

We use cylindrical polar coordinates r, ϕ, z, with the z-axis along the disclination axis. As already noted, the distribution **n**(**r**) is independent of z. It also cannot depend on r, since in the problem as formulated (a disclination in an infinite medium) there are no parameters having the dimensions of length from which a dimensionless function of r, such as **n**(**r**) is, could be constructed. The required distribution therefore depends only on the angle variable: $\mathbf{n} = \mathbf{n}(\phi)$.

Let ψ be the angle between **n** and the position vector in the plane z = constant through a given point (Fig. 30); the components of the two-dimensional (in this plane) vector **n** are

$$n_r = \cos\psi, \quad n_\phi = \sin\psi.$$

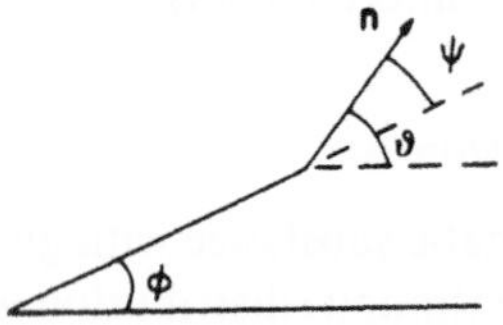

FIG. 30

The polar angle ϕ is measured from some chosen polar axis in the plane. We shall use also the angle ϑ between **n** and the polar axis; evidently $\vartheta = \phi + \psi$.

The required solution is given by the function $\psi(\phi)$. It must satisfy the condition of physical uniqueness: when ϕ changes by 2π (that is, in a passage round the origin), **n** must be unchanged apart from sign (a change of sign is permissible, since the directions **n** and $-\mathbf{n}$ are physically equivalent). This means that we must have

$$\vartheta(\phi+2\pi) = \vartheta(\phi)+2\pi n,$$

where n is a positive or negative integer or half-integer; $n = 0$ corresponds to the "undeformed" state $\mathbf{n}$ = constant. Hence $\psi(\phi) = \vartheta-\phi$ is such that

$$\psi(\phi+2\pi) = 2\pi(n-1)+\psi(\phi). \tag{38.1}$$

The number n is called the *Frank index* of the disclination.

The equation of equilibrium (to be written out below) determines the derivative $d\psi/d\phi$:

$$d\psi/d\phi = 1/f(\psi); \tag{38.2}$$

the right-hand side does not contain the independent variable ϕ, because the equation must be invariant under any rotation of the whole nematic system about the z-axis (i.e. a transformation $\phi \to \phi+\phi_0$). The function $f(\psi)$ is periodic, with period π, since the values ψ and $\psi+\pi$ are physically identical. Hence

$$\phi = \int_0^{\psi} f(x)\,dx, \tag{38.3}$$

the constant of integration being chosen so that $\psi = 0$ when $\phi = 0$. Substitution in (38.1) gives

$$\bar{f} \equiv \frac{1}{\pi}\int_0^{\pi} f(x)\,dx = \frac{1}{n-1}, \tag{38.4}$$

if $n \neq 1$; the bar denotes averaging over the period of the function.

From this, we can draw an important conclusion as to the symmetry of the disclination: when the whole picture is rotated through $\phi_0 = 2\pi/2(n-1)$ about the z-axis, the angle ψ changes by π, the distribution thus remaining unchanged: when the periodicity of $f(\psi)$ is taken into account, this transformation gives the identity

$$\phi+\frac{\pi}{n-1} = \int_0^{\psi+\pi} f(x)\,dx = \int_0^{\psi} f(x)\,dx + \int_{\psi}^{\psi+\pi} f(x)\,dx = \phi+\bar{f}\pi.$$

Thus, simply from the condition of uniqueness, the z-axis is necessarily a symmetry axis C_m whose order is

$$m = 2|n-1|, \quad n \neq 1. \tag{38.5}$$

The "streamlines" of the director are such that at each point the element of length $d\mathbf{l}$ ($dl_r = dr$, $dl_\phi = r\,d\phi$) is parallel to **n**. The differential equation of these lines is

$$dl_\phi/dl_r = n_\phi/n_r,$$

or

$$d\phi/d\log r = \tan\psi. \tag{38.6}$$

From this we see, in particular, that the streamlines include some that are straight, with $\psi = p\pi$ and p an integer. These are $2|n-1|$ radii

$$\phi = \frac{\pi}{|n-1|}p \equiv \phi_p, \quad \psi = p\pi, \quad p = 0, 1, 2, \ldots, m-1. \tag{38.7}$$

The cross-section plane of the disclination is divided by these radii into m equal sectors.

Let us now go on to the specific solution for a nematic whose deformation energy is given by (37.1).†

For a two-dimensional distribution,

$$\operatorname{div} \mathbf{n} = \frac{1}{r}\frac{dn_\phi}{d\phi} + \frac{n_r}{r} = \frac{1}{r}\cos\psi \cdot (1 + \psi'),$$

$$\operatorname{curl}_z \mathbf{n} = -\frac{1}{r}\frac{dn_\phi}{d\phi} + \frac{n_\phi}{r} = \frac{1}{r}\sin\psi \cdot (1 + \psi'),$$

$$\mathbf{n} \cdot \mathbf{curl}\, \mathbf{n} = 0,$$

with $\psi' \equiv d\psi/d\phi$. In the free energy only the K_1 and K_3 terms remain:‡

$$\int F_d r\, dr d\phi = \tfrac{1}{4}(K_1 + K_3) \int (1 - \alpha \cos 2\psi)(1 + \psi'^2)\, d\phi\, dr/r,$$

$$\alpha = (K_3 - K_1)/(K_3 + K_1).$$

The integral with respect to r is logarithmically divergent. In actual problems it is cut off above at some distance R that is of the order of the dimensions of the sample, and below at distances of the order of the molecular dimensions a, where the macroscopic theory ceases to be valid. To determine the required solution at distances $a \ll r \ll R$, we can take the factor

$$L = \int dr/r \simeq \log(R/a)$$

to be a constant simply, so that the equilibrium distribution $\psi(\phi)$ is found by minimizing the functional

$$\int_0^{2\pi} (1 - \alpha \cos 2\psi)(1 + \psi'^2)\, d\phi = \text{minimum}. \tag{38.8}$$

Euler's equation for this variational problem is

$$(1 - \alpha \cos 2\psi)\, \psi'' = \alpha \sin 2\psi\, (1 - \psi'^2). \tag{38.9}$$

This has, first of all, the two obvious solutions

$$\psi = 0 \tag{38.10}$$

and

$$\psi = \tfrac{1}{2}\pi. \tag{38.11}$$

These are axially symmetrical solutions corresponding to Figs. 29a and 29b respectively.§ They are single-valued; that is, the Frank index for these disclinations is $n = 1$ (cf. (38.1)).

† This problem was solved by C. W. Oseen (1933) and F. C. Frank (1958) for the particular case of a nematic with $K_1 = K_3$. The general solution given below is due to I. E. Dzyaloshinskiĭ (1970).

‡ The integrand does not include the total derivative $(1 - \alpha \cos 2\psi) 2\psi' = (2\psi - \alpha \sin 2\psi)'$; this does not affect the formulation of the variational problem. We shall here derive the equation of equilibrium afresh, without reverting to the general equations (37.7), (37.8), which would in practice call for a more laborious calculation.

§ In the "degenerate" case $K_1 = K_3$, $\alpha = 0$, there are solutions with ψ = any constant.

To find solutions with $n \neq 1$, we note that (38.9) has a first integral†

$$(1-\alpha\cos 2\psi)(\psi'^2-1) = \text{constant} \equiv \frac{1}{q^2}-1. \tag{38.12}$$

From this, the solution has the form (38.3), with

$$f(x) = q\left[\frac{1-\alpha\cos 2\psi}{1-\alpha q^2\cos 2\psi}\right]^{\frac{1}{2}}. \tag{38.13}$$

The constant q is found from the condition (38.4):

$$(n-1)q\int_0^\pi\left[\frac{1-\alpha\cos 2\psi}{1-\alpha q^2\cos 2\psi}\right]^{\frac{1}{2}}d\psi = \pi; \tag{38.14}$$

here, we must have $|\alpha|q^2 < 1$. These formulae give the required solution. The solution is unique for each n; since the left-hand side of (38.14) increases monotonically with q, the equation is satisfied by only one value of q. Since $f(x)$ is even, $\phi(\psi)$ is odd. The plane $\phi = 0$ is therefore a plane of symmetry for the distribution; since there is a symmetry axis C_m, there are consequently another $m-1$ planes of symmetry passing through the z-axis. Lastly, $z = 0$ is evidently a plane of symmetry. Thus a disclination with index n has the full symmetry of the point group $\boldsymbol{D}_{mh}$.

When $n = 2$, it is obvious from (38.14) that $q \doteq 1$, and the corresponding solution is simply

$$\psi = \phi = \tfrac{1}{2}\vartheta. \tag{38.15}$$

To determine the qualitative nature of the solutions found, let us examine the behaviour of the streamlines near the radii $\phi = \phi_p$ (38.7). On these radii, $\psi = p\pi$, and near them $\psi \cong p\pi$; the function (38.13) becomes a constant:

$$\frac{d\phi}{d\psi} = f(\psi) \cong q\left(\frac{1-\alpha}{1-\alpha q^2}\right)^{\frac{1}{2}} \equiv \lambda. \tag{38.16}$$

Hence

$$\psi - \pi p \cong (\phi-\phi_p)/\lambda.$$

The differential equation of the streamlines is

$$\frac{d\log r}{d\phi} = \cot\psi \cong \frac{1}{\psi-\psi_p} \cong \frac{\lambda}{\phi-\phi_p},$$

from which we find the streamlines near the radius:

$$r = \text{constant}\ \times|\phi-\phi_p|^\lambda. \tag{38.17}$$

With Cartesian coordinates and the x-axis along the radius, we have near the latter $r \cong x$, $\phi-\phi_p \cong y/x$; the streamline equation becomes

$$y = \text{constant}\times x^{1+1/\lambda}. \tag{38.18}$$

† If the integrand in (38.8) is regarded as the Lagrangian of a one-dimensional mechanical system (with ψ as the generalized coordinate and ϕ as the time), then (38.12) is the energy integral.

Various cases have now to be considered. When $n \geqslant \frac{3}{2}$, $n-1>0$, and from (38.14) $q > 0$, so that $\lambda > 0$. In this case, the streamlines start from the origin and the radius is a tangent to them.

When $n = \frac{1}{2}$, $q < 0$, and so $\lambda < 0$. A numerical analysis of (38.14) shows that $q^2 > 1$ and therefore $|\lambda| > 1$. From (38.18), y increases with x. The region near the origin cannot be dealt with in this way, since according to (38.17), when $\lambda < 0$, small values of $\phi - \phi_p$ correspond to large r.

Lastly, when $n < 0$, $-1 < \lambda < 0$, and from (38.18) $y \to 0$ as $x \to \infty$; the streamlines approach the radii asymptotically.

Figure 31 shows schematically the streamlines for disclinations with $n = \frac{3}{2}, \frac{1}{2}$ and $-\frac{1}{2}$.

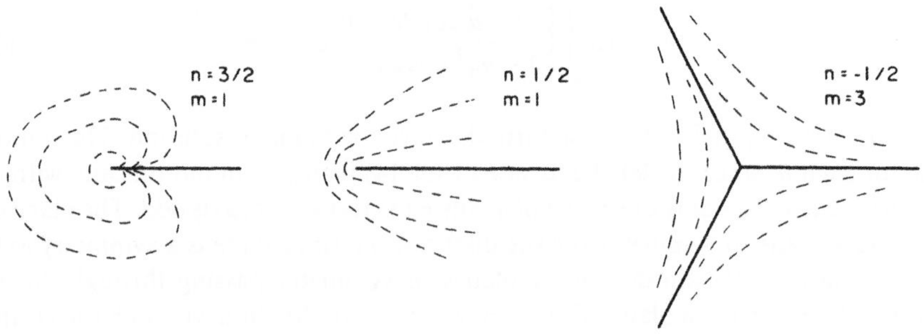

FIG. 31

§39. Non-singular axially symmetrical solution of the equilibrium equations for a nematic

The axially symmetrical deformations (38.10), (38.11) (Fig. 29) are disclinations with Frank index $n = 1$ and are exact solutions of the equations of equilibrium for a nematic medium with specified boundary conditions at the walls of the container. They are not, however, the only solutions of such problems. They are unique only as two-dimensional solutions. If we abandon the hypothesis that the vectors **n** are everywhere in planes transverse to the axis of the vessel, other solutions are possible which do not have a singularity on the axis. For example, if the boundary conditions are such that **n** must be perpendicular to the wall, the director streamlines in such a singularity-free solution are in meridional planes as shown in Fig. 32. At the wall, they leave at right angles, then bend towards the axis $r = 0$, on which **n** therefore has a quite definite direction. Moreover, we shall see that the absence of singularities from such a solution makes it thermodynamically more favourable (the total elastic free energy is less) than one with a singularity on the axis (P. E. Cladis and M. Kléman 1972). Let us now proceed to construct this solution.

We shall seek a solution that is axially symmetrical and uniform along the z-axis in cylindrical polar coordinates:

$$n_r = \cos\chi(r), \quad n_\phi = 0, \quad n_z = \sin\chi(r); \tag{39.1}$$

the angle χ is as shown in Fig. 32. The boundary condition at the wall is

$$\chi = 0 \quad \text{for} \quad r = R \tag{39.2}$$

where R is the radius of the cylindrical vessel, and on the axis we impose the condition

$$\chi = \tfrac{1}{2}\pi \quad \text{for} \quad r = 0, \tag{39.3}$$

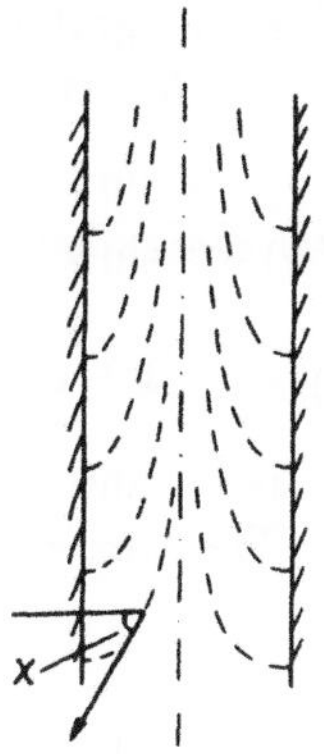

FIG. 32

which, as already shown, corresponds to the absence of any singularity. We have

$$\mathrm{curl}_\phi \mathbf{n} = -\mathrm{d}n_z/\mathrm{d}r = -\cos\chi\cdot\mathrm{d}\chi/\mathrm{d}r,$$

$$\mathrm{div}\,\mathbf{n} = \frac{1}{r}\frac{\mathrm{d}(rn_r)}{\mathrm{d}r} = -\sin\chi\frac{\mathrm{d}\chi}{\mathrm{d}r} + \frac{\cos\chi}{r}.$$

The free energy of the deformation per unit length along the z-axis is

$$\int_0^R F_d\cdot 2\pi r\,\mathrm{d}r = \pi\int_{-\infty}^{\log R}\{(K_1\sin^2\chi + K_3\cos^2\chi)\chi'^2 + K_1\cos^2\chi - K_1\sin 2\chi\cdot\chi'\}\mathrm{d}\xi, \tag{39.4}$$

where the prime denotes differentiation with respect to the variable $\xi = \log r$.†

The first integral of the equilibrium equation (that is, Euler's equation for the variational problem of finding the minimum of the functional (39.4)) is

$$(K_1\sin^2\chi + K_3\cos^2\chi)\chi'^2 - K_1\cos^2\chi = \text{constant}. \tag{39.5}$$

According to the condition (39.3), we must have $\chi\to\frac{1}{2}\pi$ and $\xi\to-\infty$. This evidently implies that $\chi'\to 0$ as $\chi\to\frac{1}{2}\pi$; the constant is therefore zero, so that

$$\chi' = -\frac{\sqrt{K_1}\cos\chi}{\sqrt{(K_1\sin^2\chi + K_3\cos^2\chi)}}.$$

From this we get the required solution, satisfying the condition (39.2), as

$$\log(R/r) = \frac{1}{\sqrt{K_1}}\int_0^\chi \frac{\sqrt{(K_1\sin^2\chi + K_3\cos^2\chi)}}{\cos\chi}\mathrm{d}\chi. \tag{39.6}$$

In contrast to the disclination (38.10), this solution is not self-similar, since it involves the

† The last term in the integrand is unimportant in stating the variational problem, but is needed when calculating the total free energy.

dimensional length parameter R. The integral (39.6) can be expressed in terms of elementary functions. The result, if $K_3 > K_1$, is

$$\left.\begin{aligned} r/R &= \sqrt{\left\{\frac{\sqrt{(1-k^2\sin^2\chi)}-k'\sin\chi}{\sqrt{(1-k^2\sin^2\chi)}+k'\sin\chi}\right\}\exp\left\{-\frac{k}{k'}\sin^{-1}(k\sin\chi)\right\}}, \\ k^2 &= (K_3-K_1)/K_3, \quad k'^2 = 1-k^2 = K_1/K_3. \end{aligned}\right\} \tag{39.7}$$

When $r \to 0$, $\frac{1}{2}\pi - \chi$ tends to zero as r, and the streamlines approach the z-axis exponentially: $r \propto \exp(\text{constant} \times z)$. The free energy associated with this solution is found to be

$$\int_0^R F_d \cdot 2\pi r\,\mathrm{d}r = \pi K_1\left\{2 + \frac{1}{kk'}\sin^{-1}k\right\}. \tag{39.8}$$

This is independent of the radius R of the vessel. The energy of the disclination in Fig. 29a (the solution (38.10)) is

$$\int_0^R F_d \cdot 2\pi r\,\mathrm{d}r = \pi K_1 L, \tag{39.9}$$

where $L = \log(R/a)$ is a large logarithm arising from the singularity on the axis. We see that the solution without a singularity is energetically more favourable than the other, unless K_1 is unusually small.

The field $\mathbf{n}(\mathbf{r})$ of this axially symmetrical non-singular solution of the equations of equilibrium can be derived from that of a disclination with $n = 1$ by a deformation that is continuous, i.e. does not involve tearing, the vectors $\mathbf{n}$ being gradually brought away from the planes z = constant. This is one case of a very general situation to be discussed in §40.

PROBLEMS

PROBLEM 1. Find the axially symmetrical solution of the equations of equilibrium for a nematic medium in a cylindrical vessel, having no singularity on the axis and corresponding to the boundary conditions in Fig. 29b.

SOLUTION. We seek the solution in the form

$$n_r = 0, \quad n_\phi = \cos\chi(r), \quad n_z = \sin\chi(r)$$

with the boundary conditions

$$\chi(R) = 0, \quad \chi(0) = \tfrac{1}{2}\pi.$$

Then

$$\begin{aligned} \operatorname{curl}_\phi \mathbf{n} &= -\cos\chi\,\mathrm{d}\chi/\mathrm{d}r, \\ \operatorname{curl}_z \mathbf{n} &= (1/r)\cos\chi - \sin\chi\,\mathrm{d}\chi/\mathrm{d}r, \\ \operatorname{div}\mathbf{n} &= 0. \end{aligned}$$

The free energy is

$$\int_0^R 2\pi r F_d\,\mathrm{d}r = \pi\int_{-\infty}^{\log R}\{K_2(\sin\chi\cos\chi - \chi')^2 + K_3\cos^4\chi\}\,\mathrm{d}\xi.$$

The first integral of the equilibrium equation is

$$K_2\chi'^2 - (K_2\sin^2\chi\cos^2\chi + K_3\cos^4\chi) = 0.$$

Integration of this gives (if $K_3 > K_2$)

$$r/R = \sqrt{\left\{\frac{\sqrt{(1-k^2\sin^2\chi)}-k'\sin\chi}{\sqrt{(1-k^2\sin^2\chi)}+k'\sin\chi}\right\}},$$

$$k^2 = (K_3 - K_2)/K_3, k'^2 = K_2/K_3.$$

As $r \to 0$, $\chi \to \frac{1}{2}\pi$ in the manner $\frac{1}{2}\pi - \chi = 2k'r/R$.

The free energy of this deformation is

$$\int_0^R F_d \cdot 2\pi r\,dr = \pi K_2\left\{2 + \frac{1}{kk'}\sin^{-1}k\right\},$$

whereas that of the two-dimensional disclination in Fig. 29b is $\pi K_3 L$.

PROBLEM 2. Examine the stability of disclinations with $n = 1$ with respect to small perturbations having the form $\delta\mathbf{n}(\phi)$ (S. I. Anisimov and I. E. Dzyaloshinskiĭ 1972).

SOLUTION. (a) The unperturbed field of a radial discliniation (Fig. 29a) is $n_r = 1, n_\phi = n_z = 0$. The perturbed field will be written as

$$n_r = \cos\Theta\cos\Phi \cong 1 - \tfrac{1}{2}(\Theta^2 + \Phi^2),$$

$$n_\phi = \cos\Theta\sin\Phi \cong \Phi,$$

$$n_z = \sin\Theta \cong \Theta.$$

where the angles Θ and Φ are functions of the angle coordinate ϕ. The energy associated with this perturbation is

$$\int F_d r\,dr\,d\phi = \tfrac{1}{4}R^2\int\{K_1\Phi'^2 + K_2\Theta'^2 + (K_3 - K_1)\Phi^2 - K_1\Theta^2\}d\phi.$$

For a general analysis, we should have to put

$$\Theta(\phi) = \sum_{s=-\infty}^{\infty}\Theta_s e^{is\phi}, \qquad \Phi(\phi) = \sum_{s=-\infty}^{\infty}\Phi_s e^{is\phi}$$

and express the energy as a function of all the Θ_s and Φ_s. It is, however, immediately obvious that the disclination in question is always unstable with respect to the perturbation Θ_0, because of the term $-K_1\Theta_0^2$ in the energy.

(b) The unperturbed field of a circular disclination (Fig. 29b) is $n_r = n_z = 0, n_\phi = 1$. We write the perturbed field as

$$n_r = \cos\Theta\cos(\tfrac{1}{2}\pi + \Phi) \cong -\Phi,$$

$$n_\phi = \cos\Theta\sin(\tfrac{1}{2}\pi + \Phi) \cong 1 - \tfrac{1}{2}(\Theta^2 + \Phi^2),$$

$$n_z = \sin\Theta \cong \Theta;$$

here, the definition of Φ is different from that in the preceding case. The corresponding energy is

$$\int F_d r\,dr\,d\phi = \tfrac{1}{4}R^2\int\{K_3(\Theta'^2 + \Phi'^2) + (K_1 - K_3)\Phi^2 + (K_2 - 2K_3)\Theta^2\}\,d\phi.$$

The most "dangerous" perturbations are Θ_0 and Φ_0; the stability conditions for these are

$$K_1 > K_3, \quad K_2 > 2K_3.$$

In itself, the result in the text and in Problem 1 that the free energy of the deformation in disclinations with $n = 1$ exceeds that of the non-singular axially symmetrical solution signifies only that these disclinations are at best metastable. We now see that the radial disclination is altogether unstable, and the circular one is stable (as regards perturbations of the type considered) only when certain relations exist between the elastic moduli.

PROBLEM 3. A nematic medium occupies the space between two parallel planes; the boundary conditions require the director to be perpendicular to one plane and parallel to the other. Determine the equilibrium configuration $\mathbf{n}(\mathbf{r})$.

SOLUTION. The equilibrium configuration is evidently two-dimensional; we take the relevant plane as the xz-plane, with the z-axis perpendicular to the boundary planes ($z = 0$ and $z = h$). We put

$$n_x = \sin\chi(z), \quad n_z = \cos\chi(z).$$

The free energy of the deformation is

$$\int F_d \, dz = \tfrac{1}{2} \int \{K_1 \sin^2 \chi + K_2 \cos^2 \chi\} \chi'^2 \, dz.$$

The first integral of the equilibrium equation is

$$(K_1 \sin^2 \chi + K_2 \cos^2 \chi)\chi'^2 = \text{constant},$$

whence, with the boundary conditions,

$$\int_0^{\chi} \sqrt{(K_1 \sin^2 \chi + K_2 \cos^2 \chi)} \, d\chi = (z/h) \int_0^{\frac{1}{2}\pi} \sqrt{(K_1 \sin^2 \chi + K_2 \cos^2 \chi)} \, d\chi,$$

or

$$z = hE(\chi, k)/E(\tfrac{1}{2}\pi, k), \quad k^2 = (K_2 - K_1)/K_1,$$

where $E(\chi, k)$ is an elliptic integral of the second kind.

§40. Topological properties of disclinations

The definition of the Frank index given in §38 depended essentially on the assumption that the disclination deformation is two-dimensional and is uniform along the disclination. We shall now show how this concept can be used in the general case of any curved disclinations in a nematic medium.

The energy of the nematic is not affected by a simultaneous arbitrary rotation of the director at every point. In this sense, we can say that the states of the nematic are degenerate with respect to the directions of the director, which are referred to as a *degeneracy parameter*. We can define *degeneracy space* as the range of variation of the degeneracy parameter that can occur without a change in energy. In the present case, this is the surface of a sphere with unit radius, each point of which corresponds to a particular direction of **n**. Here, however, we must also take into account that states of a nematic that differ by a change in the sign of **n** are physically identical. That is, diametrically opposite points on the sphere are physically equivalent. The degeneracy space of the nematic is therefore a sphere on which every pair of opposite points are regarded as equivalent.†

Let us imagine that, in the physical volume of the nematic, we pass along a closed contour γ round a disclination line. We trace this passage in terms of the direction of **n**. The point representing it in the spherical degeneracy space describes another closed contour Γ. Two cases are to be distinguished here.

In one case, Γ is literally closed. In returning to its original position, the point describes an integral number n of loops (for instance, $n = 1$ and 2 for the contours Γ_1 and Γ_2 in Fig. 33). This number is the integral Frank index.

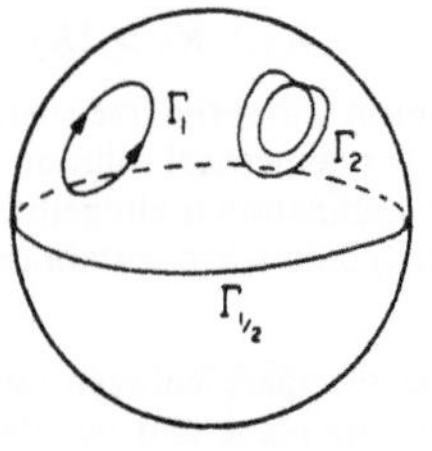

FIG. 33

† In topology, this geometrical picture corresponds to what is called a *projective plane*.

In the other case, Γ starts from a point on the sphere and ends at the diametrically opposite point. Such a contour also is to be regarded as closed, since diametrically opposite points are equivalent. The Frank index is defined as the half-integral number of "loops" then described by the point (for instance, $n = \frac{1}{2}$ for the semicircle $\Gamma_{\frac{1}{2}}$).

Any closed contour on a sphere can be transformed into any other by a continuous deformation (i.e. one that does not break the contour). Moreover, any closed contour can be continuously shrunk to a point.†

It is also possible to transform into one another any contours which begin and end at diametrically opposite points on the sphere. Such contours cannot be shrunk to a point, however: when deformation occurs, the ends of the contour may move, but must remain at the two ends of the some diameter of the sphere.

The Frank index is therefore not a topological invariant. Only its being integral or half-integral has this property.

It follows from the above that all disclinations in a nematic medium fall into two categories, each containing topologically equivalent disclinations which can be converted into one another by a continuous deformation of the field **n**(**r**) (S. I. Anisimov and I. E. Dzyaloshinskiĭ 1972). One category includes disclinations with integral Frank indices, which are topologically unstable and can be removed by a continuous deformation. Disclinations with integral index may terminate within the nematic.

The other category consists of disclinations with half-integral indices. These disclinations are not removable and are topologically stable.

The question of which of the topologically equivalent structures will in fact occur under any specified conditions depends on the relative thermodynamic favourability of these structures and is therefore outside the range of a topological analysis.

There can be point singularities in a nematic medium, as well as the linear disclination singularities. The simplest example is a point from which the vectors **n** radiate in all directions (a "hedgehog").

To determine the topological classification of point singularities, we again use the mapping on a unit sphere as degeneracy space. In the physical space occupied by the nematic, we take two points A and B joined by a contour γ surrounding the singularity O (Fig. 34). The contour γ corresponds to a certain contour Γ on the unit sphere, If now γ is rotated about the straight line AB, it describes a closed surface σ in physical space during a complete rotation back to its original position. The image Σ of σ, described by Γ, covers the unit sphere, possibly more than once. The number N of times it does so is a topological characteristic of the singular point. It is possible to regard Σ as a closed film drawn over the sphere, which evidently cannot, without cutting it in some way, be shrunk to a point; this

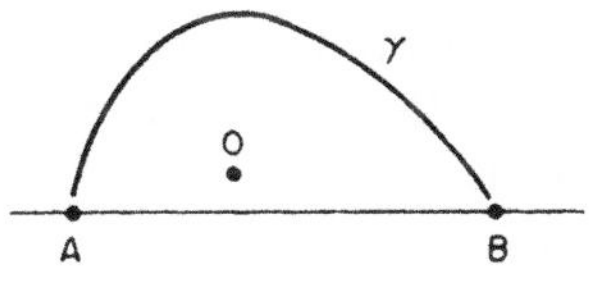

FIG. 34

† A deformation of the contour may represent either a change in the contour γ in physical space or a change in the field **n**(**r**) itself.

corresponds to the irremovability of the singularity. The value $N = 0$ (incomplete covering) corresponds to an absent or removable singularity: a film that does not completely cover the sphere can be shrunk to a point on the sphere. For singular points in a nematic, the sign of N has no significance; if the sign is changed, the directions of $\mathbf{n}$ are simply reversed in all space, and this does not affect the state of the nematic.

The number N corresponding to a point singularity can only be integral. It is easy to see that a half-integral N would in fact signify the presence of an irremovable line, not point, singularity. If Σ covers half the sphere ($N = \frac{1}{2}$), this means that, if we follow any one point on γ, we find that its image describes a contour on the sphere like $\Gamma_{\frac{1}{2}}$ in Fig. 33, which would indicate the presence of an irremovable disclination with Frank index $n = \frac{1}{2}$.†

In connection with this discussion of the topological properties of singularities in nematics, let us briefly consider the topological interpretation of dislocations, i.e. line singularities in crystal lattices. We take an infinite lattice with the x_1, x_2, x_3 axes along the three basic lattice periods a_1, a_2, a_3. The lattice energy is unchanged by parallel displacements through any distances along the axes. The ranges of variation of the degeneracy parameters (amounts of displacement) are segments with length a_1, a_2, a_3, for each of which the two end-points are regarded as equivalent, because a displacement by one period leaves the lattice in the same position and therefore in identically the same state. A segment with equivalent ends is topologically the same as a circle. Thus the degeneracy space of the lattice is a three-dimensional region based on three circles. This region can be regarded as a cube with pairs of opposite faces equivalent or, alternatively, as the three-dimensional surface of a torus in four-dimensional space.‡ On such a torus there are contours Γ that cannot be shrunk to a point, each of which is described by three integral topological invariants n_1, n_2, n_3, the numbers of passages around the three circular generators of the torus. If Γ is the image of a contour γ which in physical space passes round a singular line (a dislocation), then its three invariants are the three components of the Burgers vector measured in units of the corresponding periods a_1, a_2, a_3. Thus dislocations are topologically stable irremovable singular lines, and their Burgers vectors are topological invariants.

§41. Equations of motion of nematics

The state of a nematic medium in motion is defined by the spatial distribution of four quantities: the director $\mathbf{n}$, the mass density ρ, the velocity $\mathbf{v}$, and the entropy density S. Accordingly, the complete system of hydrodynamic equations of motion of a nematic consists of four equations giving the time derivatives of these (J. L. Ericksen 1960, F. M. Leslie 1966, T. C. Lubensky, P. C. Martin, J. Swift and P. S. Pershan 1971).

Let us begin with the equation for the director. If the nematic is in equilibrium, so that $\mathbf{h} = 0$, and moves as a whole with a velocity constant in space, then this equation simply expresses the fact that the values of $\mathbf{n}$ are transported in space at the same velocity. That is, each liquid particle moves in space with its own fixed $\mathbf{n}$. This is expressed by equating to zero the total or substantial time derivative:

$$\frac{d\mathbf{n}}{dt} = \frac{\partial \mathbf{n}}{\partial t} + (\mathbf{v} \cdot \nabla)\mathbf{n} = 0. \tag{41.1}$$

† A corresponding conclusion does not follow from similar arguments when N is integral, since a disclination with integral index is removable, and the image with integral N corresponds to an irremovable singularity.

‡ Just as a square with pairs of opposite sides equivalent is topologically the same as the two-dimensional surface of a torus in three-dimensional space.

In the general case of any motion, the right-hand side contains terms which depend on **h** and on the space derivatives of the velocity; in the first non-vanishing hydrodynamic approximation, we take just the terms linear in these quantities. The derivatives $\partial v_i/\partial x_k$ form a tensor, which may be divided into symmetrical and antisymmetrical parts:

$$v_{ik} = \tfrac{1}{2}(\partial_i v_k + \partial_k v_i), \quad \Omega_{ik} = \tfrac{1}{2}(\partial_i v_k - \partial_k v_i). \tag{41.2}$$

To determine the dependence on $\mathbf{\Omega}_{ik}$, it is sufficient to note that, in uniform rotation of the nematic as a whole with angular velocity $\mathbf{\Omega}$, the entire field $\mathbf{n}(\mathbf{r})$ rotates with that velocity. Such a rotation is represented by

$$\mathrm{d}\mathbf{n}/\mathrm{d}t = \tfrac{1}{2}\,\mathbf{curl}\,\mathbf{v}\times\mathbf{n} \quad \text{or} \quad \mathrm{d}n_i/\mathrm{d}t = \Omega_{ki} n_k:$$

the velocity of points in a body rotating as a whole is $\mathbf{v} = \mathbf{\Omega}\times\mathbf{r}$, and so $\mathbf{curl}\,\mathbf{v} = 2\mathbf{\Omega}$, and the rate of change of the director is given by a similar expression, $\mathrm{d}\mathbf{n}/\mathrm{d}t = \mathbf{\Omega}\times\mathbf{n}$. The terms depending on v_{ik} are subject to the condition $\mathbf{n}\cdot\mathrm{d}\mathbf{n}/\mathrm{d}t = 0$, since $\mathbf{n}^2 = 1$ is constant. We thus arrive at the following general form of the equation of motion of the director:

$$\mathrm{d}n_i/\mathrm{d}t = \Omega_{ki} n_k + \lambda(\delta_{il} - n_i n_l) n_k v_{kl} + N_i, \tag{41.3}$$

where†

$$\mathbf{N} = \mathbf{h}/\gamma. \tag{41.4}$$

The term **N** represents the relaxation of the director towards equilibrium under the action of the molecular field; the second term in (41.3) gives the orienting effect of the velocity gradient on the director. The coefficients γ, with the dimensions of viscosity, and λ, dimensionless, in these terms are kinetic, not thermodynamic, coefficients.‡

The equation for the time derivative of the liquid density is the continuity equation

$$\partial\rho/\partial t + \mathrm{div}\,(\rho\mathbf{v}) = 0. \tag{41.5}$$

This essentially determines the hydrodynamic velocity as the material flux density per unit mass.

The equation for the time derivative of the velocity is the dynamical equation

$$\rho\,\mathrm{d}\mathbf{v}/\mathrm{d}t = \mathbf{F}, \tag{41.6}$$

where **F** is the force on unit volume. In accordance with the general arguments in §2, body forces can be written as a tensor divergence:

$$F_i = \partial_k \sigma_{ik},$$

where σ_{ik} is the stress tensor. The dynamical equation then becomes

$$\rho\frac{\mathrm{d}v_i}{\mathrm{d}t} = \rho\left[\frac{\partial v_i}{\partial t} + (\mathbf{v}\cdot\nabla)v_i\right] = \partial_k \sigma_{ik}. \tag{41.7}$$

The form of the stress tensor will be established later.

Lastly, there is an equation for the entropy. In the absence of dissipative processes, the motion of the liquid would be adiabatic, and would be so in each volume element, which

† The notation **N** is used to show more clearly the structure of some of the following formulae, and with a view to further generalizations in §44.

‡ The absence of terms containing the gradients of density and entropy (or temperature) on the right of (41.3) is due to the necessary invariance under spatial inversion and under a change in the sign of **n**. This is further discussed in §44.

would move with a constant entropy. The entropy conservation equation would be simply the entropy continuity equation:

$$\partial S/\partial t + \operatorname{div}(S\mathbf{v}) = 0,$$

where S is the entropy per unit volume and $S\mathbf{v}$ the entropy flux density.† When dissipative processes are included, the entropy equation becomes

$$\frac{\partial S}{\partial t} + \operatorname{div}(S\mathbf{v} + \mathbf{q}/T) = 2R/T. \tag{41.8}$$

Here R is the dissipative function, and $2R/T$ gives‡ the rate of increase of entropy; it is a quadratic form in the components of the tensor v_{ik}, the vector $\mathbf{h}$, and the temperature gradient vector ∇T. The vector $\mathbf{q}$ is the heat flux density, related to the thermal conductivity. The components of this vector are linear functions of the temperature gradient components:

$$q_i = -\kappa_{ik}\partial_k T. \tag{41.9}$$

In a nematic medium, the thermal conductivity coefficient tensor κ_{ik} has two independent components and may be put in the form

$$\kappa_{ik} = \kappa_{\parallel} n_i n_k + \kappa_{\perp}(\delta_{ik} - n_i n_k), \tag{41.10}$$

where $\kappa_{\parallel}$ and $\kappa_{\perp}$ describe the thermal conductivity in the directions longitudinal and transverse relative to $\mathbf{n}$.

The energy conservation law in hydrodynamics is expressed by

$$\frac{\partial}{\partial t}(\tfrac{1}{2}\rho v^2 + E) + \operatorname{div}\mathbf{Q} = 0, \tag{41.11}$$

where E is the internal energy density and $\mathbf{Q}$ the energy flux density. The energy density is $E = E_0 + E_d$, where $E_0(\rho, S)$ relates to the undeformed homogeneous medium and E_d is due to the distortion of the field $\mathbf{n}(\mathbf{r})$. According to the remark at the end of §37, E_d is the same as the free energy F_d (37.1), except that the elastic moduli K_1, K_2, K_3 are supposed to be expressed in terms of the density and the entropy, not the temperature.

The energy conservation law is, of course, contained in the equations of motion. We shall use it to establish the relation between the function R, the tensor σ_{ik} and the vector $\mathbf{N}$ defined above.

We expand the time derivative in (41.11), using the thermodynamic relations

$$(\partial E/\partial S)_{\rho,\mathbf{n}} = T, \quad (\partial E/\partial \rho)_{S,\mathbf{n}} = \mu,$$

† This equation can be put in the equivalent form

$$\frac{\mathrm{d}}{\mathrm{d}t}(S/\rho) = \frac{\partial}{\partial t}(S/\rho) + (\mathbf{v}\cdot\nabla)(S/\rho) = 0,$$

which expresses the constancy of the entropy per unit mass transported by the liquid particles.

‡ $2R$ itself is, as in §34, the rate of dissipation of mechanical energy (*FM*, §79).

where μ is the chemical potential.† Then

$$\frac{\partial}{\partial t}(\tfrac{1}{2}\rho v^2+E)=\tfrac{1}{2}v^2\frac{\partial \rho}{\partial t}+\rho\mathbf{v}\cdot\frac{\partial \mathbf{v}}{\partial t}+\mu\frac{\partial \rho}{\partial t}+T\frac{\partial S}{\partial t}+\left(\frac{\partial E_d}{\partial t}\right)_{\rho,S} \tag{41.12}$$

Let us consider separately the last term. With Π_{ki} from (37.6), we write

$$\begin{aligned}\left(\frac{\partial E_d}{\partial t}\right)_{\rho,S}&=\left(\frac{\partial E_d}{\partial n_i}\right)_{\rho,S}\frac{\partial n_i}{\partial t}+\Pi_{ki}\partial_k\frac{\partial n_i}{\partial t}\\&=\left\{\frac{\partial E_d}{\partial n_i}-\partial_k\Pi_{ki}\right\}\frac{\partial n_i}{\partial t}+\partial_k\left(\Pi_{ki}\frac{\partial n_i}{\partial t}\right)\\&=-\mathbf{h}\cdot\partial\mathbf{n}/\partial t+\partial_k(\Pi_{ki}\partial n_i/\partial t);\end{aligned}$$

here we replace $\mathbf{H}$ by $\mathbf{h}$, since the longitudinal part of $\mathbf{H}$ disappears at once by virtue of the equation $\mathbf{n}\cdot\partial\mathbf{n}/\partial t=0$. Substituting $\partial\mathbf{n}/\partial t$ from (41.3), we write

$$\left(\frac{\partial E_d}{\partial t}\right)_{\rho,S}=(v_k\partial_k n_i+\Omega_{ik}n_k-\lambda v_{ik}n_k)h_i-\mathbf{N}\cdot\mathbf{h}+\operatorname{div}(\ldots)$$

and, separating another total divergence,

$$\left(\frac{\partial E_d}{\partial t}\right)_{\rho,S}=-\mathbf{G}\cdot\mathbf{v}-h^2/\gamma+\operatorname{div}(\ldots), \tag{41.13}$$

where

$$G_i=-h_k\partial_i n_k+\tfrac{1}{2}\partial_k(n_ih_k-n_kh_i)-\tfrac{1}{2}\lambda\partial_k(n_ih_k+n_kh_i). \tag{41.14}$$

Here and henceforward, the total divergences are not written in full, so as to make the formulae less complicated; these terms are not important in solving the problem stated, though we shall return to them at the end of the section.

The expression (41.14) can be written as

$$G_i=\partial_k\sigma_{ik}{}^{(r)}+(\partial_i E_d)_{\rho,S}, \tag{41.15}$$

where

$$\sigma_{ik}{}^{(r)}=-\Pi_{kl}\partial_i n_l-\tfrac{1}{2}\lambda(n_ih_k+n_kh_i)+\tfrac{1}{2}(n_ih_k-n_kh_i). \tag{41.16}$$

The transformation makes use of the equation

$$(\partial_i E_d)_{\rho,S}=(\partial E_d/\partial n_k)\partial_i n_k+\Pi_{lk}\partial_i\partial_l n_k.$$

The definition of the tensor $\sigma_{ik}{}^{(r)}$ is not unique: the expression (41.15) is unchanged if we add to $\sigma_{ik}{}^{(r)}$ any term $\partial_l\chi_{ilk}$, where χ_{ilk} is any tensor antisymmetrical in the last pair of suffixes ($\chi_{ilk}=-\chi_{ikl}$). Although the tensor (41.16) is not symmetrical, it can be made so by adding a term of this form with an appropriate choice of the tensor χ_{ilk}. The practical execution of this quite laborious operation will be left till the end of the section. Here, we will continue with the derivation of the equations of motion, assuming that $\sigma_{ik}{}^{(r)}$ has already been symmetrized.

---

† It should be emphasized that E relates to a specified (unit) volume; the number N of particles (molecules) in that volume is variable. In *SP* 1, the chemical potential is everywhere relative to one particle, i.e. is defined as $\mu=\partial E/\partial N$. Since $N=\rho/m$, where m is the mass of one molecule, the definition used here differs from that in *SP*1 only by a factor m. To avoid misunderstanding when comparing with the thermodynamic identity (3.2a), note that E here is the internal energy per unit volume in the strict sense, whereas in §3 $\mathscr{E}$ was defined as the energy of the matter in unit volume of the undeformed body.

Substituting (41.15) in (41.13) and separating in one term a total divergence (using the symmetry of $\sigma_{ik}^{(r)}$), we get†

$$\left(\frac{\partial E_d}{\partial t}\right)_{\rho, S} = -\mathbf{N}\cdot\mathbf{h} + \sigma_{ik}^{(r)} v_{ik} - (\partial_i E)_{\rho, S} v_i + \operatorname{div}(\ldots). \tag{41.17}$$

Lastly, substituting in (41.12) the time derivatives from (41.5), (41.7), (41.8) and (41.17), and expressing the partial derivative (with ρ and S constant) of E in terms of the total derivative by

$$\partial_i E = (\partial_i E)_{\rho, S} + \mu \partial_i \rho + T \partial_i S,$$

we find after some algebra (separating total divergences)

$$\frac{\partial}{\partial t}(\tfrac{1}{2}\rho \mathbf{v}^2 + E) = -\sigma_{ik}' v_{ik} - \mathbf{N}\cdot\mathbf{h} + (1/T)\mathbf{q}\cdot\nabla T + 2R + \operatorname{div}(\ldots), \tag{41.18}$$

where σ_{ik}' is related to σ_{ik} by

$$\sigma_{ik} = -p\delta_{ik} + \sigma_{ik}^{(r)} + \sigma_{ik}', \tag{41.19}$$

and the pressure is thermodynamically defined:

$$p = \rho\mu - E + TS; \tag{41.20}$$

$\rho\mu = \Phi$ is the thermodynamic potential (Gibbs free energy) of unit volume of material and determines, as it should, the isotropic part of the stress tensor.

Comparison of (41.18) with the energy conservation equation (41.10) shows that

$$2R = \sigma_{ik}' v_{ik} + \mathbf{N}\cdot\mathbf{h} - (1/T)\mathbf{q}\cdot\nabla T. \tag{41.21}$$

This function determines the entropy increase due to dissipative processes. It is therefore clear that the tensor σ_{ik}' in (41.19) is the dissipative (viscous) part of the stress tensor. The tensor $\sigma_{ik}^{(r)}$ does not appear in (41.21); it is the non-dissipative (additional to the pressure-dependent) part of the stress tensor‡ and is specific to a nematic (as opposed to an ordinary) liquid.

It should also be noted that the coefficient λ does not appear in the dissipative function. Although the effect represented by this dimensionless coefficient is clearly a transport and not a thermodynamic effect, it is not dissipative.§

The density of body forces in a moving nematic medium is

$$F_i = -\partial_i p + \partial_k \sigma_{ik}^{(r)} + \partial_k \sigma_{ik}' \equiv -\partial_i p + F_i^{(r)} + F_i'.$$

In a medium that is at rest in equilibrium (even if deformed), $\mathbf{F}' = 0$, and according to the equilibrium condition (37.7) $\mathbf{h} = 0$ also. According to (41.14) and (41.15), in this case

$$\mathbf{F}^{(r)} = -(\nabla E_d)_{\rho, S}, \quad \mathbf{F} = -\nabla p - (\nabla E_d)_{\rho, S}.$$

If we assume the elastic moduli to be constants independent of ρ and S, then

† Since $E_0 = E_0(\rho, S)$, $(\partial_i E_d)_{\rho, S} = (\partial_i E)_{\rho, S}$.

‡ Sometimes called the *reactive part*, whence the superscript (r).

§ This situation is paralleled, for example, by the Hall effect in the electrodynamics of conductors, which is likewise not accompanied by dissipation.

$(\nabla E_d)_{\rho,S} = \nabla E_d$, and $\mathbf{F} = -\nabla(p+E_d)$.† In equilibrium, we must also have $\mathbf{F} = 0$. It follows that (on the assumption stated) the pressure distribution in a nematic medium in equilibrium is

$$p = \text{constant} - E_d. \tag{41.22}$$

Let us now carry out explicitly the above-mentioned operation of symmetrizing the tensor $\sigma_{ik}^{(r)}$. First, we calculate explicitly the antisymmetrical part of this tensor. To calculate the difference $\sigma_{ik}^{(r)} - \sigma_{ki}^{(r)}$, we have to use the fact that

$$B_{ik} = \frac{\partial E_d}{\partial n_i} n_k + \Pi_{li}\,\partial_l n_k - \Pi_{kl}\,\partial_i n_l$$

is symmetrical in the suffixes i and k. It is not easy to verify this symmetry directly. A simpler method is an indirect one using the fact that the energy E_d is a scalar and therefore invariant under any rotations of the coordinates. With an infinitesimal rotation through an angle $\delta\boldsymbol{\phi}$, the coordinates are transformed according to

$$\mathbf{r}' = \mathbf{r} + \delta\mathbf{r}, \quad \delta\mathbf{r} = \delta\boldsymbol{\phi}\times\mathbf{r},$$

that is

$$\delta x_i = \varepsilon_{ik} x_k, \quad \varepsilon_{ik} = e_i e_k\,\delta\phi_l = -\varepsilon_{ki}.$$

The changes in the vector $\mathbf{n}$ and the tensor $\partial_k n_i$ are correspondingly

$$\delta n_i = \varepsilon_{il} n_l, \quad \delta(\partial_k n_i) = \varepsilon_{il}\,\partial_k n_l + \varepsilon_{kl}\,\partial_l n_i.$$

The invariance of E_d under this rotation signifies that $B_{ik}\varepsilon_{ik} = 0$. Since ε_{ik} is any antisymmetrical tensor, it follows that B_{ik} is symmetrical.

We can then easily bring the antisymmetrical part of the tensor $\sigma_{ik}^{(r)}$ to the form (2.11), with

$$\phi_{ikl} = n_i \Pi_{lk} - n_k \Pi_{li}.$$

The symmetrized tensor $\sigma_{ik}^{(r)}$ is then found immediately from (2.13). After some simplification, the result is

$$\sigma_{ik}^{(r)} = -\tfrac{1}{2}\lambda(n_i h_k + n_k h_i) - \tfrac{1}{2}(\Pi_{kl}\partial_i n_l + \Pi_{il}\partial_k n_l) -$$
$$-\tfrac{1}{2}\partial_l[(\Pi_{ik} + \Pi_{ki})n_l - \Pi_{kl} n_i - \Pi_{il} n_k]. \tag{41.23}$$

This actually involves only the transverse (relative to the subscript k) components of the tensor Π_{ik}. If the latter is written as

$$\Pi_{ik} = \Pi_{ik}^{(t)} + \Pi_{il} n_k n_l,$$

so that $\Pi_{ik}^{(t)} n_k = 0$, only the $\Pi_{ik}^{(t)}$ terms remain in (41.23).

Lastly, let us consider the total divergence terms, which so far have not been written out. Comparison of (41.18) and (41.11) shows that the argument of div in all these terms determines the energy flux density. The final result thus obtained is

$$Q_i = (W + \tfrac{1}{2}v^2)v_i - \Pi_{ik}\{-v_l\partial_l n_k + \Omega_{li} n_l + \lambda n_l(v_{kl} - n_k n_m v_{lm})\} +$$
$$+\tfrac{1}{2}(n_i h_k - n_k h_i)v_k + \tfrac{1}{2}\lambda(n_i h_k + n_k h_i)v_k - \sigma_{ik}' v_k - \kappa_{ik}\partial_k T, \tag{41.24}$$

where $W = p + E$ is the heat function. The first term is the same as the energy flux in ordinary fluid dynamics.

† If these assumptions are not made, the force $\mathbf{F}$ at constant temperature can be written as $\mathbf{F} = -\rho\nabla\mu$, and the equilibrium condition thus reduces to the usual μ = constant: by differentiating the expression (41.20) for the pressure and using the thermodynamic relation $dE = T\,dS + \mu\,d\rho + (dE_d)_{\rho,S}$, we find $-\nabla p = -\rho\nabla\mu - S\nabla T + (\nabla E_d)_{\rho,S}$, whence, if T = constant, we get the above expression for $\mathbf{F}$.

§42. Dissipative coefficients of nematics

The **N** and σ_{ik}' terms in the equations of motion represent relaxation processes arising from the departure of the medium from thermodynamic equilibrium, which causes **h** and v_{ik} to have non-zero values. In the ordinary hydrodynamic approximation, the departure from equilibrium is assumed to be weak; that is, **h** and v_{ik} are in some sense small. Then the σ_{ik}' are linear functions of them.

However, with the above form of the equations of motion, the terms in σ_{ik}' that depend on **h** need not be written out. The reason is that such terms from the components of **h** and **n** would have the form constant $\times\,(n_i h_k + n_k h_i)$. Such a term is already present in the non-dissipative part of the stress tensor $\sigma_{ik}^{(r)}$ (41.18); the addition of a similar term in σ_{ik}' would therefore simply amount to redefining the coefficient λ.

The general form of the linear dependence of σ_{ik}' on v_{ik} is

$$\sigma_{ik}' = \eta_{iklm} v_{lm}. \tag{42.1}$$

where the rank-four tensor η_{iklm} has the obvious symmetry properties (resulting from the symmetry of σ_{ik}' and v_{ik})

$$\eta_{iklm} = \eta_{kilm} = \eta_{ikml}. \tag{42.2}$$

This tensor also has a deeper symmetry which results from Onsager's general principle of the symmetry of the kinetic coefficients (see *SP* 1, §120; as in §33, in the rest of this section we shall formulate this principle as in *FM*, §59, and use x_a and X_a as defined there). The expression R/T for the rate of increase of entropy shows that, if the $\dot{x}_a$ are taken to be the components of the tensor σ_{ik}', then the X_a thermodynamically conjugate to them are the components of the tensor $-v_{lm}/T$.† The components of the tensor η_{iklm} act as the kinetic coefficients γ_{ab}. Onsager's principle requires that $\gamma_{ab} = \gamma_{ba}$, i.e.

$$\eta_{iklm} = \eta_{lmik}. \tag{42.3}$$

The tensor η_{iklm} has to be constructed from only the unit tensor δ_{ik} and the vector **n**, taking account of the symmetry properties mentioned. There are only five linearly independent combinations of this kind:

$$n_i n_k n_l n_m, \quad n_i n_k \delta_{lm} + n_l n_m \delta_{ik},$$
$$n_i n_l \delta_{km} + n_k n_l \delta_{im} + n_i n_m \delta_{kl} + n_k n_m \delta_{il},$$
$$\delta_{ik}\delta_{lm}, \quad \delta_{il}\delta_{km} + \delta_{kl}\delta_{im}.$$

Accordingly, η_{iklm} has five independent components; the stress tensor formed from it may be written as‡

$$\begin{aligned}\sigma_{ik}' = {}& 2\eta_1 v_{ik} + (\eta_2 - \eta_1)\delta_{ik} v_{ll} + (\eta_4 + \eta_1 - \eta_2)(\delta_{ik} n_l n_m v_{lm} + n_i n_k v_{ll}) + \\ &+ (\eta_3 - 2\eta_1)(n_i n_l v_{kl} + n_k n_l v_{il}) + (\eta_5 + \eta_1 + \eta_2 - 2\eta_3 - 2\eta_4) n_i n_k n_l n_m v_{lm}.\end{aligned} \tag{42.4}$$

The suitability of this definition of the various coefficients is seen from the following expression for the dissipative function when the z-axis is taken parallel to **n**:

$$\begin{aligned}2R = {}& 2\eta_1 (v_{\alpha\beta} - \tfrac{1}{2}\delta_{\alpha\beta} v_{\gamma\gamma})^2 + \eta_2 v_{\alpha\alpha}^2 + 2\eta_3 v_{\alpha z}^2 + \\ &+ 2\eta_4 v_{zz} v_{\alpha\alpha} + \eta_5 v_{zz}^2 + \frac{1}{T}\{\kappa_\parallel (\partial_z T)^2 + \kappa_\perp (\partial_\alpha T)^2\} + h^2/\gamma,\end{aligned} \tag{42.5}$$

---

† In the literature, the $\dot{x}_a$ and X_a are often called *thermodynamic fluxes* and *thermodynamic forces* respectively.

‡ The dissipative coefficients of nematics were introduced (in a different form) by F. M. Leslie (1966) and O. Parodi (1970). The choice of definitions of the viscosity coefficients of nematics seems to be not yet agreed upon in the literature.

where the suffixes, α, β, γ take the values x and y. Since we must have $R > 0$ (the entropy increasing), the coefficients η_1, η_2, η_3, η_5, $\kappa_\parallel$, $\kappa_\perp$, and γ are positive, and

$$\eta_2\eta_5 > \eta_4^2. \tag{42.6}$$

A nematic medium thus has a total of nine kinetic coefficients: five viscosity coefficients, two thermal conductivities, the coefficient γ (which also has the dimensions of viscosity), and the non-dissipative dimensionless coefficient λ.

The number of viscosity coefficients which appear in the equations of motion is smaller in the important case where the fluid in motion may be regarded as incompressible; for this, its velocity must be much less than that of sound. The equation of continuity for an incompressible fluid is just div $\mathbf{v} \equiv v_{ll} = 0$. The second term in the stress tensor (42.4) disappears, and the third becomes constant $\times\, \delta_{ik}(n_l n_m v_{lm})$. The latter term makes no contribution to the dissipative function, since it gives zero in the product $\sigma_{ik}'\, v_{ik}$ because of the resulting factor $v_{ik}\delta_{ik} = v_{kk} = 0$; it has the same tensor structure as $-p\delta_{ik}$ in the complete stress tensor σ_{ik}. On the other hand, in incompressible fluid dynamics the pressure appears (like the velocity) as just one of the unknown functions of coordinates and time, determined by solving the equations of motion; it is here not a thermodynamic quantity related to other similar ones by the equation of state. The terms $-p\delta_{ik}$ and constant $\times\, \delta_{ik}(n_l n_m v_{lm})$ in the stress tensor can therefore be combined; this simply amounts to redefining the pressure. The viscous stress tensor for an incompressible nematic fluid therefore reduces to

$$\sigma_{ik}' = 2\eta_1 v_{ik} + (\eta_3 - 2\eta_1)(n_i n_l v_{kl} + n_k n_l v_{il}) + (\bar{\eta}_2 + \eta_1 - 2\eta_3) n_i n_k n_l n_m v_{lm}, \tag{42.7}$$

where $\bar{\eta}_2 = \eta_2 + \eta_5 - 2\eta_4$; it contains only three independent viscosity coefficients. The corresponding dissipative function is (with the z-axis along $\mathbf{n}$)

$$2R = 2\eta_1(v_{\alpha\beta} - \tfrac{1}{2}\delta_{\alpha\beta}v_{\gamma\gamma})^2 + \bar{\eta}_2 v_{zz}^2 + 2\eta_3 v_{\alpha z}^2 + \\ + \frac{1}{T}\{\kappa_\parallel(\partial_z T)^2 + \kappa_\perp(\partial_\alpha T)^2\} + h^2/\gamma, \tag{42.8}$$

since $v_{\alpha\alpha} + v_{zz} = 0$; the inequality (42.6) makes the coefficient $\bar{\eta}_2$ positive.

PROBLEM

Determine the force on a straight disclination (with Frank index $n = 1$) moving transverse to itself (H. Imura and K. Okano 1973).

SOLUTION. Let us consider the disclination in coordinates for which it is at rest and along the z-axis, while the liquid moves at a constant speed v in the x-direction. The distribution $\mathbf{n}(\mathbf{r})$ in the disclination in these coordinates is steady, and is given (for a disclination with radial director streamlines, Fig. 29a) by

$$n_x = \cos\phi, \quad n_y = \sin\phi,$$

where the polar angle $\phi = \tan^{-1}(y/x)$. In equation (41.3), we have $\partial\mathbf{n}/\partial t = 0$ and $v_{ik} = 0$ (the flow being uniform), leaving

$$v\,\partial\mathbf{n}/\partial x = \mathbf{h}/\gamma.$$

This gives for the weak molecular field resulting from the motion

$$\mathbf{h} = \gamma v\, \boldsymbol{\nu} \times \mathbf{n}\, \partial\phi/\partial x,$$

where $\boldsymbol{\nu}$ is a unit vector in the z-direction; in the absence of motion, $\mathbf{h} = 0$, since a disclination at rest is an equilibrium state of the medium. The dissipative function is

$$2R = h^2/\gamma = \gamma v^2(\partial\phi/\partial x)^2 = \gamma v^2 y^2/(x^2 + y^2)^2.$$

The energy dissipated per unit time and per unit length of the disclination is represented by the integral

$$\int 2R\, dx\, dy = \pi\gamma v^2 L, \quad L = \log(R/a),$$

where R is the transverse dimension of the region of motion and a the molecular dimensions. This dissipation must be compensated by the work vf done by the force $\mathbf{f}$ acting on the disclination. Hence

$$f = \pi\gamma v L.$$

A similar result is obtained for a disclination with circular streamlines (Fig. 29b).

§43. Propagation of small oscillations in nematics

The complete system of exact equations for the hydrodynamics of nematics is very complicated, but it is simpler for small oscillations, where the equations can be linearized.

In considering the propagation of small oscillations in nematic media, let us first recall the oscillation types (modes) that exist in ordinary liquids. Firstly, there are ordinary sound waves, for which the dispersion relation between the frequency ω and the wave vector $\mathbf{k}$ is $\omega = ck$ and the propagation speed is

$$c = \sqrt{(\partial p/\partial \rho)_s}. \tag{43.1}$$

The oscillations in a sound wave are longitudinal (see *FM*, §64).

Next, there are strongly damped viscous waves, with dispersion relation

$$i\omega = \eta k^2/\rho, \tag{43.2}$$

where η is the viscosity coefficient (see *FM*, §24). These waves are transverse (the velocity $\mathbf{v}$ is perpendicular to $\mathbf{k}$), and are therefore often called *shear waves*. They can have two independent directions of polarization; the dispersion relation does not depend on these.

Lastly, in a liquid at rest, small oscillations of temperature (and entropy) are propagated as waves, likewise strongly damped, with dispersion relation

$$i\omega = \chi k^2, \tag{43.3}$$

where χ is the thermometric conductivity of the medium (see *FM*, §52).

Analogous types of wave exist in nematic media, but the presence of an additional dynamical variable, the director $\mathbf{n}$, gives rise to further types peculiar to these media.

Let us begin with ordinary sound in nematics. It is easy to see that, in the limit of sufficiently long waves (i.e. sufficiently small k), the corrections to the speed of sound that are due to the presence of the additional dynamical variable are slight, so that the speed of sound is again given by the simple formula (43.1). We can write the director in the oscillating medium as $\mathbf{n} = \mathbf{n}_0 + \delta\mathbf{n}$, where $\mathbf{n}_0$ is the unperturbed value, constant throughout the medium, and $\delta\mathbf{n}$ is a small variable part; since $\mathbf{n}^2 = \mathbf{n}_0{}^2 = 1$, $\mathbf{n}_0 \cdot \delta\mathbf{n} = 0$. Comparison of the left-hand side of (41.3) with the first two terms on the right shows that $\omega\delta n \sim kv$, or $\delta n \sim v/c$; the term $\mathbf{N} = \mathbf{h}/\gamma$ is a higher-order small quantity, since by (37.9) the molecular field $h \propto k^2$. The term E_d in the energy density of the liquid is therefore

$$E_d \sim K(k\delta n)^2 \sim K(kv/c)^2,$$

i.e. is of the order of k^2 relative to the leading term $\sim \rho v^2$. In the approximation considered, this energy may therefore be neglected, and this proves the above statement regarding the speed of sound.

In the next approximation with respect to k, there is absorption of sound arising from dissipative processes. The specific feature of a nematic, as compared with an ordinary liquid, is that this absorption is anisotropic, depending on the direction of propagation of the sound wave; see Problem 1.

The remaining types of oscillation in nematics have a dispersion relation similar to (43.2) and (43.3): $\omega \propto k^2$. This means that, for sufficiently small k, we always have $\omega \ll ck$.

In turn, it follows that in such oscillations the fluid may be regarded as incompressible.† The continuity equation then reduces to div $\mathbf{v} = 0$ or, for a plane wave, $\mathbf{k}\cdot\mathbf{v} = 0$. The oscillations considered are therefore transverse shear oscillations relative to the velocity oscillations.

To investigate these various oscillations, we linearize the equations of motion, putting $\mathbf{n} = \mathbf{n}_0 + \delta\mathbf{n}$, $p = p_0 + \delta p$. In the first approximation, the molecular field is linear in the derivatives of $\mathbf{n}$, and therefore linear in $\delta\mathbf{n}$:

$$\mathbf{H} = K_1 \nabla \operatorname{div} \delta\mathbf{n} - K_2 \,\mathbf{curl}\,[\mathbf{n}_0(\mathbf{n}_0 \cdot \mathbf{curl}\,\delta\mathbf{n})] + K_3\,\mathbf{curl}\,[\mathbf{n}_0 \times (\mathbf{n}_0 \times \mathbf{curl}\,\delta\mathbf{n})]. \quad (43.4)$$

The first term in the "reactive" part of the stress tensor (41.16) is quadratic in $\delta\mathbf{n}$ and is therefore to be omitted. We have also to omit the quadratic terms which arise in taking the tensor divergence $\partial_k \sigma_{ik}^{(r)}$ in (41.7) and $(\mathbf{v}\cdot\nabla)\mathbf{v}$ on the left-hand side. This equation thus becomes

$$\rho\partial v_i/\partial t = -\partial_i \delta p + \tfrac{1}{2}(n_{0i}\partial_k h_k - n_{0k}\partial_k h_i) - \tfrac{1}{2}\lambda(n_{0i}\partial_k h_k + n_{0k}\partial_k h_i) + \partial_k \sigma_{ik}. \quad (43.5)$$

In (41.3), it is sufficient to replace $\mathbf{n}$ by $\mathbf{n}_0$ in the first two terms on the right and omit $(\mathbf{v}\cdot\nabla)\delta\mathbf{n}$ on the left:

$$\partial \delta n_i/\partial t = \Omega_{ki} n_{0k} + \lambda(\delta_{il} - n_{0i}n_{0l}) n_{0k} v_{kl} + h_i/\gamma. \quad (43.6)$$

Because $\mathbf{n}_0 \cdot \delta\mathbf{n} = 0$ and $\mathbf{v}\cdot\mathbf{k} = 0$, the vectors $\delta\mathbf{n}$ and $\mathbf{v}$ have only two independent components each. The equations (43.5) and (43.6) thus form a set of four linear equations. They define four oscillation modes, in each of which the velocity and the director undergo coupled oscillations. Usually, however, the situation is considerably simplified by the fact that the dimensionless ratio

$$\mu = K\rho/\eta^2 \quad (43.7)$$

is small, $\sim 10^{-2} - 10^{-4}$; here K and η denote the order of magnitude of the elastic moduli of the nematic and its viscosity coefficients, $\eta_1, \bar{\eta}_2, \eta_3, \gamma$. It will be shown below that we can then distinguish two substantially different types of oscillation, for each of which (43.5) and (43.6) allow certain simplifications.

In one type, the frequency is related to the wave number by

$$i\omega \sim \eta k^2/\rho, \quad (43.8)$$

which is similar to (43.2); for a reason to be explained below, these are called *fast shear* oscillations. In both equations (43.5) and (43.6), we can then neglect all terms containing $\mathbf{h}$: it is seen from (43.8) that

$$\delta n \sim kv/\omega \sim \rho v/\eta k,$$

and so the molecular field is

$$h \sim K k^2 \delta n \sim \rho v k K/\eta.$$

Using these estimates, we can readily verify that the terms in $\mathbf{h}$ in the equations are small in comparison with those in v_{ik}, their ratio being $\sim \mu$. The equations of fast shear oscillations thus reduce to

$$\rho\partial v_i/\partial t = \partial_k \sigma_{ik}' - \partial_i \delta p, \quad (43.9)$$

$$\partial \delta n_i/\partial t = \Omega_{ki} n_{0k} + \lambda(\delta_{il} - n_{0i}n_{0l}) n_{0k} v_{kl}. \quad (43.10)$$

The first equation does not involve $\delta\mathbf{n}$, and determines the velocity oscillations and the dispersion relation, after which the second equation immediately gives the accompanying oscillations of the director (see Problem 2).

---

† It may be recalled (see *FM*, §10) that a fluid in non-steady motion may be regarded as incompressible if $v \ll c$ and $\tau \gg l/c$, where τ and l are the times and distances over which the velocity changes appreciably. For oscillatory motion, the first condition is always satisfied at sufficiently low oscillation amplitudes, while the second implies that $\omega/k \ll c$.

Let us now turn to the second type of shear oscillations with the condition $\mu \ll 1$, the *slow* director oscillations that are specific to nematics. In these, the order of magnitude of the alternating part of the director is determined by the balance between the derivative $\partial\delta\mathbf{n}/\partial t$ on the left of (43.6) and the term $\mathbf{h}/\gamma$ on the right: $\omega\delta n \sim h/\gamma$, and, since $h \sim Kk^2\,\delta n$, the dispersion relation for these oscillations is qualitatively

$$i\omega \sim Kk^2/\gamma. \qquad (43.11)$$

Evidently, the derivative $\rho\partial\mathbf{v}/\partial t \sim \rho v\omega$ on the left of (43.5) is then small compared with the terms $\partial_k\sigma_{ik}' \sim \eta vk^2$ on the right, and may therefore be omitted. The equation

$$-\partial_i\,\delta p + \tfrac{1}{2}(n_{0i}\,\partial_k h_k - n_{0k}\,\partial_k h_i) - \tfrac{1}{2}\lambda(n_{0k}\,\partial_k h_k + n_{0k}\,\partial_k h_i) + \partial_k\,\sigma_{ik}' = 0 \qquad (43.12)$$

gives the relation between the velocity and director oscillations, and the dispersion relation is then found from (43.6); see Problem 3.

Note that the ratio of the frequencies (43.11) and (43.8) is $\omega_s/\omega_f \sim \mu$. Thus, for a given k, ω_s is much less than ω_f, and this is the reason for calling the oscillations slow and fast respectively.

Lastly, the temperature oscillations in a nematic medium at rest differ from the corresponding ones in an ordinary liquid only by the anisotropy in the dispersion relation, which is similar to (43.3); see Problem 4.

PROBLEMS

PROBLEM 1. Determine the absorption coefficient of sound in a nematic medium.

SOLUTION. The absorption coefficient† is calculated as the ratio

$$\Gamma = \overline{R}/c\rho\overline{v^2}$$

(see §35); the dissipative function is given by (42.5), in which the term $\mathbf{h}^2/\gamma$ may be omitted: as already mentioned, the molecular field $h \propto k^2$, and therefore $h^2/\gamma \propto k^4$, whereas the other terms in R are proportional to k^2, a lower power of the wave number. A simple calculation gives‡

$$\Gamma = \frac{\omega^2}{2\rho c^3}\Big\{(\eta_1+\eta_2) + 2(\eta_3+\eta_4-\eta_1-\eta_2)\cos^2\theta + (\eta_1+\eta_2+\eta_5-2\eta_3-2\eta_4)\cos^4\theta + $$
$$+ [\kappa_\perp + (\kappa_\parallel - \kappa_\perp)\cos^2\theta]\left(\frac{1}{c_v}-\frac{1}{c_p}\right)\Big\},$$

where θ is the angle between $\mathbf{k}$ (and therefore $\mathbf{v}$) and $\mathbf{n}$. The calculation of the thermal-conduction part of the absorption is entirely analogous to the similar one for an ordinary liquid (*FM*, §79); c_p and c_v are the specific heats per unit mass of material.

PROBLEM 2. Find the dispersion relation for fast shear oscillations.

SOLUTION. For a plane wave, with $\mathbf{v} \propto \exp(i\mathbf{k}\cdot\mathbf{r} - i\omega t)$, (43.9) becomes

$$-i\rho\omega v_i = -ik_i\delta p + ik_k\sigma_{ik}'.$$

For an incompressible nematic, the viscous stress tensor is given by (42.7), and a simple calculation (using the fact that $\mathbf{v}$ is transverse, $\mathbf{v}\cdot\mathbf{k} = 0$) brings the equation to the form§

$$i\rho\omega\mathbf{v} = i\mathbf{k}\delta p + a_1k^2\mathbf{v} + a_2k^2\mathbf{n}(\mathbf{n}\cdot\mathbf{v}) + a_3k\mathbf{k}(\mathbf{n}\cdot\mathbf{v}), \qquad (1)$$

where

$$a_1 = \eta_1 + \tfrac{1}{2}(\eta_3 - 2\eta_1)\cos^2\theta,$$
$$a_2 = \tfrac{1}{2}(\eta_3 - 2\eta_1) + (\eta_2 + \eta_1 - 2\eta_3)\cos^2\theta,$$
$$a_3 = \tfrac{1}{2}(\eta_3 - 2\eta_1)\cos\theta,$$

† Here we denote this quantity by Γ to avoid confusion with the dissipative coefficient γ.

‡ In calculating the quadratic terms, all oscillatory quantities must of course be written as real, their dependence on t and $\mathbf{r}$ being given by factors cos $(\mathbf{k}\cdot\mathbf{r} - \omega t)$.

§ To simplify the formulae, the suffix 0 is omitted from $\mathbf{n}_0$ in the remainder of the Problems.

with θ the angle between $\mathbf{k}$ and $\mathbf{n}$. Multiplying (1) by $\mathbf{k}$, we get an expression for the pressure oscillations in terms of the velocity oscillations:

$$\delta p = ik(\mathbf{n}\cdot\mathbf{v})(a_3 + a_2\cos\theta). \tag{2}$$

The required dispersion relation is given by the transverse components of (1). Multiplying this by $\mathbf{n}\times\mathbf{k}$, we get

$$\begin{aligned} i\omega_\perp &= k^2 a_1(\theta)/\rho \\ &= k^2(\eta_1 \sin^2\theta + \tfrac{1}{2}\eta_3\cos^2\theta)/\rho, \end{aligned}$$

corresponding to oscillations of $\mathbf{v}$ at right angles to the plane through $\mathbf{k}$ and $\mathbf{n}$. The dispersion relation for oscillations polarized in this plane is found by multiplying (1) by $\mathbf{n}$ and eliminating δp by means of (2):

$$\begin{aligned} i\omega_\parallel &= k^2\{a_1(\theta) + \sin^2\theta\, a_2(\theta)\}/\rho \\ &= k^2\{\tfrac{1}{4}(\eta_1 + \bar\eta_2)\sin^2 2\theta + \tfrac{1}{2}\eta_3\cos^2 2\theta\}/\rho, \end{aligned}$$

where $\bar\eta_2$ is as in (42.7).

Both dispersion relations agree, of course, with the qualitative estimate (43.8).

PROBLEM 3. Find the dispersion relation for slow shear oscillations.

SOLUTION. For a plane wave, $\delta\mathbf{n} \propto \exp(i\mathbf{k}\cdot\mathbf{r} - i\omega t)$, the linearized molecular field is

$$\begin{aligned} \mathbf{h} &= \mathbf{H} - \mathbf{n}(\mathbf{n}\cdot\mathbf{H}) \\ &= -K_1\{\mathbf{k} - \mathbf{n}(\mathbf{n}\cdot\mathbf{k})\}\mathbf{k}\cdot\delta\mathbf{n} - K_2\mathbf{v}(\mathbf{v}\cdot\delta\mathbf{n}) - K_3(\mathbf{k}\cdot\mathbf{n})^2\delta\mathbf{n}, \end{aligned}$$

where $\mathbf{v} = \mathbf{n}\times\mathbf{k}$ and $\mathbf{v}^2 = k^2\sin^2\theta$. Equation (43.12), with σ'_{ik} from (42.7), becomes

$$\begin{aligned} &-i\mathbf{k}\delta p - a_1 k^2\mathbf{v} - a_2 k^2\mathbf{n}(\mathbf{n}\cdot\mathbf{v}) - a_3 k\mathbf{k}(\mathbf{n}\cdot\mathbf{v}) + \\ &\quad + \tfrac{1}{2}i(1-\lambda)\mathbf{n}(\mathbf{h}\cdot\mathbf{k}) - \tfrac{1}{2}i(1+\lambda)\mathbf{h}(\mathbf{n}\cdot\mathbf{k}) = 0; \end{aligned} \tag{1}$$

the functions $a_1(\theta)$ and $a_2(\theta)$ have been found in Problem 2. Multiplying this by $\mathbf{v}$, we find the relation between the oscillations of $\mathbf{v}$ and $\delta\mathbf{n}$ polarized perpendicularly to the plane of $\mathbf{k}$ and $\mathbf{n}$:

$$a_1(\mathbf{v}\cdot\mathbf{v}) = -i\frac{1+\lambda}{2k^2}(\mathbf{n}\cdot\mathbf{k})(\mathbf{h}\cdot\mathbf{v}) = \tfrac{1}{2}i(1+\lambda)(\mathbf{n}\cdot\mathbf{k})K_\perp(\mathbf{v}\cdot\delta\mathbf{n}), \tag{2}$$

where

$$K_\perp = K_2\sin^2\theta + K_3\cos^2\theta.$$

Equation (43.6) multiplied by $\mathbf{v}$ is

$$-i\omega(\mathbf{v}\cdot\delta\mathbf{n}) = \tfrac{1}{2}i(1+\lambda)(\mathbf{n}\cdot\mathbf{k})(\mathbf{v}\cdot\mathbf{v}) - k^2K_\perp(\mathbf{v}\cdot\delta\mathbf{n})/\gamma.$$

Eliminating $(\mathbf{v}\cdot\mathbf{v})$ by means of (2), we find the dispersion relation for oscillations polarized perpendicularly to the plane of $\mathbf{k}$ and $\mathbf{n}$:

$$\omega_\perp = k^2K_\perp\left\{\frac{1}{\gamma} + \frac{(1+\lambda)^2\cos^2\theta}{4a_1}\right\}.$$

To find the dispersion relation for oscillations polarized in the plane of $\mathbf{k}$ and $\mathbf{n}$, we take the component of (1) in the direction perpendicular to $\mathbf{k}$ in that plane, and multiply by $\mathbf{n}$, obtaining

$$(\mathbf{n}\cdot\mathbf{v})(a_1 + a_2\sin^2\theta) = -\tfrac{1}{2}i(1+\lambda\cos 2\theta)K_\parallel(\mathbf{k}\cdot\delta\mathbf{n}),$$

where

$$K_\parallel = K_1\sin^2\theta + K_3\cos^2\theta.$$

Similar operations on (43.6) give

$$i\omega(\mathbf{k}\cdot\delta\mathbf{n}) = \tfrac{1}{2}ik^2(1+\lambda\cos 2\theta)(\mathbf{n}\cdot\mathbf{v}) + k^2K_\parallel(\mathbf{k}\cdot\delta\mathbf{n})/\gamma.$$

Elimination of $\mathbf{n}\cdot\mathbf{v}$ from these two equations gives the dispersion relation

$$i\omega_\parallel = k^2K_\parallel\left\{\frac{1}{\gamma} + \frac{(1+\lambda\cos 2\theta)^2}{4(a_1 + a_2\sin^2\theta)}\right\}.$$

Both relations are in agreement with the qualitative estimate (43.11).†

PROBLEM 4. Find the dispersion relation for temperature oscillations in a nematic at rest.

---

† When k is real, the real quantity $i\omega$ must be positive, and the oscillations are damped in the course of time, not spontaneously amplified. All the dispersion relations found in Problems 2 and 3 have this property.

SOLUTION. The transformation of (41.8) for an incompressible nematic is made in exactly the same way as for an ordinary liquid (see *FM*, §50), and the result is

$$\partial T/\partial t = \chi_{ik}\partial_i\partial_k T,$$

$$\chi_{ik} = \kappa_{ik}/\rho c_p = \chi_\parallel n_i n_k + \chi_\perp(\delta_{ik} - n_i n_k),$$

with κ_{ik} from (41.10). For oscillations with $\delta T \propto \exp(i\mathbf{k}\cdot\mathbf{r} - i\omega t)$, we find the dispersion relation

$$i\omega = k^2(\chi_\parallel \cos^2\theta + \chi_\perp \sin^2\theta).$$

§44. Mechanics of cholesterics

Cholesteric liquid crystals or *cholesterics* differ from nematics in that there is no centre of inversion among their symmetry elements. The directions $\mathbf{n}$ and $-\mathbf{n}$ of the director remain equivalent; see *SP* 1, §140.

The absence of a centre of symmetry has the result that the free energy of a deformation may contain a term linear in the derivatives—the pseudoscalar $\mathbf{n}\cdot\mathbf{curl}\,\mathbf{n}$. The general form of the free energy may be written as

$$F_d = \tfrac{1}{2}K_1(\operatorname{div}\mathbf{n})^2 + \tfrac{1}{2}K_2(\mathbf{n}\cdot\mathbf{curl}\,\mathbf{n} + q)^2 + \tfrac{1}{2}K_3(\mathbf{n}\times\mathbf{curl}\,\mathbf{n})^2, \tag{44.1}$$

where q is a parameter having the dimensions of reciprocal length. This difference causes a fundamental change in the nature of the equilibrium state of the medium (in the absence of external interactions): it is no longer uniform in space ($\mathbf{n}$ = constant) as in nematics.

The equilibrium state of a cholesteric corresponds to a distribution of directions of $\mathbf{n}$ for which

$$\operatorname{div}\mathbf{n} = 0, \quad \mathbf{n}\cdot\mathbf{curl}\,\mathbf{n} = -q, \quad \mathbf{n}\times\mathbf{curl}\,\mathbf{n} = 0; \tag{44.2}$$

the free energy (44.1) then has its minimum value of zero. The solution of these equations is

$$n_x = \cos qz, \quad n_y = \sin qz, \quad n_z = 0. \tag{44.3}$$

This *helicoidal* structure can be regarded as the result of twisting about the z-axis a nematic medium originally oriented with $\mathbf{n}$ = constant in one direction in the xy-plane. The orientational symmetry of a cholesteric is periodic in one direction in space (the z-axis). The vector $\mathbf{n}$ returns to its previous value after every interval $2\pi/q$ in the z-direction; since $\mathbf{n}$ and $-\mathbf{n}$ have equivalent directions, however, the true period of repetition of the structure is half this, or π/q. Of course, the macroscopic description of the helicoidal structure of a cholesteric by the equations (44.3) is meaningful only if the pitch of the structure is much greater than molecular dimensions. In actual cholesterics this condition is satisfied ($\pi/q \sim 10^{-5}$ cm).

In deriving the equations of equilibrium and motion of nematics, no use was made of their possession of a centre of inversion. The same general equations are therefore valid for cholesterics. There are, however, a number of differences. First, there is a change in the expression for F_d with which the molecular field $\mathbf{h}$ is to be calculated from the definition (37.5). Next, the presence of a term in the free energy that is linear in the derivatives causes a difference between the isothermal and adiabatic values of K_2; cf. the end of §37. In the hydrodynamic equations as formulated in §§41 and 42, the basic thermodynamic variables are the density and entropy. Accordingly, the adiabatic elastic moduli (as functions of ρ and S) are to be used.

Lastly, there is a substantial change in the hydrodynamic equations of cholesterics, as compared with those for nematics, in that further terms appear in the dissipative parts of the equations, namely in the stress tensor σ_{ik}', the heat flux $\mathbf{q}$, and the quantity $\mathbf{N}$ on the right of (41.3) (F. M. Leslie 1968):

$$\left.\begin{aligned} \sigma_{ik}' &= (\sigma_{ik}')_{\text{nem}} + \mu(n_i e_{klm} + n_k e_{ilm}) n_m \partial_l T, \\ N_i &= (N_i)_{\text{nem}} + \nu e_{ikl} n_k \partial_l T, \\ q_l &= (q_l)_{\text{nem}} + \nu_1 e_{lki} n_k h_i + \mu_1 (e_{lmi} n_k + e_{lmk} n_i) n_m v_{ik}; \end{aligned}\right\} \tag{44.4}$$

the terms with the suffix nem denote the expressions given by the hydrodynamics of nematics. The additional terms in these relations are a pseudotensor and pseudovectors, not a true tensor and vectors. This removes the symmetry under spatial inversion, and for that reason the terms do not appear in nematic hydrodynamics. Note that the construction of similar terms that are true tensors or vectors is not possible, owing to the requirement that the equations are invariant under a change in the sign of $\mathbf{n}$. For example, a term in σ_{ik}' having the form constant $\times$ $(n_i \partial_k T + n_k \partial_i T)$ or a term in $\mathbf{q}$ having the form constant $\times$ $\mathbf{h}$ would change sign with $\mathbf{n}$, whereas the stress tensor and the heat flux must be invariant under this transformation. Similarly, a term constant $\times$ ∇T in $\mathbf{N}$ is impossible, since it is invariant under a change in the sign of $\mathbf{n}$, whereas $\mathbf{N}$ (which determines the derivative $d\mathbf{n}/dt$) would have to change sign.

The coefficients in (44.4) are connected by relations which follow from Onsager's principle. In applying this principle (cf. §42), we choose as the $\dot{x}_a$ (the thermodynamic fluxes) the quantities σ_{ik}', q_i and N_i. The form of the dissipative function (41.21), or more precisely that of $2R/T$, which determines the increase of entropy, shows that the corresponding thermodynamic forces X_a are the quantities $-v_{ik}/T$, $\partial_i T/T^2$ and $-h_i/T$. It must also be noted that the σ_{ik}' are even and the q_i and N_i odd under time reversal, as is seen from their positions in (41.3), (41.7) and (41.8). If the x_a and x_b have the same parity under this transformation, then the corresponding kinetic coefficients are related by $\gamma_{ab} = \gamma_{ba}$; if they have opposite parities, then $\gamma_{ab} = -\gamma_{ba}$. Now, comparing the cross coefficients in (44.4),† we find

$$\nu_1 = \nu T, \quad \mu_1 = \mu T.$$

We can thus write (44.4) in the final form

$$\begin{aligned} \sigma_{ik}' &= (\sigma_{ik}')_{\text{nem}} - \mu[n_i(\mathbf{n}\times\nabla T)_k + n_k(\mathbf{n}\times\nabla T)_i], \\ \mathbf{N} &= \mathbf{N}_{\text{nem}} + \nu\mathbf{n}\times\nabla T, \\ \mathbf{q} &= \mathbf{q}_{\text{nem}} + \nu T\mathbf{n}\times\mathbf{h} + 2\mu T\mathbf{n}\times(v\mathbf{n}), \end{aligned} \tag{44.5}$$

where $(v\mathbf{n})$ denotes the vector with components $v_{ik} n_k$.

In the mechanics of cholesterics, there is thus a dependence of the stress tensor and the vector $\mathbf{N}$ on the temperature gradient.‡ The form of this dependence (the vector product $\mathbf{n}\times\nabla T$) signifies that the temperature gradient gives rise to twisting moments acting on the director and on the mass of the liquid. The molecular field which accompanies a rotation of the director relative to the liquid, and the liquid velocity gradients, cause heat fluxes.

† When comparing, note carefully the order of suffixes in the factor e_{ikl}.

‡ The presence of terms containing the gradient of a second independent thermodynamic quantity, such as the pressure, among the dissipative terms in the equations of motion is forbidden (*FM*, §49) by the law of increase of entropy. The presence of such terms would lead to terms in the dissipative function which contain the products $\nabla p \cdot \nabla T$ and $\mathbf{h} \cdot \nabla p$, and these, in the absence of terms containing $(\nabla p)^2$, would make it impossible for R to be positive definite.

One hydrodynamic effect peculiar to cholesterics may be illustratively described as the percolation of a liquid through a helicoidal structure at rest (W. Helfrich 1972). It is as follows.

Let us imagine a cholesteric medium whose helicoidal structure is fixed in space, for example by some form of adhesion to the boundary walls of the medium. We shall show that there can then exist a uniform flow of the liquid along the axis of the structure (the z-axis).

Since the structure (44.3) corresponds to the equilibrium state of the medium, it makes the molecular field zero: $\mathbf{h} = 0$. The presence of the percolating flux causes some distortion of the structure and accordingly a small molecular field (together with the flow velocity v). This field can be determined from the equation of motion (41.3) of the director. Since the field $\mathbf{n}(\mathbf{r})$ is at rest in the zero-order approximation with respect to the velocity, $\partial\mathbf{n}/\partial t = 0$, and, since the liquid flow is assumed uniform ($v_z = v = \text{constant}$), $v_{ik} = \Omega_{ik} = 0$. The equation thus becomes

$$(\mathbf{v}\cdot\nabla)\mathbf{n} = v d\mathbf{n}/dz = \mathbf{h}/\gamma.$$

With $\mathbf{n}(z)$ from (44.3), we then find

$$\mathbf{h} = \gamma v \mathbf{q}\times\mathbf{n}, \tag{44.6}$$

where the vector $\mathbf{q}$, with magnitude q, is in the z-direction. Under the conditions considered, the expression (41.21) for the dissipative function becomes $2R = \mathbf{h}^2/\gamma$ and, with $\mathbf{h}$ from (44.6),

$$2R = \gamma v^2 q^2. \tag{44.7}$$

This gives the energy dissipated per unit time and per unit volume of the liquid. In steady motion, it is balanced by the work done by the external sources that maintain the pressure gradient $p' \equiv dp/dz$ acting along the z-axis. The body force density in the medium is given by just the gradient $-\nabla p$; the work done by these forces per unit time and per unit volume is $-p'v$, and on equating this to $2R$, we find the percolation velocity

$$v = |p'|/\gamma q^2. \tag{44.8}$$

The director $\mathbf{n}$ rotates with angular velocity vq relative to a liquid particle percolating through the helicoidal structure. This rotation is accompanied by "friction" described by the coefficient γ, which determines the velocity of the flow.

Under actual conditions, the velocity cannot be constant over the whole width of the flow: it must be zero at the walls of the containing tube. The velocity decreases in a layer having a thickness δ. Now the only parameter of length for the motion in question is $1/q$. If we suppose that all the viscosity coefficients of a cholesteric are of the same order of magnitude, there are also no dimensionless parameters other than ~ 1. Under these conditions, evidently only $\delta \sim 1/q$ is possible. Thus, for flow in a tube whose radius is much greater than $1/q$, (44.8) is valid everywhere except in a very thin layer at the wall, with thickness of the order of the pitch of the helicoidal structure.

§45. Elastic properties of smectics

According to the accepted terminology, *smectic liquid crystals* or *smectics* comprise anisotropic liquids with various layer structures. At least some of these have a microscopic molecular density function that depends on only one coordinate (z, say) and is periodic in that coordinate: $\rho = \rho(z)$. It may be recalled (see *SP* 1, §128) that the density function determines the probability distribution of various positions of particles in the body; in this case, such positions can be treated as a whole, that is, $\rho\,dV$ is the probability for the centre of mass of an individual molecule to be in the volume dV. A body with density function $\rho(z)$ may be regarded as consisting of equidistant plane layers with free relative movement. In each layer, the molecular centres of mass are arranged randomly, and in this sense each is a two-dimensional liquid, but the liquid layers may be either isotropic or not. This difference may be due to the nature of the ordered orientation of molecules in the layers. In the simplest case, the anisotropy of the orientation distribution is specified only by the direction of **n**, say the direction of the longest axis of the molecule. If this direction is at right angles to the plane of the layers, then the latter are isotropic, so that the z-axis is an axis of symmetry in the body; this appears to be the structure of what are called *smectics A*. If the direction of **n** is oblique to the xy-plane, that plane contains a preferred direction, and there is no axial symmetry; this appears to be the structure of what are called *smectics C*.

In the following, we shall discuss only the simpler smectics A, and call them just "smectics". In all known smectics A, as well as the axial symmetry about the z-axis, there is equivalence of the two directions of the z-axis. If the smectic has also a centre of inversion, its macroscopic symmetry (the point symmetry group) is the same as in nematics; the microscopic symmetry, and therefore the mechanical properties, are of course quite different.

There is a very important reservation concerning what has been said so far. The existence of a structure in which the density varies within the body presupposes that the displacements caused in small regions of the body by the thermal fluctuations are sufficiently small. However, for a structure with $\rho = \rho(z)$ these fluctuational displacements increase without limit as the body becomes larger; see *SP* 1, §137. Strictly speaking, this means that there cannot exist a one-dimensional periodic structure in an infinite medium. In practice, however, this statement has only a highly conventional significance, because the fluctuations increase only slowly (logarithmically) as the body becomes larger. Estimates using typical values for the material constants show that the one-dimensional periodic structure could be lost only for enormous sizes impossible in practice, and so the $\rho(z)$ structure is feasible in any realistic problem.

It should be emphasized, at the same time, that the medium does not become an ordinary liquid when the $\rho(z)$ structure is disturbed by fluctuations and ρ = constant. The fundamental difference from a liquid lies in the properties of the density fluctuation correlation function between different points, $\langle\delta\rho(\mathbf{r}_1)\,\delta\rho(\mathbf{r}_2)\rangle$. In an ordinary liquid, this function is isotropic, and decreases exponentially as $r = |\mathbf{r}_2 - \mathbf{r}_1| \to \infty$; see *SP* 1, §116. In a system with $\rho = \rho(z)$, the correlation function remains anisotropic, and as $r \to \infty$ it decreases only slowly, as a power function, and more slowly as the temperature falls; see *SP* 1, §138.

In going on to construct a mechanics of smectics, we have to begin by finding an expression for the deformation free energy density. Because of the microscopic homogeneity of the medium in the xy-plane, the displacements of its points in that plane are related to the change in energy only in so far as they change the density of the substance. We therefore choose as the fundamental hydrodynamic variables (in addition to the temperature, which is assumed constant throughout the medium) the density ρ and the displacement $u_z \equiv u$ of the points in the medium along the z-axis. The deformation energy depends on the density change $\rho - \rho_0$ (where ρ_0 is the density of the undeformed medium) and on the derivatives of the displacement u with respect to the coordinates. The first derivatives $\partial u/\partial x$, $\partial u/\partial y$ cannot occur in the second-order terms in the free energy: if the body is rotated rigidly about the x or y axis, these derivatives change, whereas the energy must obviously remain constant.†

As always in elasticity, the spatial variation of all quantities will be assumed to be sufficiently slow, so that the deformation energy is determined by the first non-vanishing terms in the expansion in powers of the spatial derivatives. We shall also, however, assume an even stronger condition: the displacements u themselves are so small that the layers everywhere remain almost parallel to the same xy-plane.‡

Under these assumptions, and using the symmetry of the medium, we find for the free energy of the deformation of the smectic

$$\left.\begin{aligned} F_d &= F - F_0(T) \\ &= \tfrac{1}{2}(A/\rho_0)(\rho-\rho_0)^2 + C(\rho-\rho_0)\partial u/\partial z + \tfrac{1}{2}B\rho_0(\partial u/\partial z)^2 + \tfrac{1}{2}K_1(\triangle_\perp u)^2, \\ \triangle_\perp &= \partial^2/\partial x^2 + \partial^2/\partial y^2. \end{aligned}\right\} \tag{45.1}$$

A term§ $(\partial u/\partial z)\triangle_\perp u$ is prohibited by the assumed equivalence of the two directions of the z-axis, i.e. by the symmetry under the transformation $u \to -u$, $z \to -z$, $x, y \to x, y$ (reflection in the xy-plane) or $u \to -u$, $z \to -z$, $y \to -y$, $x \to x$ (rotation about the horizontal second-order axis, the x-axis); for the same reason, there is no term $(\rho - \rho_0)\triangle_\perp u$. Including the first term of the expansion in second derivatives (which does not appear in the elasticity theory for solids) is necessary since F_d does not contain first derivatives with respect to x and y. The stability conditions for the undeformed state, i.e. the conditions for the energy (45.1) to be positive, are

$$A > 0, \quad B > 0, \quad AB > C^2. \tag{45.2}$$

The use of the notation K_1 in (45.1), as in (37.1), is deliberate. A deformation of a layer structure of smectics can be described by a distribution $\mathbf{n}(\mathbf{r})$ of the director, regarded as the normal to the deformed layers specified by the equations $u(\mathbf{r})$ = constant. For a small distortion of the layers,

$$n_x \cong \partial u/\partial x, \quad n_y \cong \partial u/\partial y, \quad n_z \cong 1, \tag{45.3}$$

and then $(\triangle_\perp u)^2 = (\operatorname{div}\mathbf{n})^2$, which is just the quantity in the corresponding term in (37.1). The coefficients B and C in (45.1), however, characterize the specific crystal nature of smectics which distinguishes them from nematics.¶

---

† These derivatives occur in the elastic energy of solids in combinations of u_{xz} and u_{yz} with the derivatives of u_x and u_y, which are unaffected by the rotation mentioned.

‡ In this sense, the range of application of the mechanics of smectics as developed here is narrower than for the nematic mechanics considered previously, which allowed director fields $\mathbf{n}(\mathbf{r})$ differing to any extent from the undeformed uniform distribution.

§ Such as occurred in *SP* 1, §137.

¶ The director $\mathbf{n}$ (regarded as the preferred direction of orientation of the molecules in the layers) is not an independent hydrodynamic variable in smectics A. With a variable $\mathbf{n}$ in nematic hydrodynamics it is characteristic that a uniform rotation of $\mathbf{n}(\mathbf{r})$ throughout the body causes no change in the energy. It is for this reason that a slow change in $\mathbf{n}$ through the body involves only a small change in the energy, which depends only on the derivatives of $\mathbf{n}$ and can be expanded in powers of these. In smectics, however, such a rotation alters the orientation relative to the layer structure and would change the energy considerably. In smectics C, where the director is at some definite angle to the normal, a uniform rotation of $\mathbf{n}$ about the normals at a constant inclination would again not affect the energy. This provides another hydrodynamic variable, namely the component of $\mathbf{n}$ in the plane of the layers.

In the approximation (45.3), $\mathbf{n}\cdot\mathbf{curl}\,\mathbf{n} \cong \mathbf{curl}_z\,\mathbf{n} = 0$. The term $\mathbf{n}\cdot\mathbf{curl}\,\mathbf{n}$ thus does not occur in the free energy of smectics, nor therefore does the cholesteric distortion of the structure (§44), whether or not the symmetry elements include a centre of inversion.

The equations of equilibrium of a smectic are found by minimizing the total free energy with respect to the variables ρ and u, with the added condition $\int \rho\, dV = \text{constant}$, expressing the constancy of the total mass of the body. Minimizing the difference

$$\int F_d\, dV - \lambda \int \rho\, dV$$

(where λ is a constant Lagrange multiplier) with respect to ρ, we find

$$A(\rho - \rho_0)/\rho_0 + C\partial u/\partial z = \lambda,$$

relating the density change to the deformation of the layers. Taking ρ_0 to be the density when $\partial u/\partial z = 0$, we have $\lambda = 0$ and

$$\rho - \rho_0 = -\rho_0 m \partial u/\partial z, \quad m = C\rho_0/A. \tag{45.4}$$

The dimensionless coefficient m is related to Poisson's ratio σ for a rod cut from the smectic in the z-direction. For

$$(\rho - \rho_0)/\rho_0 = -(V - V_0)/V_0 = -(u_{xx} + u_{yy} + u_{zz})$$

(see (1.6)), where $u_{zz} = \partial u/\partial z$ and u_{xx}, u_{yy} are the strain tensor components in the xy-plane. Putting $u_{xx} = u_{yy}$, we have

$$u_{xx} = -\tfrac{1}{2}(1-m)u_{zz},$$

and comparison with (5.4) shows that

$$\sigma = \tfrac{1}{2}(1-m). \tag{45.5}$$

When $m = 0$, $\sigma = \frac{1}{2}$, the value for a liquid.

Eliminating the density change from (45.1) and (45.4) gives the free energy in terms of u only:

$$F_d = \tfrac{1}{2}\rho_0 B'(\partial u/\partial z)^2 + \tfrac{1}{2}K_1(\triangle_\perp u)^2, \tag{45.6}$$

where

$$B' = B - C^2/A. \tag{45.7}$$

Variation of the total free energy with respect to u now gives, after some integrations by parts,

$$\delta \int F_d\, dV = -\int F_z \delta u\, dV, \tag{45.8}$$

where

$$F_z = \rho_0 B' \partial^2 u/\partial z^2 - K_1 \triangle_\perp{}^2 u. \tag{45.9}$$

Evidently F_z is the force per unit volume acting in the z-direction in the deformed smectic if the density change is not "adjusted" to the deformation.

In equilibrium, $F_z = 0$, and the displacement u satisfies the linear differential equation

$$\rho_0 B' \partial^2 u/\partial z^2 - K_1 \triangle_\perp{}^2 u = 0. \tag{45.10}$$

If the body is also subjected to externally applied body forces, these must be included on the left-hand side; cf. (2.8).

The ratio $\sqrt{(K_1/\rho_0 B')}$ has the dimensions of length, and a rough estimate of it is $\sqrt{(K_1/\rho_0 B')} \sim a$, where a is the period of the one-dimensional structure, i.e. the distance between layers. If the smectic is subjected to a deformation that varies considerably in the xy-plane over distances $\sim l_\perp \gg a$, then it follows from (45.10) that in the z-direction the deformation varies considerably only over distances $l_\parallel \sim l_\perp^2/a \gg l_\perp$.

As an example, let us find the Green's function for (45.10), i.e. the displacement $u = G_{zz}(\mathbf{r}) \equiv G(\mathbf{r})$ at a variable point $\mathbf{r}$ due to a single concentrated force applied at $\mathbf{r} = 0$ and acting in the z-direction; cf. §8, Problem. This function satisfies the equation

$$\rho_0 B \partial^2 G/\partial z^2 - K_1 \triangle_\perp^2 G + \delta(\mathbf{r}) = 0. \tag{45.11}$$

Taking the Fourier transform of this equation (i.e. multiplying it by $e^{-i\mathbf{k}\cdot\mathbf{r}}$ and integrating over d^3x), we find for the Fourier components of $G(\mathbf{r})$

$$G_\mathbf{k} = [\rho_0 B' k_z^2 + K_1 k_\perp^4]^{-1},$$

where $k_\perp^2 = k_x^2 + k_y^2$. The inverse Fourier transformation gives the function sought, as the integral

$$G(\mathbf{r}) = \int \frac{e^{-i\mathbf{k}\cdot\mathbf{r}}}{\rho_0 B' k_z^2 + K_1 k_\perp^4} \frac{d^3k}{(2\pi)^3}. \tag{45.12}$$

This integral is logarithmically divergent as $\mathbf{k} \to 0$. To give it a definite significance, we have to eliminate the motion of the body as a whole, assuming some arbitrarily chosen point $\mathbf{r} = \mathbf{r}_0$ in the body to be fixed; the numerator of the integrand then becomes $e^{i\mathbf{k}\cdot\mathbf{r}} - e^{i\mathbf{k}\cdot\mathbf{r}_0}$, and the divergence is eliminated.

Let us now return to the influence of thermal fluctuations on the properties of smectics, and consider their elastic properties. The problem can be formulated most definitely as follows: how do the fluctuations affect the deformation due to a concentrated force applied to the body, i.e., how does the Green's function $G(\mathbf{r})$ vary? It is found that the change is given by replacing k_z^2 and $k_\perp^4$ in (45.12) by $k_z^2[\log(1/ak_z)]^{-4/5}$ and $k_\perp^4[\log(1/ak_\perp)]^{2/5}$ respectively, a being of the order of the structure period.† In turn, this change can be intuitively interpreted as a change in the effective values of the elastic moduli B' and K_1 when the characteristic wave number of the deformation decreases, and so its extent ($\sim 1/k$) increases. We see that B'_{eff} decreases as $[\log(1/ak_z)]^{-4/5}$ when $k_z \to 0$, and $K_{1\text{eff}}$ increases as $[\log(1/ak_\perp)]^{2/5}$ when $k_\perp \to 0$. In practice, however, such effects could become significant only for unrealistically large dimensions.

To conclude this section, we shall show that the expression (45.6) for the elastic energy of a smectic can be somewhat generalized by including some higher-order terms, though without bringing in further coefficients.

To do so, we note that the energy contribution given by the first term in (45.6) is physically due to the change in the distance a between the layers; the derivative $\partial u/\partial z$ is equal to the relative change in this distance under a displacement $u_z = u$, and the term may therefore be written as $\frac{1}{2}\rho_0 B'(\delta a/a)^2$. The distance between the layers may, however, change because of the dependence of u on x and y as well as that of z. This is easily seen by imagining all the layers to be simultaneously rotated through an angle θ about the y-axis, say, in such a way that the period of the structure in the z-direction remains equal to a; the

† See G. Grinstein and R. A. Pelcovits, *Physical Review Letters* **47**, 856, 1981; *Physical Review* A **26**, 915 (1982); E. I. Kats, *Soviet Physics JETP* **56**, 791, 1983. It is necessary in the analysis to include the terms of the third and fourth order in u in the expansion of the free energy.

distance between the layers, measured along the normal to them, becomes $a\cos\theta$. For small θ, the change in the distance between the layers is

$$\delta a = a(\cos\theta - 1) \cong -\tfrac{1}{2}a\theta^2.$$

Since at the same time the displacement in the rotation considered is u = constant $+ x\tan\theta \cong$ constant $+ x\theta$, we have

$$\delta a/a = -\tfrac{1}{2}(\partial u/\partial x)^2.$$

In this form the expression is valid for any dependence of u on x; if u depends on y also, $(\partial u/\partial x)^2$ becomes $(\nabla_\perp u)^2$.

Taking this effect into account, we must write the free energy (45.6) as

$$F_d = \tfrac{1}{2}\rho_0 B'\left[\frac{\partial u}{\partial z} - \tfrac{1}{2}\left(\frac{\partial u}{\partial x}\right)^2 - \tfrac{1}{2}\left(\frac{\partial u}{\partial y}\right)^2\right]^2 + \tfrac{1}{2}K_1(\triangle_\perp u)^2. \qquad (45.13)$$

This expression will be used in the Problem.

PROBLEM

A layer of smectic with thickness h and plane boundaries parallel to the layer structure planes is uniformly stretched in the z-direction perpendicular to the layer. Find the critical tension beyond which the layer structure becomes unstable with respect to transverse perturbations (W. Helfrich 1971).†

SOLUTION. A uniform stretching is a deformation $u = \gamma z$, where the constant $\gamma > 0$. To investigate the stability, we put $u = \gamma z + \delta u(x, z)$, where δu is a small perturbation which satisfies the boundary conditions $\delta u = 0$ for $z = \pm\frac{1}{2}h$ (the xy-plane being taken in the middle of the layer). As far as the second-order terms, the total elastic energy of the perturbation, per unit length in the y-direction, is

$$\int \delta F_d \, dx\, dz = \tfrac{1}{2}\int \{\rho_0 B'(\partial\delta u/\partial z)^2 - \rho_0 B'\gamma(\partial\delta u/\partial x)^2 + K_1(\partial^2\delta u/\partial x^2)\}\, dx\, dz; \qquad (1)$$

the term in $\gamma\partial\delta u/\partial z$ disappears on integration over dz, because of the boundary conditions.

We shall consider perturbations having the form

$$\delta u = \text{constant} \times \cos k_z z \cos k_x x, \quad k_z = n\pi/h, \quad n = 1, 2, \ldots,$$

i.e. a transverse modulation of the layer structure. The condition for the structure to be stable is that the energy (1) be positive. Replacing all $\sin^2$ and $\cos^2$ factors in the integrand by their mean values $\frac{1}{2}$, we obtain this condition in the form

$$\rho_0 B'(k_z^2 - \gamma k_x^2) + K_1 k_x^4 > 0.$$

The limit of stability as γ increases is determined by the occurrence of a real root k_x^2 of the trinomial on the left of this inequality; complex k_x do not satisfy the condition that the perturbation be finite throughout the xy-plane. The first such root appears for the perturbation with $n = 1$, and gives the critical value of γ with the corresponding $k_x = k_{cr}$:‡

$$\gamma_{cr} = (2\pi/h)(K_1/\rho_0 B')^{\frac{1}{2}}, \quad k_{cr} = (\pi/h)(\rho_0 B'/K_1)^{\frac{1}{4}}.$$

§46. Dislocations in smectics

The concept of a dislocation in a smectic has the same significance as in an ordinary crystal. The only difference is that, since the microscopic structure of smectics has one-dimensional periodicity (in the z-direction), the Burgers vector of a dislocation is always

† This instability is analogous to that of a straight rod under compression (§21).

‡ The value of k_{cr} determines only the wave number of the perturbation in the xy-plane, not the whole symmetry of the deformation that occurs. To find the latter, it is necessary to go beyond the approximation of equilibrium equations linear in δn; the situation here is similar to that of convective instability in a plane-parallel layer of liquid (see *FM*, §57, and J. M. Delrieu, *Journal of Chemical Physics* **60**, 1081, 1974).

along the z-axis and its magnitude is always an integral multiple of the period a of the structure.

Bearing this in mind, we find that the deformation around a dislocation in a smectic is described by the same formula (27.10), with an appropriate definition of the elastic modulus tensor λ_{iklm}. For this purpose, we define the stress tensor σ_{ik} in the smectic in accordance with the usual relationship

$$F_z = \partial_k \sigma_{zk}, \tag{46.1}$$

where F_z is the "internal stress" body force (45.9). We also use the strain tensor corresponding to the displacement $u_z = u$; its non-zero components are

$$u_{zz} = \partial u/\partial z, \quad u_{xz} = \tfrac{1}{2}\partial u/\partial x, \quad u_{yz} = \tfrac{1}{2}\partial u/\partial y. \tag{46.2}$$

The force (45.9) can be put in the form (46.1) if we express the stress tensor in terms of the strain tensor by $\sigma_{ik} = \lambda_{iklm} u_{lm}$, with†

$$\lambda_{zzzz} = \rho_0 B', \lambda_{zxzx} = \lambda_{zyzy} = -K_1 \triangle_\perp, \lambda_{zxzy} = \lambda_{zxzz} = \lambda_{zyzz} = 0; \tag{46.3}$$

some of these are operators.

Formula (27.10) for the displacement $u_z = u$ becomes

$$u(\mathbf{r}) = -\lambda_{zklz} b \int_{S_D} n_l \frac{\partial}{\partial x_k} G(\mathbf{r}-\mathbf{r}')\,df' \tag{46.4}$$

where $G \equiv G_{zz}$ is the function (45.12).

Let us consider two particular cases: straight screw and edge dislocations. In the first case, the dislocation axis is parallel to the Burgers vector (the z-axis). This case requires no further calculations. It is evident *a priori* that the deformation u will depend only on the coordinates x and y. The medium is isotropic in the xy-plane. We can therefore apply immediately the results of §27, Problem 2, according to which

$$u = b\phi/2\pi, \tag{46.5}$$

where ϕ is the polar angle of the position vector in the xy-plane.

The edge dislocation case is more complicated (P. G. de Gennes 1972). Here the dislocation axis is at right angles to the Burgers vector; suppose it to be along the y-axis. Then the surface S_D in the integral (46.4) can be taken as the right-hand half of the xy-plane, and the vector $\mathbf{n}$ normal to it will be along the negative z-axis. The only non-zero component λ_{zkzz} is $\lambda_{zzzz} = \rho_0 B'$, so that (46.4) becomes

$$u(\mathbf{r}) = b\rho_0 B' \int_{-\infty}^{\infty}\int_{0}^{\infty} \frac{\partial G(\mathbf{r}-\mathbf{r}')}{\partial z}\,dx'\,dy'.$$

We substitute G from (45.12); the differentiation with respect to z gives a factor ik_z, and the integration with respect to y gives $2\pi\delta(k_y)$; the delta function is then eliminated by integration with respect to k_y. In the integral

$$\int_0^\infty e^{-ik_x x'}\,dx',$$

† The remaining components λ_{iklm} can be chosen so that $F_x = F_y = 0$; these components do not occur in (46.4).

in order to ensure convergence, we must treat k_x as $k_x - i0$. The result of integrating with respect to x', y' and k_y is then

$$u(\mathbf{r}) = -b \int_{-\infty}^{\infty} \frac{\exp(ik_x x)}{k_x - i0} I(k_x, z) \frac{dk_x}{2\pi},$$

where

$$I(k_x, z) = \int_{-\infty}^{\infty} \frac{k_z \exp(ik_z z)\, dk_z}{k_z^2 + \lambda^2 k_x^4\, 2\pi}, \quad \lambda^2 = K_1/\rho_0 B'.$$

This last integral is calculated by closing the contour of integration with an infinite semicircle in the upper or lower half-plane (for $z > 0$ and $z < 0$ respectively) of the complex variable k_z and taking the residue at the pole $k_z = i\lambda k_x^2$ or $k_z = -i\lambda k_x^2$:

$$I = \pm \tfrac{1}{2} i \exp(-\lambda k_x^2 |z|),$$

where the upper and lower signs correspond to $z > 0$ and $z < 0$. The displacement is thus

$$u(x, z) = \pm \frac{b}{4\pi i} \int_{-\infty}^{\infty} \exp\{-\lambda k_x^2 |z| + ik_x x\} \frac{dk_x}{k_x - i0}. \tag{46.6}$$

The spatial derivatives of this are, however, more interesting than the displacement itself. The derivative $\partial u/\partial x$ is

$$\frac{\partial u}{\partial x} = \pm \frac{b}{4\pi} \int_{-\infty}^{\infty} \exp\{-\lambda k_x^2 |z| + ik_x x\}\, dk_x$$

$$= \pm \frac{b}{4\sqrt{(\pi\lambda |z|)}} \exp\{-x^2/4\lambda |z|\}. \tag{46.7}$$

According to (46.6), the derivatives with respect to z and x are related by

$$\partial u/\partial z = \pm \lambda \partial^2 u/\partial x^2,$$

whence

$$\frac{\partial u}{\partial z} = -\frac{bx}{8\sqrt{(\pi\lambda |z|^3)}} \exp\{-x^2/4\lambda |z|\}. \tag{46.8}$$

The deformation tends to zero exponentially as $|x| \to \infty$, but much more slowly (by a power law) as $|z| \to \infty$.

§47. Equations of motion of smectics

The mechanics of smectics has in common with that of nematics the fact that both involve hydrodynamics with extra variables in comparison with an ordinary liquid. For nematics, the variable concerned is the director **n**; for smectics it is the displacement u of the layers (P. Martin, O. Parodi and P. S. Pershan 1972). The latter point needs elucidation. The velocity is defined in hydrodynamics as the momentum of unit mass of matter. Its component v_z need not, in the present case, be equal to $\partial u/\partial t$. In a smectic, mass transfer (in the z-direction) can take place not only by the deformation of layers but also by the percolation of matter through a one-dimensional structure at rest, as described for cholesterics in §44. This phenomenon is not specific to liquid crystals; a similar effect can occur in solid crystals, where it is due to diffusion of defects (see the first footnote to §22). In smectics, however, it cannot in principle be eliminated by increased blurring of the periodic structure with a large number of defects (vacancies) and a greater mobility of the molecules.

In adiabatic motion, each element of the liquid transfers its constant entropy s (per unit mass); if at some initial instant s is constant throughout the medium, it remains so. Since the condition of constant s relates to the entropy per unit mass, it will be convenient to use the internal energy of the medium per unit mass also; this will be denoted by ε. For a deformed smectic, ε is given by a formula analogous to (45.1):

$$\begin{aligned}\varepsilon_d &= \varepsilon - \varepsilon_0(s)\\ &= \tfrac{1}{2}(A/\rho_0{}^2)(\rho-\rho_0)^2 + (C/\rho_0)(\rho-\rho_0)\partial u/\partial z + \tfrac{1}{2}B(\partial u/\partial z)^2 + \tfrac{1}{2}(K_1/\rho_0)(\triangle_\perp u)^2, \quad (47.1)\end{aligned}$$

where ρ_0 is the density of the undeformed medium; the coefficients A, B, C here are not the same as in (45.1), being now the adiabatic values of the elastic moduli (assumed to be expressed as functions of s), not the isothermal ones as in (45.1). The isothermal and adiabatic values of K are equal, for the same reasons as in nematics; see the end of §37.†

The volume of unit mass is $1/\rho$. The thermodynamic relation for the energy differential is therefore

$$\begin{aligned}d\varepsilon &= T\,ds - p\,dV\\ &= T\,ds + p\,d\rho/\rho^2.\end{aligned}$$

The pressure in the medium can therefore be found by differentiating the expression (47.1):

$$p = \rho^2(\partial\varepsilon/\partial\rho)_s \cong A(\rho-\rho_0) + \rho_0 C\partial u/\partial z. \quad (47.2)$$

The sequence of operations in constructing the equations of motion of smectics is then very similar to the derivation of those of nematics in §41. To emphasize this analogy, we shall, as in §41, use the energy $E = \rho\varepsilon$ and the entropy $S = \rho s$ per unit volume.

The equation of continuity has the usual form‡

$$\partial\rho/\partial t + \operatorname{div}(\rho\mathbf{v}) = 0. \quad (47.3)$$

The dynamical equation for the velocity must have the form

$$\rho dv_i/dt = \partial_k\sigma_{ik}; \quad (47.4)$$

cf. (41.7). The form of the stress tensor will be established later.

One further equation arises from the presence of the additional variable, and expresses the difference between v_z and $\partial u/\partial t$:

$$\partial u/\partial t - v_z = N. \quad (47.5)$$

The quantity $-N$ gives the rate of percolation, i.e. the velocity of the liquid relative to the

---

† Strictly speaking, $\partial u/\partial z$ in (47.1) should be written as $\partial u/\partial z - \delta_0(s)$, where $\delta_0(s)$ is the value of $\partial u/\partial z$ for entropy s in the absence of external forces. Considering the motion with a given s, we can take as the undeformed state this particular state and put $\delta_0(s) = 0$. It should be emphasized, however, that we then cannot, for example, differentiate the expression (47.1) with respect to s in order to determine the temperature from $T = (\partial\varepsilon/\partial s)_\rho$.

‡ Although we are ultimately concerned only with the linearized equations of motion, the linearization will not be performed at every stage of the derivation, since this would complicate the formulae.

one-dimensional lattice; it is a transport quantity, and an expression for it will be derived later.

Lastly, the entropy equation, taking account of dissipative processes in the medium, has the form (41.8):

$$\frac{\partial S}{\partial t} + \operatorname{div}(S\mathbf{v} + \mathbf{q}/T) = 2R/T. \tag{47.6}$$

As in §41, we calculate the time derivative of the total energy per unit volume of the medium, which appears in the energy conservation equation (41.11). The only difference is in the form of the last term in (41.12): we now have†

$$\begin{aligned}\left(\frac{\partial E_d}{\partial t}\right)_{\rho,S} &= \left(\frac{\partial E_d}{\partial(\partial_z u)}\right)_{\rho,S}\frac{\partial}{\partial z}\frac{\partial u}{\partial t} + K_1(\triangle_\perp u)\left(\triangle_\perp\frac{\partial u}{\partial t}\right)\\ &= -h\frac{\partial u}{\partial t} + \operatorname{div}\{\ldots\};\end{aligned} \tag{47.7}$$

as in §41, the total divergence terms are not written out. The notation here is

$$\begin{aligned}h &= \frac{\partial}{\partial z}\left(\frac{\partial E_d}{\partial(\partial_z u)}\right)_{\rho,S} - K_1\triangle_\perp{}^2 u\\ &= \rho_0 B\frac{\partial^2 u}{\partial z^2} + C\frac{\partial(\rho-\rho_0)}{\partial z} - K_1\triangle_\perp{}^2 u.\end{aligned} \tag{47.8}$$

If h is regarded as the z-component of a vector $\mathbf{h} = \mathbf{n}h$ (where $\mathbf{n}$ is a unit vector in the z-direction), we can easily see that this vector may be written as a divergence:

$$h_i = \partial_k \sigma_{ik}{}^{(r)}, \tag{47.9}$$

where the symmetrical tensor $\sigma_{ik}{}^{(r)}$ has the components

$$\left.\begin{aligned}&\sigma_{xx}{}^{(r)} = \sigma_{yy}{}^{(r)} = K_1\triangle_\perp \partial u/\partial z, \quad \sigma_{zz}{}^{(r)} = \rho_0 B\partial u/\partial z + C(\rho-\rho_0),\\ &\sigma_{xz}{}^{(r)} = -K_1\triangle_\perp \partial u/\partial x, \quad \sigma_{yz}{}^{(r)} = -K_1\triangle_\perp\partial u/\partial y, \quad \sigma_{xy}{}^{(r)} = 0.\end{aligned}\right\} \tag{47.10}$$

Substituting in (47.7) $\partial u/\partial t$ from (47.5) and again separating a total divergence in one term, we can write

$$\begin{aligned}(\partial E_d/\partial t)_{\rho,S} &= -hN - v_i\partial_k\sigma_{ik}{}^{(r)} + \operatorname{div}\{\ldots\}\\ &= -hN + v_{ik}\sigma_{ik}{}^{(r)} + \operatorname{div}\{\ldots\}.\end{aligned}$$

This expression differs from (41.17) only in the significance of h and N.‡ Proceeding as in §41, we obtain the same expression (41.21) for the dissipative function:

$$2R = \sigma_{ik}'v_{ik} + Nh - (1/T)\mathbf{q}\cdot\nabla T, \tag{47.11}$$

where σ_{ik}' is the viscous part of the stress tensor:

$$\sigma_{ik} = -p\delta_{ik} + \sigma_{ik}{}^{(r)} + \sigma_{ik}'. \tag{47.12}$$

† Here and henceforward, we neglect the change in the elastic moduli within the medium.

‡ And in the absence of the term $v_i(\partial_i E)_{\rho,S}$. Such a term would, however, occur in this case also as a third-order small quantity, negligible in comparison with the second-order ones.

The dynamical equation (47.4) with this stress tensor is, after linearization, omitting the term $(\mathbf{v}\cdot\nabla)\mathbf{v}$,

$$\rho_0 \partial v_i/\partial t = -\partial_i p + h_i + \partial_k \sigma_{ik}', \tag{47.13}$$

the vector $\mathbf{h} = \mathbf{n}h$ being defined by (47.8).

The viscous part σ_{ik}' of the stress tensor, the heat flux $\mathbf{q}$ and the percolation rate N (thermodynamic fluxes) are, as usual, given by expressions linear in the thermodynamic forces $-v_{ik}/T$, $(1/T^2)\partial_i T$ and $-h/T$; the coefficients in these expressions satisfy relations which follow from Onsager's principle. We shall not repeat the derivation (cf. §§42 and 44), but simply give the result, assuming that (as is usually the case) the smectic has a centre of inversion; this has not so far been assumed.

The viscous part of the stress tensor is then given by the same expression (42.4) as for nematics, with $\mathbf{n}$ in the z-direction. The heat flux and the percolation rate are

$$\left.\begin{aligned} q_z = -\kappa_{\parallel}\partial T/\partial z + \mu h, \quad \mathbf{q}_{\perp} = -\kappa_{\perp}\nabla_{\perp}T,\\ N = \lambda_p h - (\mu/T)\partial T/\partial z; \end{aligned}\right\} \tag{47.14}$$

since the dissipative function is positive, we must have

$$\kappa_{\parallel},\ \kappa_{\perp},\ \lambda_p > 0, \quad \mu^2 < T\lambda_p\kappa_{\parallel}. \tag{47.15}$$

Percolation makes possible in smectics an effect similar to that described for cholesterics at the end of §44. If the periodic structure of the smectic is in some way fixed in space, there can be a uniform steady flow in the z-direction. It follows from (47.13) that for such a flow $dp/dz = h$, and from (47.5), with N from (47.14),

$$v_z = -\lambda_p h = -\lambda_p dp/dz. \tag{47.16}$$

There is one important remark to be made regarding the above discussion of the kinetic coefficients in smectics. The divergence of the fluctuations in smectics (§46) has a particularly marked effect in transport phenomena and may substantially alter their nature.†

§48. Sound in smectics

In ordinary liquids, and in nematic liquid crystals, there is only one branch of weakly damped acoustic vibrations, namely longitudinal sound waves. In solid crystals and amorphous solids, there are three acoustic branches of the linear dispersion relation (§§22, 23). One-dimensional crystals, i.e. smectics, occupy once again an intermediate position, having two acoustic branches (P. G. de Gennes 1969).

The attenuation coefficients of these waves are of no interest in the present discussion, and in order to determine just their speed of propagation we shall neglect all dissipative terms in the equations of motion. The complete set of linearized equations comprises: the continuity equation

$$\partial\rho'/\partial t + \rho\,\mathrm{div}\,\mathbf{v} = 0 \tag{48.1}$$

(here and henceforward, we omit the subscript zero in ρ_0; ρ' and p' are the variable parts of the density and pressure); equation (47.16), which reduces to

$$v_z = \partial u/\partial t, \tag{48.2}$$

† See E. I. Kats and V. V. Lebedev, *Soviet Physics JETP* **58**, 1172, 1984.

there being no percolation; and the dynamical equation (47.13),

$$\rho\,\partial \mathbf{v}/\partial t = -\nabla p' + \mathbf{n}h, \tag{48.3}$$

where, according to (47.2),

$$p' = A\rho' + \rho\,C\,\partial u/\partial z. \tag{48.4}$$

In (47.8) for h, the term $K_1\Delta_\perp{}^2 u$, which contains higher-order derivatives, is to be omitted, since it is of too high an order in the wave number k, which in acoustic waves is to be regarded as a small quantity:

$$h = \rho B\partial^2 u/\partial z^2 + C\partial\rho'/\partial z. \tag{48.5}$$

In actual smectics, B and C are usually small in comparison with A. Under these conditions, which will be assumed to hold, the nature of the two acoustic branches in smectics is more readily perceived.

If we neglect in the equations of motion all terms containing the small coefficients B and C, they reduce to those of an ordinary liquid with the equation of state $p' = A\rho'$, i.e. with compressibility $(\partial p'/\partial\rho')_s = A$. The corresponding vibrations are ordinary sound waves—longitudinal compression and rarefaction waves in the medium. Their speed of propagation is

$$c_1 = \sqrt{A} \tag{48.6}$$

and is, in the approximation considered, independent of the direction.

The speed c_2 of propagation of waves in the second acoustic branch is, as we shall see, much less than c_1: $\omega/k = c_2 \ll c_1$. As regards these vibrations, therefore, the medium may be treated as incompressible; see the first footnote to §43. The continuity equation then reduces to the incompressibility condition $\operatorname{div}\mathbf{v} = 0$; in (48.5), we omit the second term, so that (48.3) becomes

$$\rho\partial\mathbf{v}/\partial t = -\nabla p' + \mathbf{n}\rho B\partial^2 u/\partial z^2. \tag{48.7}$$

Differentiating the z-component of this equation with respect to z and substituting $v_z = \partial u/\partial t$ gives

$$\rho\partial^2\delta/\partial t^2 = -\partial^2 p'/\partial z^2 + \rho B\partial^2\delta/\partial z^2,$$

where $\delta = \partial u/\partial z$. Taking the divergence of (48.7) gives, with the incompressibility condition,

$$\triangle p' = \rho B\partial^2\delta/\partial z^2.$$

Lastly, eliminating p' from these two equations, we obtain one equation for δ:

$$\partial^2\triangle\delta/\partial t^2 = B\{-\partial^4\delta/\partial z^4 + \partial^2\triangle\delta/\partial z^2\}. \tag{48.8}$$

The dependence of the displacement u on the coordinate z means that the distances a between adjacent layers vary: $\delta a = a\partial u/\partial z$, and the relative change in a is given by $\delta = \partial u/\partial z$. Thus (48.8) describes the propagation of a transverse ($\mathbf{k}\cdot\mathbf{v} = 0$) wave in which the distances between the layers oscillate at constant density. For a plane wave, in which $\delta \propto \exp(i\mathbf{k}\cdot\mathbf{r} - i\omega t)$, (48.8) gives

$$\omega^2 k^2 = Bk_\perp{}^2 k_z{}^2$$

and hence the velocity of propagation

$$c_2 = \sqrt{B}\sin\theta\cos\theta, \tag{48.9}$$

where θ is the angle between $\mathbf{k}$ and the z-axis. The velocity is anisotropic, and is zero for propagation either parallel to the z-axis ($\theta = 0$) or in the xy-plane ($\theta = \pi/2$).

PROBLEM

Find the speed of propagation of acoustic waves in smectics for any relation between the moduli A, B and C.

SOLUTION. Differentiating (48.3) with respect to t and eliminating $\partial\rho'/\partial t$ and $\partial u/\partial t$ by means of (48.1) and (48.2), we find the equation

$$\partial^2\mathbf{v}/\partial t^2 = A\nabla \operatorname{div}\mathbf{v} - C\nabla\partial v_z/\partial z + \mathbf{n}[-C\partial \operatorname{div}\mathbf{v}/\partial z + B\partial^2 v_z/\partial z^2].$$

For a plane wave, in which $\mathbf{v} \propto \exp(i\mathbf{k}\cdot\mathbf{r} - i\omega t)$, this becomes

$$-\omega^2\mathbf{v} = -A\mathbf{k}(\mathbf{k}\cdot\mathbf{v}) + C\mathbf{k}k_z v_z + \mathbf{n}[Ck_z(\mathbf{k}\cdot\mathbf{v}) - Bk_z^2 v_z]. \tag{1}$$

Let the wave vector $\mathbf{k}$ be in the xz-plane. Then it follows from (1) that $\mathbf{v}$ is in the same plane, its x and z components being given by the two equations

$$v_z[c^2 - (A+B-2C)\cos^2\theta] + v_x(C-A)\sin\theta\cos\theta = 0,$$
$$v_z(C-A)\sin\theta\cos\theta + v_x[c^2 - A\sin^2\theta] = 0,$$

where $c = \omega/k$ is the speed of propagation of the wave, and θ the angle between $\mathbf{k}$ and the z-axis. Equating to zero the determinant of this system gives the dispersion relation

$$c^4 - c^2[A + (B-2C)\cos^2\theta] + (AB - C^2)\sin^2\theta\cos^2\theta = 0.$$

The larger and smaller roots of this quadratic in c^2 give the speeds c_1 and c_2. In particular,

$$c_1 = \sqrt{A} \quad \text{when} \quad \theta = \tfrac{1}{2}\pi,$$
$$c_1 = \sqrt{(A+B-2C)} \quad \text{when} \quad \theta = 0.$$

The speed c_2 in these directions is zero.

INDEX

Printed and bound by CPI Group (UK) Ltd, Croydon, CR0 4YY
03/10/2024
01040334-0013